CHARLES WADSWORTH CAMP AND WORLD WAR I

War's Dark Frame and History of the 305th Field Artillery

CHARLES WADSWORTH CAMP

Introduction by
JONATHAN D. BRATTEN

Edited and with an Afterword by
CHARLOTTE JONES VOIKLIS

Tesser Well Books
Goshen, Connecticut

Introduction © 2023 by Jonathan D. Bratten

Afterword © 2023 by Charlotte Jones Voiklis

War's Dark Frame originally published in 1917 by Dodd, Mead and Company

History of the 305th Field Artillery originally published in 1919 by The Country Life Press

Cover design by Renata DiBiase

E-book ISBN: 979-8-9872623-2-0
Print ISBN: 979-8-9872623-0-6
LCCN: 2023930060

Note on the Text

Because these two books are in the public domain, original facsimiles and other electronic editions are easily available. This edition of *War's Dark Frame* and *A History of the 305th Field Artillery* makes changes to the original text with the aim of giving ease to the modern reader. American spelling has been used consistently, grammar and punctuation has been modernized, and the use of foreign words has also been made consistent with modern usage, following both *Chicago Manual of Style*, 17th edition, and *Merriam-Webster Dictionary*, online. There are instances of dialogue where the author has used transliterated spelling to indicate an accent. Occasionally words have been added or sentences rewritten to make it easy for the modern reader. Sometimes spelling or capitalization or accents were incorrect or missing, and when a proper name was wrong, it was corrected to today's spelling. If a place name was unidentifiable, the original was left in place. Abbreviations were written out in their first instance. When clarification was necessary, brackets were used in some, but not all, instances of added text; sometimes it was added in parentheses. Again, this is to give the reader the most seamless experience possible, without sacrificing too much of the original flavor. There are also occasions where I have not been sure what the meaning of a slang word of phrase is, and have left it. There are plenty of passages and sentences where ambiguity remains.

There were occasional slang terms and references that if not obviously derogatory at the time, certainly are now. These have been deleted. They were not many, but I wanted this edition to strike a balance between ease of reading for the modern reader and historical accuracy. Again, the original is readily available for any scholars who want to read an unmediated version. I did keep the original use and spelling of the word "Esquimaux" because although modern usage would disallow it, the combination of archaic spelling and its location in the text made its retention meaningful. I retained the occasional derogatory use of "Jerry" and "Hun" for German.

After much deliberation, I decided to not include a wealth of ancillary material included in the original of *History of the 305th Field Artillery*. Illustrations and lists are difficult to render consistently and clearly in e-book format, and although it pained me not to include the lists of deaths and promotions, I decided that the purpose of this specific volume was to focus on Charles' voice. The originals in their full glory are available elsewhere, and pdfs are available for free on www.madeleinelengle.com

— Charlotte Jones Voiklis

Introduction

CHARLES WADSWORTH CAMP, WORLD WAR I, AND
THE CREATION OF A NATIONAL ARMY

Jonathan D. Bratten

Historians often muse over the question of how noteworthy people develop in their formative years. Much is written about famous leaders, but what of notable authors? One might at first think that tracing a line from the much beloved author Madeleine L'Engle to the First World War would be a stretch. After all, since she was born weeks after the Armistice in 1918, the war could hardly have had a great impact on her. And yet, we must consider how a parent's experience plays into how they raise their child. And for Madeleine's father, Charles Wadsworth Camp, the Great War played an outsized role in his life.

You might wonder why republishing the World War I memoirs of a man long deceased would be a matter of importance in a world now one hundred years removed from that great epoch. We live in a world shaped and still smarting from the first half-century of conflict in the 1900s. The first world conflagration did little to dampen the fires of enmity burning in Europe, causing a Second World War with the rise of fascist governments in Europe. From that war grew another fifty years of animosity and combat, as the great powers of the world were locked in the struggle of the Cold War, which often flared into life in places like Korea, Vietnam, and the Middle East. Out of those conflicts grew new nations, new governments, new social movements, and new ways of thinking. While World War I lays in the past, its effects are still very much with us today. Camp's experiences in that war—borne out in his writings—bring us a unique lens into that era.

Born in 1879, Charles Wadsworth Camp was a writer and journalist well before 1914, writing for magazines such as *Colliers* and *The Century*. He also wrote fiction, penning several mystery novels prior to the outbreak of war. But it was the war that seemed to seize his interest most passionately. As a journalist, he covered the war's beginning in 1914 and traveled to Ireland to report on the Easter Rising in 1916. All this at a time when the rest of the United States seemed bent on ignoring the war as much as possible, except as a financial pursuit. Indeed, President Woodrow Wilson won a second term in 1916 on the slogan: "He kept us out of the war." America was set on isolation rather than intervention. It was a nation very aware of the affairs tearing the world apart, but not yet keen on becoming involved in them.

For those like Camp who felt as though America had a duty to fulfill in stopping the bloodletting in Europe, this national feeling must have been infuriating. To see the world torn apart before their very eyes and yet have business persist as usual in the United States. Millions fell between the Somme and Verdun in 1916, while the war also ravaged the Balkans, Eastern Europe, and Russia, and moved across Africa, into the Arabian Peninsula, and into the Dardanelles. There was no end in sight. And yet, most Americans continued to act as though this world of violence and suffering had nothing to do with them. For those who felt a feeling of kinship to Great Britain and France, this attitude was incredibly frustrating. It is in this light that we need to see Camp's 1917 publication *War's Dark Frame*, written after his visits to England, France, and Belgium in late 1916. It is far from a glorification of war. There is no gung-ho call for armed legions of Americans to enter the fray. What Camp emphasized, instead, was just how horrible the war itself was for people rather than nations.

His goal was to give his readers an idea of just what the French,

British, Belgians, and their allies had been contending with. "For it isn't easy to understand war in America," he stated. Camp painted the picture of the impact of war on individuals—beginning with his experiences on the transport ship over to England, where he met a young woman who had married a British soldier, who was wounded shortly thereafter. She spent the whole passage wondering if she would see him, with every part of her longing for him. Camp describes the joy of the young woman's meeting with her husband— only to find out he is off to the front again.

This was far from Charles Wadsworth Camp's first time to Europe, as he had traveled extensively before the war to London and Paris, as well as to Cairo and Shanghai. He and his wife, also named Madeleine (referred to as Mado in this volume for clarity), had loved to travel together. As mentioned above, he had reported on the war in 1914, and then on the Easter Rising in 1916. But seeing wartime Europe for the first time came as something of a shock to him. As he stated, "The world was different and wrong." In particular, he found England transformed entirely by the war; in reporting on it, he gives his first jab at American isolationism: "All Britain is heart and soul in the war. Even then it was hard to accept as real the brilliant, careless complacency of our own country."

But it was war-scarred France that brought Camp to drop much of his neutral writing, for he was overcome with sympathy for the suffering of the British and the French, as opposed to the "iron bound German system." He used the systematic destruction of French towns as a medium to show how terrible this war was from those in the past and to paint the image of Germans as something beyond the limits of civilized people. A very willing French government brought him to the town of Gerbéviller, where he was shown the devastation of the town and told about the rape and murder of French civilians by Bavarian troops. This particular episode fits into a part of the "atrocity narrative" that was common at that time—although there is historical evidence for the events at Gerbéviller actually transpiring. The result of this visit was probably exactly what the French had hoped it would be. "We went out of Lorraine with a sense of flight before a sinister invasion perilous to the entire world," Camp relayed in his reporting, "of unusual and ruthless creatures, suddenly unmasked by the tearing claws of war." (80) The hint is there in Camp's writing: sure, America, you may be at peace now, but do you think the Germans will stop when they have Europe?

Camp visits Paris, Lorraine, the Marne, Champagne, Flanders, Arras—in short, all the places that the French and British would have liked an American journalist to see in order to inflame US sensibilities concerning the war. He visits coal miners building tunnels under No Man's Land into the German lines and describes a heart-pounding narrative of life in the British sector, where he dodges artillery shells and then has tea with British officers. His small snapshots of the war are vivid and filled with color. In the end, he keeps bringing his readers back to the same problem: what is the United States doing about all of this?

In one such editorial, Camp delivers to his readers a very stark challenge, one that was on the minds of many in the United States at the time: Perhaps Americans were no longer able to take up a challenge like the Great War because multiculturalism had deteriorated the American will. Camp explained: "I gathered, not particularly from this conversation, rather everywhere in England and France, that a belief had grown since the beginning of the war in our lack of homogeneity. We were, it was suspected, incapable of direct and concerted action. In those days the men who were actually treading the exhausting mill frequently placed upon us—whether justly, who can tell?—the taint of many races, the incoherence of too vast a variety of creeds and desires and antipathies."

This was the great question of the era: "Would the United States be able to fight when nearly 15 percent of its population was foreign-born?" Waves of immigration from Europe in the late nineteenth and early twentieth century brought millions of people to the United States. Many wondered if this meant a loss of homogeneity and a weakening of national will. And so, when Camp returned to the United States, this was very much on his mind. He reported: "Then one afternoon we steamed into New York harbor, and I saw a city that seemed proud of an incomprehensible ignorance of the meaning of war." (203)

As it turned out, Camp did not have long to wait for this ignorance to end. In early 1917, the Germans resumed their strategy of unrestricted submarine warfare, sinking any vessel suspected of aiding the Entente powers. American newspapers flooded with human-interest stories, including those of crews whose ships had been sunk while investors watched nervously as millions of tons of their cargo went to the bottom. In March, the British exposed the Zimmerman Telegram that revealed how Germany was encouraging Mexico to attack the United States. After a month of frenzied debate, President Wilson requested and received a declaration of war from Congress and, all of a sudden, the United States was involved in World War I.

That great question of what the country could do remained, however. As it entered the war, the United States had a miniscule army and very little manufacturing prepared for war production. It would take time to man, train, and equip an army that could fight at the scale of war that the battlefields of France and Belgium demanded. The nation's professional Regular Army was constabulary, built for manning the outposts of the American empire rather than fighting war on an industrial scale. The National Guard—consisting of part-time citizen-soldiers—still carried with it much of the militia tradition that had characterized it since 1636. Regular Army and National Guard units were formed into divisions capable of fighting in this new type of war, but they were still not enough.

To fully man this force, the nation would need to use conscription, something it had not done since the American Civil War. But it would have to be on a whole new scale unlike anything seen before. For the first time in a war involving the United States, the full powers of the national government were turned towards war-making: all transportation was nationalized, industry responded to War Department orders, journalism was restricted, war bonds were announced, and the people of the United States began to be "called to the colors" by the millions. By the end of 1918 there would be four million Americans in uniform, with two million across the ocean in France.

For Camp and many others, the question still lingered over whether a cosmopolitan nation could unite to fight a war. How could all these varied races, creeds, and nationalities fit into one army? That question was surely nowhere more on peoples' minds than with the creation of the 77th Division. A draftee division, the 77th was created in the zone encompassed mostly by New York City. While most divisions could look to common ties of statehood to bind its men together, here was a division that would have to pull from what was then the most polyglot, multicultural city in the entire world.

Even the division concept was novel to the US Army. While as a tactical organization the "division"—a collection of brigades that contained the individual regiments of infantry, artillery, and other arms—went back to the American Revolution, "permanent divisions" were a product of this newly industrialized war. The numbering sequence reflected this industrial shift: divisions numbered one to twenty-five would be reserved as nomenclature for the Regular Army; divisions twenty-six to seventy-five for the National Guard; and divisions seventy-six and above were reserved for the National Army. Inside each of these divisions—with 28,000 men per division—were four infantry regiments, three field artillery regiments, an engineer regiment, three machine gun battalions, and various supporting supply, medical, signal, and military police units.

The numbering of regiments followed this scheme, with infantry regiments in the New England National Guard's 26th Division beginning at 101, 102, 103, and 104 and the artillery— officers of which would help train Camp's unit once in France—had the same sequence. New York's 27th Division continued the sequence with 105, 106, 107, and 108, and so on through the remainder of the eighteen other National Guard divisions. Thus, the 77th Division would eventually field the 305th, 306th, 307th, and 308th Infantry Regiments and the 304th, 305th, and 306th Field Artillery Regiments.

Charles Wadsworth Camp's history of his own 305th Artillery Regiment inside the 77th Division, written just after the Armistice of

1918, gives us a unique glimpse into this great experimental organization. Camp had been a member of the New York National Guard's famous 7th Infantry Regiment from 1902 to 1908. Known as the "silk-stocking regiment" in the Civil War, it was half military unit and half a "who's who" of New York high society, with their armory located on Park Avenue. It had become the 107th Infantry Regiment of the 27th Division in the new US Army-wide reorganization, as the old state units were mashed together to form the new divisions. The National Guard elements were all tied to localities and tended to reflect those communities. Camp's old cosmopolitan, white-collar 7th Infantry Regiment of New York differed greatly from the more rural 2nd New York—now the 105th Infantry—consisting of upstate farmers, mechanics, and millworkers. This was reflective of National Guard units across the country: some urban, some rural. In an ever-expanding army, military experience of any kind was valued greatly and, therefore, Camp with his experience was made one of the officers of the regiment. And what a regiment it was.

Like most of the National Army (draftee) divisions, the 77th was created out of nearly nothing. Thus, each division experienced much of the same: arrive at camp, wait for uniforms to arrive, wait for officers to be assigned, wait for more soldiers to arrive, wait for equipment to arrive, wait for training to begin. In essence, a lot of waiting. And with it, a vast quantity of paperwork, which Camp noted, with chagrin, seemed to be the main enemy they faced. Like most writers of this era who were in the American Expeditionary Force (AEF)—the name given to the combined Army, Navy, and Air Service units bound for Europe—Camp had a keen eye for the humorous and his sardonic tone was similar to those of his generation.

Those early days of building a fighting division from scratch are portrayed via Camp's tongue-in-cheek remarks on the utter farce that it could often be. Officers would show up, be appointed to batteries and staff positions, and then just as quickly be transferred off to some other unit. It took months for basic things to arrive, such as tack for the battery horses, let alone the heavy artillery guns themselves. Camp's description of the time it took for the regiment to get even one old field gun to train on would have left a military-preparedness proponent like Teddy Roosevelt in tears. Camp further captures the inherent chaos of this new national war and the inhumanity of being part of the American Army, a massive machine that ballooned to nearly four million by war's end. Soldiers were torn away from their friends and shipped off to other outfits, to fulfill the needs of the army. And the 305th was luckier than most regiments in those hectic times, as their executive officer was Henry Stimson, who served as secretary of war from 1911 to 1913, and was the future secretary of state and then secretary of war for Franklin D. Roosevelt throughout World War II.

The overall question remained of whether a democracy like the United States could produce a fighting force that could go toe-to-toe with the autocracy of Germany. Camp felt this keenly, possibly remembering his conversations with French and British officers during his trip in 1916. This informs the pride that one can see in his writing, as he describes how the division began to come together in the fall of 1917. When the first battery of the 305th Field Artillery received its guns, Camp wrote, "The crowd cheered that single battery as it crunched through the snow past the reviewing stand, little Wing, the Chinaman, on one of the lead horses, pointing with unconscious pride [at] the democratic, the universal power of our army." (54)

However, Camp's reporting is not all a philosophical discussion on the merits of a democratic military. It is also a chronicle of young men on an adventure, young men entering something greater than themselves. In this, Camp falls into line with most of the American writers of World War I memoirs. For the young men—and women—of the nation, this was a grand adventure. And yes, women saw service as well in World War I. For the first time, women were formally allowed to serve in uniform—a sign of changing times and the desperate need that the war brought for unique skills. As well as serving in the Army Nurse Corps, women also served as telephone operators controlling the vast network of communications for the AEF. The US Navy brought in women to serve as clerks and typists, who were labeled "Yeomanettes." Women also served in aid organizations such as the Red Cross, Salvation Army, and Young Men's Christian Association. Barred from combat roles, women still saw the frontlines as ambulance drivers, mechanics, nurses, and in a myriad of other roles. Women would brave German artillery fire to drive the wounded to hospitals or to serve out coffee and doughnuts to dough-boys on the frontlines. This is one aspect of World War I that is missing in Camp's narrative, though it is often included in many other wartime reminiscences. The sight of an American woman did much to hearten soldiers on the front and may have contributed to the eventual approval of women's suffrage in the United States. For many of these young Americans, it was the first time leaving their state, let alone the country. Everything was new and exciting, a feeling captured in the slew of books written by veterans, from 1919 to 1924. However, after that point, such memoirs written by Americans became incredibly scarce. Indeed, it seemed almost as if an entire generation had gone entirely silent. Perhaps as a result of the Great Depression, or a result of a second world war being needed, these veterans simply stopped telling their stories.

While writing in this same genre, Camp managed to express sentiments without the overly cocky bravado typical of many American memoirists. He provides a realistic face to the war in a matter-of-fact voice. Like War's Dark Frame, his history of the 305th is not written in a way that glorifies war. Instead, he explores how the war mattered for the men of the regiment, hoping in some way to also show why the war was so important for the United States.

One reason it was important was because the US Army established a new way of creating an army. As discussed previously, while a draft had been used during the Civil War, the draftees at that time went to the volunteer regiments of the various states. Contrary to this, during World War I, the massive 28,000-men divisions were formed from the "whole cloth" in the service of the whole nation, rather than representing just their state. In the process, the US Army brought a universal military experience to about four million Americans who would eventually serve in the war. At the same time, these citizen-soldiers forced the army to change. It could no longer be the same rigidly conservative institution it had been before World War I, for progressive ideas from the era had begun to seep in. In this symbiotic way, both society and the military were changed by the war.

With the 77th Division being the first National Army unit on the frontlines, Camp was very aware of the significance of their performance. Prior to their entrance into combat, he mused, "The National Army was a good deal of an experiment. It contained every type, race,[1] and temperament. Had its brief training fused these uncongenial elements into a serviceable whole?"

Camp details the unit's training and eventual deployment to the first zones of operation

in France, where they are ultimately tested in battle. The artillery was crucial to victory in World War I—it was the artillery's job to keep the enemy guns silent with long-range, counter-battery fire, to destroy attackers with a rain of high-explosive shells, and to break up enemy defenses with immense barrages. Camp's writing is vivid and pulse-pounding as he describes his regiment's efforts to hide their 75mm field guns from prying German aircraft and probing fire from enemy batteries. The lack of American air superiority might seem odd to so many who take it for granted these days. Camp provides an excellent example of what it was like to have to fight when someone might always be watching from above. And as a member of an artillery regiment, his depictions of the war vary from those written by infantrymen, for Camp was able to take in more of the countryside and overall situation rather than just the view of a foxhole or trench line. As a result, he provides a different perspective than most memoirs of this type written at the time.

In the end, Camp's concerns about this "experiment," proved to be mostly for naught. While National Army divisions often struggled in combat, so did their Regular Army and National Guard counterparts. The war was a steep learning curve for the US Army, which still maintained circa 1914-era ideas about how war should be fought. The army had to learn four- years-worth of hard-won lessons in as many months. It tended to be the units that arrived in Europe earliest that did the best, mainly because they had more time to learn and adapt. The 1st Division arrived in the late summer of 1917, the 26th Division in October, and a handful of other divisions followed. Arriving in April of 1918, the 77th Division was able to accrue the most experience of any of the National Army divisions. Still, that did not stop the debacle of what has been called the "Lost Battalion" of October 1918, when an infantry battalion from the 77th was cut off by the Germans and nearly annihilated. But far from breaking, this battalion managed to hold out long enough to be rescued, even after receiving heavy casualties. The concept of a democratic army, it seemed, was a success.

The war changed everything for the United States. Formerly a backwater power in the Americas, the country was now a global power that could project force across the globe. In fact, all the way to Russia, as some National Army members found themselves in Siberia in 1919. What was this new country to look like? And who should have a share in it? These questions reverberated through American society in the years and decades after the war, even as the United States pulled back into isolation again. But the lid was off Pandora's box. Americans had gotten a taste of the world stage and something profound had changed within the country. It would take time for that "something" to distill into Roosevelt's "Four Freedoms" and the idea of the United States as a protector of self-determination.

This was the world that Madeleine L'Engle was born into, just weeks after the Armistice. Though she could not know it, she was entering an entirely different world than had existed just one year prior. And not all of it was positive. For the veterans, too, the world had changed in a dramatic way. They had seen some of the worst of humanity, and the best—and many could not forget some of the horrors they had seen. Those that returned home, carrying visible and invisible wounds, came home to a population that cared only briefly about what these war veterans had seen and done. The nation had been devastated by the 1918 Influenza Epidemic, which killed around 675,000 people in a matter of a few months, at the same time as the AEF was incurring its highest casualties in the Meuse-Argonne

Offensive at the end of the war. Few wanted to remember so much death. The Roaring Twenties and the Great Depression that followed left the World War I generation behind, to slip into a painful obscurity.

Charles Wadsworth Camp seems to have been among those veterans who never could recover their youth after their experiences in the Great War. Mentally, he entered depressive states, a common occurrence in veterans of that war. Physically, he also carried another trait unique to World War I veterans: respiratory issues due to poison gas, probably exposed to along the Vesle River or in the Forest of Argonne. Because these ailments impacted his health and ability to write, Charles and Mado moved to France in 1930, where their daughter Madeleine was put into a Swiss boarding school. Like so many of his generation, Charles never recovered from the war, mentally or physically. It is not known exactly how many US veterans of the war carried mental health injuries back home with them, but the number is probably substantial, based on the information that is available. The Great Depression struck them an even harder blow than most, as evidenced by the Bonus March in 1932, when World War I veterans held a massive protest march in Washington, DC, to demand financial assistance. It was the war that would take Charles's life in 1936, when he died of pneumonia that took over his gas-damaged lungs.

The woman who would someday capture the hearts of children and adults alike with her depiction of a daughter's journey through space and time to save her father had her own father stolen from her by the Great War. This makes the writings of Charles Wadsworth Camp all the more important. Not only do they offer a unique perspective on the war, they are the clues he left behind as we work towards a world where daughters do not lose their fathers to war.

---

1. The army was in fact still segregated, the 92nd and 93rd Divisions were African American.

War's Dark Frame

by Charles Wadsworth Camp

Contents

TO
MY WIFE
AND
MY MOTHER

Chapter One

THE SUBMARINE ZONE

"Can't be a submarine. We're too far out!" a passenger shouted.

"Keep quiet, it's all right!" said another, in response.

"Don't get excited!" cried another.

Exclamations came in men's voices, unnaturally suppressed. From the women arose one or two half-choked cries. Feet hastened along the decks. Apprehensive but without panic we poured through the companionway. You admired the women in that moment, because they had an appearance of steeling themselves against dreadful inevitabilities. And the sea was appropriately sullen and unquiet.

Many of us, I think, foresaw what we would find at the forward rail—a view of the crew, with purposeful faces, in an emergency drill. Yet the necessity for that exercise, the wisdom of shocking us from our Sabbath somnolence by the raucous alarm of the ship's bell, reminded us of how closely we had approached the incredible spectacle of a civilization in arms against itself. "What would the next day bring . . . or the next?" we wondered.

Abruptly, we realized that war for the individual has the quality of a perpetual and tragic disaster. Later, in the cities of Europe, in the devastated districts, in the towns under bombardment, in the front-line trenches, that truth was forced upon me. So I have remembered chiefly the human incidents and impressions that will have a real meaning for the individual, who has had the foresight to visualize himself, his family, and his friends entangled in the struggle.

For it isn't easy to understand war in America. The entrance to the pier in New York Harbor teaches you that. Beyond comes a mental alteration as pronounced as the change from brilliant sunshine to the somber obscurity of the shed. It is accented by the tight line before the gangway, by the suspicious examination of passports and luggage, by the unstudied talk among the inspectors of bombs, of spies, of the possibility of submarines. And the gangway is the threshold of war.

On all boats bound for Europe these days there is an atmosphere of difficult partings and a reluctance to discuss the future. There are, moreover, people who bring war home to you. That afternoon of the drill, for instance, a boy, not yet twenty, watched and reassured two women who counted the hours before we would be off the Irish coast. All along he had interested us in a sorrowful fashion, because he had been wounded in the head at Ypres, and a disability had remained that meant he was no longer of value in battle to his country. Always he seemed older than the old men, as if he could never forget and be young again. A tall, straight, ruddy-faced man, nearly at middle age, joined him. The newcomer, following his custom, wore no hat. We gathered around him because, since he was on his way to the front from Canada, whatever he said seemed to possess a special eloquence.

"Funny time for fire drill!" said the newcomer. "Splendid nerve tonic though. You know, I wouldn't be surprised if the Huns took a shot at us. It's about due."

"I want to die with my boots off and without fame," another man said plaintively.

We laughed, returning to our cards, our reading, or our naps. The boy who had fought at Ypres demanded a game of deck tennis. He had no difficulty finding three other players, for the growing tenseness was unfriendly to reserve—already everyone knew everyone else.

An elderly gentleman from the South wandered restlessly across the smoke room and interrupted the bridge game. "They say this boat's loaded so heavily," he said.

"I bid a heart," responded one of the players, "we all know she's got a big freight manifest, Mister . . ."

"Think she has! Go down like a shot!" continued the man from the South. "I've been talking to one of the officers. Says there's no way to avoid floating mines. No respecters of neutrals. You've heard of the . . ." And then he proceeded to list half a dozen boats recently injured or sunk by mines. The card player who had spoken before grew impatient.

"Your lead to a heart," he said.

The elderly Southerner turned away, muttering with a prideful air, "Just the same, since I got on this boat I've never ceased thanking God I'm a powerful swimmer—a right powerful swimmer, sir."

The incident was funny because nobody laughed. We glanced at each other and took up the game. But, perhaps, the one who brought war closest was a pretty American girl bound for England with her mother. We understood she was married to a Scotch officer. We wondered why she had been in America and where her husband was, for she didn't wear mourning.

"The girl has a story," one woman after another commented.

To realize it you only had to look at her eyes and at the convalescent pallor of her face, as striking as that of the boy wounded at Ypres. She wanted, moreover, to talk about her experience. That, too, was in her eyes. Because of the past, possibly because of something she approached, she desired to tell her story. The last evening, as we crept up the channel, she yielded to the growing tenseness that fought reserve. She sat with her mother on deck, staring at the boats that had been swung out, listening to talk of the extra life belts that had been distributed—mere italics for possibilities of which the women were, patently, trying not to think.

The sun sank behind a low, brown mass on the horizon—the coast of Ireland. We reviewed the crimes and the tragedies it had witnessed since the commencement of the war. We imagined the round backs of indifferent submarines and black specks of humanity struggling in the yellowish, menacing water. A multitude of fishing trawlers pitched and reeled drunkenly. It was difficult to realize that their only game was submersibles, their only task the protection of such craft as ours.

Groups of people still lined the rails, scanning the dusky water. All afternoon they had seen periscopes. Each piece of driftwood in the forbidden zone had attained an importance never dreamed of in the scheme of things. That night the moon appeared and quiet men cursed it.

"They get us against it in silhouette and we're gone," one and then another commented.

The prow parted the transformed water almost reluctantly. It was as if the elderly Southerner had impressed on the boat itself his aphorism concerning floating mines. As we went on, feeling our way, with a sense of dodging unseen and treacherous obstacles, the pretty girl told her story—a brutal one that brought the war closer.

The first chapter, just a year old, was her marriage in Nice to an invalided officer of a Highland regiment. Before his complete recovery, he had been unexpectedly recalled to active service. The uncertainties of waiting to marry had appalled them, so they married hurriedly, in spite of her mother (now watchful and lounging in her steamer chair) who was panicked and shocked by their actions. The couple's honeymoon was a swift journey to the military base at Rouen. The girl's voice was fearful, rather than reminiscent, as she spoke about it.

"He left me at a queer hotel on the main street while he went to report," she began. "He didn't know exactly what his orders would be —whether he would stay at Rouen for a while, or whether they would hurry him to the trenches with new troops. The room they gave me had six doors and none of them possessed a key. It may sound silly, but it was late and I was afraid, afraid of everything. I wasn't sure he would come back at all, and if he didn't I knew I might never see him again. Strange sounds drifted in from the dark street: I heard soldiers marching; queer songs in French and English; and far off, a bugle. I was lonely, and homesick, and unhappy. I knew he wouldn't come back and all those doors frightened me. I tried to barricade them, but I couldn't find enough chairs. Then he ran in, and he laughed at my barricade, which he had had to tumble over. He had to go that night, and I walked through the dark streets with him, although he said I'd better not, because it would only make it harder for both of us. But I went, and at the military station there were soldiers everywhere, and confusion, and a train—that waited. I didn't dare look at it, but I knew when it started that it was his train, for he said goodbye."

"I looked then and saw him climb into a carriage filled with soldiers," she added. "He waved his hand, shouting to an officer he knew to see that I got back to the hotel and later to Paris where my mother would be waiting."

Her mother, good-humored and middle-aged, laughed resentfully. "Instead of that she dragged me to Rouen," she said, then added, "You need another wrap, my dear."

The girl shook her head. "So I went back," she continued, "crying through the dark streets with that strange officer. Halfway I stopped, remembering I didn't have a cent. My husband hadn't given me any money. You see we had been married such a little while. We hadn't learned to think of such things."

She then spoke of her interminable days of waiting in Rouen. She had been at the point of winning a staff appointment for her husband, with lighter dangers, when word of him—hourly expected —had been delivered to her.

"Oh, quite brutally," she said. "I didn't know what it meant, death or a wound. I only knew I must go, so I persuaded a high officer to give me a pass for a military train. I spent a lifetime on that train. Over many hours it crawled only a little way. Finally, they told me to get out. They drove me to a small hospital back of the lines. The odor of it! And he lay there, a sister bending over him. She said I mustn't cry. It was hard, because he didn't know me, because he seemed like one already dead."

Her voice dwindled, her mother stirred, and then, as if to spare the girl, the mother explained how her daughter had drawn her husband from the black valley through months of nursing in France and England. She had broken down—so the doctors ordered her to America, away from the hospital odor and the perpetual reminders of war.

"She's going back too soon," her mother said.

"Naturally," the girl answered, "because he writes he is on light duty again, and he's trying to persuade them he's fit to return to the trenches. I won't have it. I couldn't stand that suspense again. But of course they won't let him. He has a piece of shrapnel shell within an inch of his heart. He's done his bit."

"You know," she went on, "I'll have to harden myself. I've grown soft in America, because it's so far from the war. You can't remain sane unless you are hard in the presence of this war."

Reviewing her story, questioning its final words, you realized how true that was. You shrank from the water flashing by, because you knew it measured your approach towards those fantastic occurrences against which men and women must harden their hearts or suffer beyond reason.

Naturally, I thought that was all I was ever to know of the young wife's history, yet the next day there was to be a sequel, read at first hand, cheerless and unexpected. We sat until late that last night, and she spoke from time to time of the approaching meeting.

"He's sure to be at Liverpool. Suppose anything should happen to this boat?" she exclaimed. But for the most part she was silent.

"We will spend the night on deck," her mother said, "in case anything happens."

In the smoke room, I heard men talking of sleeping on the lounges there. An elderly and morose commercial traveler heightened their misgivings with stories of his escape from the torpedoed *Arabic*.

"She went down in ten minutes," explained the man. "Five minutes would see the last of this boat as she's loaded. If you were caught below decks, good night! Talk about rats in a trap!"

"Oh, forget it!" another man said, under his breath. "I've heard that old fool sink the *Arabic* a dozen times in the last half hour. Once is enough for any boat."

But the morose traveler had spoken to the women about his premonitions. They wandered restlessly, or stared across the cold and troubled water, rehearsing his warnings. This one man had sown the seeds of panic. The women didn't want to go to bed. Then a squad of sailors came by with hose, pails, and swabs. They went to work with quiet confidence. One of them spoke good-naturedly:

"Better be off to bed, ladies. If you don't, you'll get wetter than if a torpedo struck us in the bloomin' witals."

Some of the women laughed then. There was nothing else to do, so off they went. And in the morning, the women weren't alone in showing signs of a sleepless night spent in bed fully clothed. A vast relief shone in the eyes of the young wife and her mother. Only a few hours away, the convalescent waited to welcome them back to England.

Indeed, to most of the passengers the brown mass of Holyhead, rising at starboard, appeared like a beacon of safety. A deck steward, who had grown communicative, grinned and said, "Just as well they think that way." Without thought for my own feelings, he assured me that the really dangerous part of the trip lay just ahead. Yet, without adventure, we raised above the sands the gigantic skeleton of the Birkenhead tower and swung in across the bar of the River Mersey.

Stretched before us were Liverpool's suburbs, uninteresting rows of buildings, as a foreground for the routine activity of a war-time seaport. Remembered steamships lay in the docks or at anchor, painted a dead gray, converted into transports or auxiliaries. One of the best known of all wore a livery of white and green with red crosses here and there. Bandaged men stared dumbly at us from the rails.

Liverpool had altered significantly. From it, the war stretched its grimy fingers to draw us closer into its lethal atmosphere. A sentry paced the landing stage. No more than a handful of people waited there. As we drew closer, we all noticed a tall, straight, young fellow in a Highland uniform. He walked up and down impatiently, swinging a little stick, glancing with anxious eyes at the crowd of us by the forward rail. The girl and her mother were near. They cried out and glanced at each other tearfully. With jerky motions, they waved their handkerchiefs. The young officer, with a piece of shrapnel near his heart, suddenly swung his stick, paused, and stared up at the tear-stained faces.

"Doesn't he look fit?" the girl cried proudly. "But not really fit—never fit for war again."

More intimate affairs grasped our attention. Sheep-like, we were herded into the dining room to face the alien [immigration] officers. While we awaited our inquisitions, the young Highlander entered, exuding a naive pride in his uniform, which had won him a permit to pass the guards, and hastened this moment of fervent greeting. He stood close to us with his wife; for a time, they spoke softly, and then all at once their voices were raised. The flushed girl exclaimed, as if she had been struck, and the husband laughed with an embarrassed indifference.

"Then my letter didn't reach New York until after you had sailed," he explained. "They are sending me back to the front. Of course I am well enough. It was good of them to give me leave to meet you," he said, pausing and glancing at her curiously. At first she didn't answer. Then, she turned with a gesture of despair and walked spiritlessly away.

Chapter Two

THE STRANGE ENGLAND

For several hours, we suffered examination before the alien officers. With a progressive severity, the European ports had made the entrance of neutrals difficult. One by one we faced the little group of officers seated at a table in the dining room, while the exit doors were carefully guarded. Some of us were questioned for only a moment or two, others were grilled uncomfortably, while the next on the list waxed impatient. They asked: "What were you going to do? Where had you come from? What parts of the kingdom did you wish to visit? What was your ultimate destination?" Your past was ransacked.

As you stood by the little table, facing the unsmiling men, you felt as if you were a suspect. You questioned how a person trying to enter with criminal intent could possibly stare back without fear, or answer the searching questions without revealing a tremor. There are spies, beyond a doubt, who did survive such ordeals jauntily. It became obvious to us, however, that by then very few of them would have had an opportunity of attempting it. For a German spy to have slipped from New York to England, through such a net, would have approached the miraculous. On this day, men and women, released after particularly extended examinations, felt themselves aggrieved.

"Do I look like a spy?" one woman demanded hysterically, as she gathered her luggage. "What do those people think nerves are? If I had been a spy I'd have screamed. I'd have asked them to arrest and shoot me, just to get it over with."

The search of baggage took scarcely less than a minute. You were made to feel again the possibility of hidden bombs, or deadly weapons, or secret documents. Although we had docked at noon, it was very nearly dark before we were collected and embarked on the special train. In the carriages, with the suspense of the trip through the submarine zone over and the irritation of the examinations done with, we lay back, anticipating a momentary peace. Instead, reminders of war crowded more thickly upon us. The guards were either very young or very old. Prominently displayed in each compartment was a sign commanding us to draw the blinds on request, as a measure of safety. While we were in the dining car, a guard came through and gave that order. The midland countryside, flat and placid in the fading light, was shut out. We turned to our meal with a realization of how different this trip was from any we had previously taken. We had a sensation of stealth, of a personal partici- pation in the deception of Zeppelins. The rumbling of the train seemed discreet. When we glanced, daringly, behind the edge of a blind, we saw clouds banked against a pallid sky. A furnace broadcast a fierce red glow. The landscape was sullen, a little frightening. The world was different and wrong.

The women looked to, as if for reassurance, the mere boys who served us. They seemed to walk on tip-toe and appeared to discourage conversation. One of the boys, however, answered a ques- tion thusly: "We're the only kind they can use. The men are doing their bit, sir."

The arrival of the train at Euston conveyed little beyond the impression of quieter days: the shed was sufficiently lighted, there was a scramble to identify luggage at the vans, the pursuit of porters, and the snaring of taxicabs. And yet, upon driving into the street the alteration sprang upon us. It makes no difference how much you may have read or heard of darkened London, the reality reaches you with a sense of shock, not wholly unpleasant. It stirs your memory and you can't guess why at first, because you have never seen anything like it. Then you understand, as you rattle through the obscurity, as you catch the trivial illumination from shrouded lamps, as you stare at the glow from shop windows—discreet, a little mysterious, more provocative than the most vivid electrical displays. It recalls what you have imagined of Elizabethan London. And the city does have an air of romance. It is very lovely, too, because everything ugly is crushed beneath the shadows and everything beautiful acquires a meaning new and is sentimental.

Under such conditions the city offers exciting contrasts. It is magic to step from the medieval romance of the streets into the glittering present of hotels, restaurants, or theaters. Inside, where many lights burn, the only material alteration is the carefully drawn blinds. As far as you can judge, many people crowd this world of pleasure. The great change inside is not only physical. The ever-present officers and soldiers point this out to you. It takes some time to grow accustomed to these splendid men in uniform. You stare at them, observe their unstudied gaiety, and are aware of a vast depression. Some are back from the front on a few days leave. Others—in blue hospital uniforms, with pallid faces or missing limbs—advertise their convalescence with a pitiful pride. The greater number, however, are men still in training, on leave from the various cantonments. One smiles at the talk of a scarcity of men without compulsion.

These fellows are the best of the nation—young, sturdy, handsome, awaiting their baptism of fire with a quiet confidence. They know, too, what that means. This war has left them no illusions. High explosives, gas, liquid fire, are common subjects in their conversations over the tea table or dinner. They face such things with a stolid determination that surprises. It is the most thrilling aspect of London, this procession of youths that have assumed the khaki—a symbol of their supreme sacrifice that they wear too easily. Yet, in reality, there is something ecstatic about their young faces—something quite beyond definition.

As the days passed, one wondered that London should be so crowded. At the popular restaurants it was always necessary to engage a table in advance. I heard acquaintances lament that they had to lunch at cheap tea houses after craving for admittance to eight or nine packed dining rooms in the neighborhood of Piccadilly Circus.

The theaters housing popular plays shared the same inflated prosperity. London had never known such a season. Yet in spite of its easy chatter, its surface cheerfulness, it was, to an extent, restrained. There was little dancing outside of private houses. Evening clothes were frowned upon. You saw them only worn among the vulgar patrons of the most blatant hotels. Khaki was the color note by day and by night. Those wearing it went about with women likewise soberly dressed. And, in nearly every bravely smiling face, you caught an appreciation of imminence. The eyes of soldiers and the eyes of wives, sweethearts, and relatives seemed strained to regard an unknown and melancholy prospect.

In a sense, you felt yourself an intruder. Civilian clothing was an anachronism in London. You realized that the soldier was responsible for this city, crowded and a trifle unreal. You wondered if all of England was like this. You felt that you must see the country districts.

For this excursion, an American acquaintance and I took advantage of one of the bank holidays. We drove first of all to Cambridge. Even from the road we could see that rural England was more thoroughly transformed than the metropolis. We passed aviation instruction grounds. We saw practice observation balloons in the air— unwieldy, misshapen objects, carrying boys ambitious to make themselves targets for German antiaircraft guns. Transport trains rumbled by. In one or two villages we saw artillery parked. Khaki-clad figures paced the sidewalks or strolled among the fields. All England, you felt, was in brown.

The alteration impressed us most, perhaps, in the two great university towns, Oxford and Cambridge. We wandered through quadrangles and halls, missing students in caps and gowns, seeking the familiar atmosphere of undergraduate activity. Instead, we found proctors who displayed their brief rolls of foreigners and the physically unfit. "The others," they explained gravely, "have gone to the war."

And war was here among the hedges, as thoroughly as it was in London. We found few caps and gowns, but khaki was plentiful. Several of the colleges had been turned into training schools for officers. Men of university age and appearance went through evolutions and studied tactics in the ancient quadrangles, preparing themselves to replace the Oxford and Cambridge men already killed or rendered unfit for service.

There were hospitals, too. Figures on crutches or grotesquely bandaged struggled about the grounds or across the commons—places that just a few years ago were noisy and active with the play of whole-bodied, careless youngsters.

It was among the convalescents in Cambridge that I found a veteran of those terrific first days—one of the few survivors of the hell of Mons and the retreat that had followed; of the Battle of the Marne; of the deadly turmoil at Ypres. We saw him standing in the entrance of a garage at Cambridge, when we drove up and paused for gasoline. His hair was grizzled and his face had many small lines that gave it an expression a trifle quizzical. His crutches and the blue band encircling the sleeve of his service overcoat stamped him as still under hospital treatment. He wore sergeant's chevrons and a Scotch cap set at an absurd angle, and a little black pipe protruded from his mouth. Discontent was clearly visible on his whimsical, middle-aged face. While he talked, I waited for an opportunity to find out the cause of his irritation. His most fervent description of the horror of the retreat was: "Oh, mon, but that was warm work." The same expression did for the Marne and Ypres.

"But when and how were you wounded?" I asked.

He flushed and puffed rapidly on his stubby black pipe. He no longer looked one straight in the eye. When he answered, his voice was low and ashamed: "Not at Mons," he said, "not at the Marne, not at Ypres." His voice thickened with revolt, as he continued, "It was on a day when there was nothing doing. You understand? As quiet as you please. I lay in a dugout, reading a letter from my bonnie girl.

Along comes a shell and explodes in the entrance—on a quiet day after all my chances. Disgusting's what I say. A fifteen months' job so far, and they took pieces of that letter out of me in France. They took them out of me on the hospital boat. They took them out of me here."

His eyes twinkled, as he said: "And I guess there's some wee paragraphs still . . . there."

But then the twinkle died; the discontent returned. This man had a grievance beyond the manner of his wounding. By chance, we learned what it was about.

My companion fumbled in his pocket and produced some small change and suggested: "Perhaps, you'd like to drink to the memory of those days."

The sergeant's discontent exploded. "A British soldier!" he cried. "A sergeant in his Majesty's army! Me drink! I'm a baby. A blessed, swaddling infant." He tapped the blue band on his arm and added, "Just because I'm under hospital treatment I can't have my glass of beer. That's gratitude for you! After what I've come through. I've learned my lesson. I've had my dose."

"You mean," I said mildly, "that when you are quite well nothing will persuade you to put yourself in the way of such ingratitude again? You won't go back to war if they need you?"

He braced himself on his crutch, took one quick puff at his pipe. "Like to see," he said gruffly, "the man that'd ask me that when I'm good and well." He raised his hand in a simple salute.

So they grumble, these veterans! When one turns from these soldiers, these men like discarded battle refuse, to the new, untried recruits, it's hard not to see their future

By chance, at nightfall, we came to a town that was the center of a vast training cantonment. Because of the restrictions on automobile lights, it was necessary to stop where darkness catches you. We watched the dusk descend over the green and rolling hills. From the distant hamlets and from the nearer cottages—picturesque, with low thatched roofs—no lights gleamed. The twilight acquired a primeval quality. It encased one as in an armor, worn against an eager and treacherous enemy.

Soldiers, too self-conscious perhaps because of this primitive projection from their surroundings, made sentimental gestures to smirking or bashful girls along the roadside. Sometimes the girls laughed, but not frequently. The change in the countryside grasped one tighter than ever because of this pursuit of romance, almost reluctant and a little appalled, through the turmoil of a dreadful reality.

We hurried on. Utter darkness caught us in the main street. We crept the few remaining yards to the hotel we had chosen. In the dining room, brown and black with khaki and the usual soberly dressed women, the proprietor greeted us. He was regretful; we could have dinner but no room.

As we ate, our feeling of intrusion increased. These women, it was clear, had left their homes to live in an uncomfortable hotel in order that they might be with their husbands, their sons, or their brothers until the order should be given, until this cantonment should break up, until these officers should leave for Flanders to face the chances of which the newspapers with thoughtless cruelty perpetually reminded them. From their bearing, you caught their appraisal of each day's value. There was little laughter. The murmuring voices created a monotone, full of misgivings, pitifully abashed.

It was a relief to go forth with a guide to seek a lodging. Just across a stone bridge we found it in a small, quaint hotel. This, too, was crowded with officers and their families. In the tiny bar you heard only military talk.

"How are your fellows doing with their patrol work?" one patron asked.

"Jolly well. It's almost a pity there aren't Huns about for them to fool," the other responded.

"What about Captain Smith, doctor?" queried another.

A laugh from the doctor, who replied: "Measles, of all things! Must have got it on leave. Fortunately, no one's been exposed."

During this time, you travel safely in England only with an identity book, furnished after investigation by the police of the district in which you live. It is required that you report yourself and have your book stamped by the local police in every town you visit in the forbidden districts. We set forth, therefore, for the police station. As soon as we had crossed the stone bridge we became hopelessly lost. I had never dreamed of such darkness. There was no moon and the sky was clouded, obscuring the stars. From no building escaped the faintest gleam of light. In the main street you could imagine yourself in a wilderness. The night was like a smothering blanket. It appeared to offer your outstretched hands a palpable resistance. People ran into you or you ran into others, laughing and making apologies.

"Where is the police station?" we asked one person.

"Heaven knows," was the response. "By daylight it ought to be a couple of blocks to the left, if you run into the church first."

"How will we know the church?" we asked.

"It is very large, and solid," a wag answered, "You'll recognize it if it stops you."

A constable, met in this obscure and abrupt fashion, kindly took us in tow. Whispering sympathetically, he stamped our books. "Now," he asked, "do you think you can find your way back? It's a long time, you know, before daybreak." He gave us detailed directions, which we followed almost wholly by the sense of touch.

It was difficult to go to sleep that night. Until very late, I listened to the perpetual shuffling of feet along the sidewalks—the tentative feet of countless young men, condemned to war, groping a course through a complete and inimical darkness.

After that, London no longer seemed black. As we drove in, its few hooded lamps seemed brazenly to invite disaster. We brought back with us one conviction: rural England is not apathetic—all of Britain's heart and soul is in the war. Even then, it was hard to accept as real the brilliant, careless complacency of our own country. That became a memory from the remote past.

Not many days after, the lesson was strengthened by the sight of heroes marching through an admiring, worshipful multitude. I hadn't realized that the war had already led to memorial days. On that morning, the few remaining Australian and New Zealand troops who had made the heroic and tragic landing at Gallipoli were gathered in

London for what will always be called, in their honor, Anzac Day. It brought the war very close to step into the Strand and to see above the bobbing heads of a nearly silent crowd the brown campaign hats with the colored bands of the New Zealanders.

There were so many spectators—women, old men, young girls, and a multitude of youths in uniform—that it was difficult at first to get close to the marchers. At the curb finally, one no longer needed to probe that silence of the great crowd—singular and a little startling. The faces of the soldiers amid the bright animation were serious and full of remembrance. The brisk, round notes of the bugles and the tapping of drums were unlike such sounds as we remember from Fifth Avenue or in the armories of a land at peace. With a lithe rhythm, the thin brown line of soldiers came on. It was a survival. With it marched ghosts, an infinite army of shadows—once such men as these—and familiar and friendly to these eyes that glanced curiously at the human river of the Strand. Mere boys, here were veterans of such fighting as the world had seldom seen. It was disquieting to forecast. In a few weeks, how many more shadows would crowd the thinned ranks?

The Australians joined the New Zealanders. They marched to Westminster Abbey, where the Queen and King came to mourn with them, to share as far as possible in their somber pride. The crowd filled the sidewalks and the streets. Men and women bent from windows, clung to railings, or sought precarious footing on the wheels of wagons and stalled omnibuses. It was as great as the crowd at a football game. It was as soundless as those who gape at the funeral procession of some imposing personage.

A group of wounded stood on the roof of a low building near the Abbey entrance. The Queen and King paced from the Abbey. The King wore a service uniform similar to the uniforms of the wounded on the low roof. As he stepped into his carriage, a single hand protruded from the mangled group, and a single voice cried out, piercingly, hysterically, as if the King must be made to hear and understand and perform a miracle: "Think of my arm! Oh, think of my arm!"

The crowd was too dense to get fingers to ears. Nor would it have been any use, for a column, whose eloquence was voiceless, deliberately emerged from the Abbey. Nurses in melancholy gray wheeled incomplete men in invalid chairs, or blanket-covered stretchers, down the footpaths between the lawns. Some crawled painfully after on crutches, or bent over like very old men who can no longer measure their strength. The comparatively sound followed, filling Parliament Square in ranks that awaited the word to march. Policemen spoke roughly, forcing stragglers into the ranks. This picture of a constable, a guardian of peace, handling a soldier, an instrument of death, created an incongruity that illustrated the whole illogical effrontery of war.

Again the bugles blared, again the brown ranks stepped quickly out—two thousand men, nearly all of whom had been wounded or had suffered from the fevers of camp life. Again the procession of handsome, purposeful young faces moved swiftly by, with groping expressions, as if missing someone. The incomplete wrecks on the stretchers and in the chairs made futile movements, attempted a fragile cheer. The sun continued its brilliancy, untroubled by the smallest cloud. It was like the phantasmagoria of nightmare beneath a heaven crowded with white tempest. One wanted to fling up one's hands and burst into tears for the dead—and for those not yet dead.

Chapter Three

BATTLE, ZEPPELINS, AND DEMOCRACY

As far from the front as London, it became obvious that the nearer one approaches this war the nearer one visualizes a vivid growth of democracy. A number of incidents, at the time apparently trivial, assume in retrospect a very real importance.

I had been interested in the women's participation. I had visited hospitals and watched the nurses at their merciful work. I had seen them, with an amusing diffidence, accomplish men's tasks on trams and busses, even at the wheels of taxicabs. I knew of their labor in the munition factories. I was not prepared, however, for the surprise an English friend—a man of wealth and influence—offered me when I visited him at his home in Surrey. I was curious when his daughter didn't appear for luncheon. "I'll show you this afternoon what she's at," he promised, mysteriously. He wouldn't say anything else.

We set off in his automobile, stopping at a soldiers' convalescent home to pick up two wounded men, for Englishmen don't like to use automobiles without sharing their luxury with the sufferers. We halted at a neat farm in a hollow. A horsey-looking individual appeared.

"Where's the young lady?" my host asked him.

The man took his pipe from his mouth and pointed it in the direction of a distant rise. "In summit field with Jerry and Jinny," he said.

"How's she coming on?" my host asked.

The horsey-looking man puffed thoughtfully at his pipe. "Better," he said grudgingly, "than I calculated. She knows something about horses."

We went on to the top of the rise. The soldiers, because of their wounds, couldn't leave the automobile, but my host led me down a lane to a broad field. A solitary, dusty figure crossed the field with long strides, calling cheerily to the raw-boned horses she drove, clinging with real skill to the handle of a plough.

"My daughter," my wealthy host announced, with real pride ringing in his voice.

She was a very pretty girl—all the handsomer, one felt, for having a thorough coat of tan. Nor could her corduroy skirt or her khaki blouse diminish the grace of her figure. It was easy to understand her father's pride. She talked pleasantly with us about her work. There was no attempt to make light of it. She didn't define it in terms of sacrifice.

"I'm sorry I missed you this morning," she said in her casual, educated voice, absurdly at variance with her occupation. "But, you see, I must be at work by six, so I leave home at five. I carry my luncheon in a basket and it's jolly good at noon, even in solitude."

"When do you get home?" I gasped.

"Usually in time for a late dinner," she told us. "You know I must cover this field tonight, or I'll have no dinner at all."

We watched her as she called to her horses and strode gracefully away.

"That's her life," her father reflected, "and, on the whole, I fancy it's better for her than teas and dances and the things girls used to do. She loves horses, so she's capable."

"But why . . ." I began.

"Don't you understand?" he said. "She releases one man more to go to war."

An airplane whirred across the downs towards France. The

wounded soldiers welcomed us back to the automobile and I gazed at their bandages and crutches. Certainly the plough girl was democratic. Yet, you couldn't help thinking it was a pity her devotion should have no more beautiful object than the release of a man to become, let us say, like one of these maimed fellows who somehow managed to enliven their invalid pallor with smiles for us.

At every turning, the sign posts of social change meets you. I remember a middle-aged woman in black who rode ahead of me one afternoon on the top of a bus. A newsboy in Hay- market burst the bounds of propriety with a strident yell. We all had a partial glimpse of the poster in his hand, announcing the sinking of a British ship. The woman who in peace times, you felt, would have been in an automobile, turned to me with a cry of fright: "Did you see the name of the ship?" she asked. "I couldn't."

I had noticed one of the posters just before mounting the bus. "It's the [...,]" I answered. "She was sunk in the Mediterranean."

The color rushed back to her face and the sharp anxiety faded from her eyes. "Thanks," she said, and turned away. After a moment she looked back, for it was evident she felt the need of an explanation. "You see," she said, "I have lost my brother in the navy and my son is with the high seas fleet. One goes about expecting news like that all the time. I ought not to be glad it was in the Mediterranean, for there are many women whose fear will grow when they hear that word. Thank you. You understand?"

"I understand," I replied.

I descended the bus, thinking, a little more than two years ago this woman would not have spoken to a stranger, no matter what her sudden doubt. So the perpetual strain, its general distribution, draws people to each other for comfort, because so many of them can say: "I understand."

Anyone will tell you that the Zeppelin raids have encouraged such a community of feeling. They have destroyed, in every portion of the population of the London and East Coast districts, the comfortable aloofness from actual warfare to which English civilians have for centuries been accustomed. The fact that noncombatants have frequently been the only victims has intensified this impression of a common outrage and a common sympathy.

The Germans, it is fair to assume, haven't bothered about who might be hurt. As a matter of fact, in proportion to the energy and ammunition expended, together with the loss of Zeppelins and their crews, the results have been nearly negligible. It is, all the more on that account, ironic that the innocent should have been the chief sufferers.

"If they go after a factory," an officer said to me, "they get a workingman's house a mile or so away. If they go for a barracks they get a farm. It's small comfort for the old men and the women and children done in that no real damage has been accomplished."

No one seems to know what the Zeppelins were after the night they dropped bombs on one of the great Inns of Court. A house of some peaceful barristers here, the shattering of some ancient carvings and glass in a chapter there, and about the lawns a few gaping holes—that was the extent of the damage. Zeppelin raids have all the casual inconsistency of a tempest.

When I was in the city, I saw that London had learned (about as thoroughly as Paris) to take care of this menace. With the decline of the moon a little nervousness was apparent, but for the most part people faced the prospect calmly. During one week, after the departure of the moon had made the heavens safer for aircraft, we had three of these visits in a row. At tea on the afternoon of the first raid, I heard a retired admiral and a famous editor discussing it as one talks of an approaching horse race or a ball game.

"Everything is quite perfect for them. The wind of the last fortnight has died away," the admiral said, rubbing his hands. "Now if you want to lay a wager..."

At the theater that night, although the audience shared this sense of anticipation, the play progressed cheerfully. When we came out after the final curtain, we saw that the heavens were torn by the groping fingers of countless searchlights. From the wide spaces of Trafalgar Square, we could watch shrapnel bursting close to the shafts of light, and we pointed out to each other what we imagined to be the minute shape of a Zeppelin flying high.

"Maybe the bloke fawncies 'es over Lunnon," a constable said. "If so, Gawd 'elp 'im when 'e tries to fly back."

"Aw, they dawn't get over the 'eart of Lunnon these days," said a cab driver, lounging nearby in the hope of a fare. "Show ayn't worth the price of stying out. 'Ome for you, gentlemen?"

Later, in a room overlooking the Embankment, a party of us watched in darkness. The fingers of light still groped, but there was no more shrapnel. A pretty young girl grasped her father's arm and cried out, her voice vibrating with disappointment, "Daddy! You promised I should see a Zep tonight."

"Never mind, my dear," the father said indulgently. "Bobby, suppose you call up . . . at the War Office and ask where the rascals have gone."

After a time, Bobby returned from the telephone. Apologetic, he said: "[War Office] says they're headed for the home fires."

Our host drew the curtain and snapped on the lights. We blinked. The pretty girl pouted. She seemed to think her father had somehow failed her. "A game of bridge," he suggested, "or is it too late?" One was rather relieved that the German Admiralty couldn't see London intimately that night. Its chagrin would have been too painful.

Sometime later I chanced to see a quotation from a Munich paper that recalled that very date and threatened London with similar "nights of terror." During that same week, I lunched with an officer of one of the guard regiments, who said: "Of course you know the Zeps were fussing about again last night." I told him I had seen the lights and had shared a little of the excitement.

"I was in the barracks at . . . ," he chatted. "I don't know how many of the things there were. One of them sailed directly over the barracks square. We were crowded in looking up. What a place for a bomb! This fellow dropped a number in some empty fields as usual. You could see their fuses twisting down. Then he showed a red light on his tail—some kind of a signal, I fancy—and swung towards the channel. I think our air guns were spoiling his evening. At least the shrapnel was bursting all about. Last we saw of him. He must have felt an awful fool, but they ought to be getting accustomed to that."

Before the moon had come again, one had nearly forgotten (like most Londoners) to be apprehensive of the great dirigibles. In such indifference lies the tragedy of Count Zeppelin.

———

IF, HOWEVER, SUCH CONSIDERATIONS AS ZEPPELINS AND ANXIETY FOR relatives at the front have accented the virtues of democracy, its faults have also fattened through the war. The French and the English appreciated that during the first few months. It challenged me during my brief trip to Ireland during the Sinn Féin rebellion. I have no intention of taking up the military or political phases of that affair; they have been sufficiently dissected and fought over.

Indeed, my chief recollection is of confusion. It began at Euston where they had no idea whether the boat would leave Holyhead or not. Haggard women wept and men ran up and down with an anxiety for which the officials had no antidote. A young fellow in the uniform of the Naval Flying Corps came along and held out an envelope and a bundle of newspapers.

"If you get through, please try to mail these in Ireland," he said to me. "I can't go and my family's in Dublin. I've heard nothing."

"If the Zeps come tonight," a bystander offered, "nobody'll get through. The train won't budge from London."

But the Zeppelins didn't come and we left, and in the train the confusion persisted. An army officer shared a compartment with another correspondent and myself. We turned out the lights, rolled ourselves in our overcoats, and tried to sleep. But we couldn't sleep. There was too much noise in the corridors and a monotonous undertone issued from the other compartments where people, full of misgivings for relatives and friends, discussed the future that they approached with uncertain steps. There were black clouds in the sky through which the moon was like a dying lamp.

"I'm used to roughing it," the officer said. "I've slept often enough less comfortably at the front. It isn't that that worries me. I've been transferred to a regiment stationed at some distance from Dublin. If they tell me at Holyhead the trains aren't running on the other side, I'll have to go back to London."

At Holyhead, the confusion sent him back to London, because nobody seemed to know anything with certainty. The boat, however, lay against the dock with steam up. During the minute examination of our papers, rain added to the shivering discomfort of those black hours before the dawn. Reluctantly, we were permitted to embark, and we tried to settle ourselves in the midst of a confusion that increased. There was wild but serious talk of a fleet of new and gigantic German submarines, which were supposed to be somewhere in the Irish Sea preparing to cooperate with the rebels.

"They're sure to give us a chase," a man whispered.

Many agreed with him. You couldn't help admiring these people who went forward in face of such a belief.

Ireland loomed out of a haze touched by the first gray light. The haze seemed a veil for sinister things. The passengers arose and stretched themselves, as if emerging from the shadow of one disaster to gather strength to elude another. And at the dock the confusion, for us at any rate, culminated. Here it had the whimsical, lovable quality of the country. An officer stopped us at the gangplank and inquired: "Where are you going?"

"Ashore for breakfast, for a lodging, to look around," I replied.

"You can't land without a pass from the provost marshal in Kingstown," the officer demanded.

"You mean," I asked, "that we will have to go, back on this boat?"

"Oh, no," he answered seriously, "because you can't leave on this boat without a pass from the provost marshal in Kingstown."

———

BY STRATEGY AND FAIR WORDS, WE GOT ASHORE AND TO THE PROVOST marshal. Of the confusion there, as I have suggested, enough has already been written. When I left on a clear, ruddy evening it occurred to me that rather too much undemocratic order was emerging from the hurly- burly, for I had to run a gauntlet: of Scotland Yard men, of the British Army, of the Irish Constabulary, of Mr. Redmond's Nationalist Volunteers. On the boat, however, the old state was in evidence. We were crowded by the first refugees from Dublin—men and women with overwrought nerves who knew of that story of the gigantic German submarines. Moreover, the barricades on the waterfront at Kingstown had seemed to give this rumor rather too much body.

We crept out of the harbor double-shrouded: canvas light shields were stretched along the sides and the portholes were closely shuttered. Only one entrance, far forward and completely dark, was left open to the lower deck. There was a dim light in the smoke room, and we counted the minutes there, while the refugees, a trifle hysterical, exchanged experiences. Suddenly, what everybody had feared seemed to spring upon us. The lights snapped out. Through a blackness nearly palpable, someone cried out.

"Submarines!" they said.

The thought of panic in this shrouded boat was more oppressive than the sudden night.

"Sit still!" said another.

Then a man spoke wistfully and saved the situation: "What are you afraid of? It couldn't be any worse than a happy Easter in Dublin."

Some of us laughed and gradually the ominous stirring subsided. Everyone sat still until by and by the lights came on and we looked at each other and smiled. A man ran in, crying breathlessly: "Holyhead light! I say, I can see Holyhead light!" A great sigh went up. We all crowded to the front deck to watch that red and friendly greeting.

Chapter Four

PARIS AND ITS WAR SPIRIT

Each trip through the submarine zone, in fact, has its own thrill until you grow, to a measure, hardened. When I was ready to leave for France, the channel crossing seemed for a number of reasons less pleasant than usual. Only one line was in operation, and that was taking the long route from Southampton to Le Havre. That the Sussex tragedy had had something to do with the choice was obvious. People spoke of the approaching excursion with misgivings. It is likely that the antidote for most of them was the extended formalities they had to accomplish before they were permitted to risk their lives at all. The police, the American consul, the French consul, local detectives, Scotland Yard agents, and French secret service men— those were some of the obstacles to dishonest travel between the continent and England.

I was amused when I drove with my baggage to the pier entrance in Southampton. I had been conducted that afternoon by the courtesy of the Admiralty through one of the great dockyards. Therefore, I didn't come down from London on the special train with the rest of the passengers. At a stated hour the gate was thrown open and I was permitted to drive in after an examination of passports. I found an elderly porter in front of the ticket office and asked him to take my luggage from the cab. Following extended and silent consideration he agreed, apparently against his better judgment. While he worked he shook his head continually. When I turned to enter the ticket office, he grasped my arm. His gesture and his face expressed a desire to spare me an indiscretion. This time he spoke.

"Where you off to, sir?" he inquired.

"To buy my ticket for France," I told him. I am convinced he was a Wesleyan. I have never seen a longer face.

"Better not do that, sir," he said mournfully, "until you find out whether you're going."

I laughed and walked on. He called after me with the effect of pursuing an erring soul. With each word his voice grew shriller: "Very often they don't go, sir. I tell you, they don't go. They stops 'em at the dock."

I was tired, so when I was aboard the boat I entered my bunk; but sleep was nearly out of the question, because of that justifiable care and severity of which the old porter had warned me. Men and women struggled through until just before dawn: at times they complained loudly or congratulated themselves in equally unrestrained voices, but the idea of sleeping occurred to few. The man who shared my cabin went to bed with his shoes on. Perhaps he was wise.

There wasn't much talk after breakfast. The passengers sat or walked about, anxiously scanning the water. The coast of France emerged from the haze. We passed the skeleton masts of several ships, sunk by submarines. We made the harbor entrance and spirits revived. Such chatter as last night's disturbed the boat again. People wondered if there would be a new ordeal at the dock. And there was, for France is as careful and suspicious as England.

It was one of those hot, brilliant days Normandy receives on occasion. The harbor, untroubled by the slightest breeze, was like a mirror for the violent sun. We were herded into a single-story shed on the waterfront. A tall, military policeman with a bristling moustache guarded the gate to the examination room. Beyond him we had glimpses of a long deal table, around which sat numerous inquisitors, in uniform and out, French and British.

Because of the crowd in the little room, it was impossible to put down one's coat and hand bags. Their weight increased by the minute, as the unclouded sun baked the flimsy roof overhead. Many of us commenced to look as if we were more in need of a physician's certificate than one of entry. Then, at a grumbled word from the inquisitors, the proceedings opened and with a commendable partiality a huge military policeman roared: "Ladies first! Step forward, and don't push abawt so. Now, lady, you got your passport ready?"

There were more women than one would have thought. Because of the increasing heat and the weight of baggage, the situation had approached the intolerable when the military policeman cried out: "No more ladies?"

Suspense! A sigh of relief as the silence persisted! Those of us who were not at the front of the line began to compute the duration of our ordeal. A groan disturbed our ranks, for the military policeman was evidently following an extended order of precedence.

"All with diplomatic passports," came his leonine voice, "kindly step forward."

And, after a number of important-appearing men had been passed through, "Are there any more with diplomatic passports?"

The case was desperate. I called over the heads of the others, "Sergeant! I have a journalistic passport."

"What?" he thundered back.

"A journalistic passport," I repeated, less hopefully. It meant nothing and I knew it.

"Let that gentleman through!" he roared.

As I struggled forward, I felt his intention was to discipline my presumption with some sharp words and a command to take the rear of the line. His frown was ominous, his bristling mustache unsympathetic.

"Let's see your passport," he growled. "What do you mean? I asked for diplomatic passports."

I handed him the much-viséd document. He glanced it over and then a more dangerous belligerency colored his tone. "You got an office in London?" he inquired.

"No," I answered meekly. "I have a sort of an office in New York."

The threat faded from his appearance and his voice. He smiled with a childish and excited interest. "New York!" he echoed, swinging the gate open. "Step right in, sir. Make yourself comfortable." As I obeyed, he asked, "Why didn't you say that in the first place? I'm from New York not two years ago. Expect to go back after the war, if I don't get killed. I used to run an elevator in the Waldorf. What's the news from Broadway? Give my regards to Times Square."

He was too friendly. I was among the last of the sufferers to be released by him into the hands of the judges.

As at Liverpool, the narrow mesh of these spy nets was made apparent. As a farther check, I am guessing, we were made to spend nearly seven hours in Le Havre waiting for the departure of the special train for Paris. I wasn't sorry, for Le Havre in itself held plenty of interest. It is the working capital of invaded Belgium; it is one of the great English bases. Consequently, the uniforms of French, Belgians, and British were everywhere in evidence, but the British, naturally, predominated.

From the waterfront, I watched transports enter and leave the harbor. On the docks the work of unloading proceeded with a precise efficiency. In the streets wagons and automobile trucks to which good-natured Tommies clung hurried tempestuously. Officers strolled here and there, swinging little canes. Their faces were rather more serious than the faces scanned in London. All at once you realized that you were actually on the soil of war-torn France, within a few miles of the grotesque and deadly battle of the trenches.

In the train the shadow of the war deepened again. As we steamed inland across a landscape that, for me, had always had an air of sedate pleasuring, we caught glimpses of tents and the intricate movements of men at battle drill. Elderly French Territorials in faded blue and red uniforms lined the railroad tracks and guarded the bridges. As our cars flashed past, they presented arms or stood at attention. We threaded through great supply trains on temporary tracks in the vicinity of Rouen. The heat was unreal in such a country. It seemed that it must be emanating from the great, nearby forge of battle. Then darkness closed over the steaming world, as if to hide from our eager eyes the elaborate machinery of war. At St. Lazare, we passed the last examination and scattered to our hotels.

Curiously, arriving at night as I did, my first impression was that Paris was more nearly normal than London. Almost at once I realized that this was due to the contrast between the few but unveiled street lamps and the unblanketed glow from buildings, compared to the darkened thoroughfares and the curtained windows of England. In addition, there was the difference in the Anglo-Saxon temperament —after all, largely our own—and the admirable Gallic intensity of temporary appreciation that even this war has been powerless to destroy.

The terrace cafés were crowded and the many soldiers, wearing their graceful steel helmets, seemed undisturbed by what they had already survived and unappalled by that which awaited them at the close of their brief permissions [leaves].

By daylight the truer values became noticeable. Nearly every woman was dressed in mourning and their white faces haunted one —because out of the eyes, in which there were no tears, stared a fierce pride that burned up grief. I talked with one of these women at a simple tea. Her history had been rapidly sketched for me. She was the widow of a colonel who had been wounded in an early battle and killed almost immediately after his return to duty. Before the war this woman had lived in a charming apartment near the Avenue du Bois de Boulogne, the most expensive quarter of Paris. Like many army officers, her husband had spent all of his income. Now with her child, a nine-year-old boy, she lived in a single garret room, sewing. By these odd jobs she strove to maintain the shadow of a home.

From the deep frame of mourning, her sorrowful face glowed with that pride that has made all Frenchwomen, to an extent, resemble each other. As if there were no other subject worth talking about, she spoke almost at once of her husband and the manner of his death.

"I was so happy when he came back with his wound for that little time, and when he went I thought the good Lord would let him return again. When they killed him he wasn't painfully hurt but, you see, the great artery in his thigh was cut. He understood, of course, but his men were in a bad place, so he had them prop him up, and he directed the defense and sent a message to me while he bled to death, knowing all the time, until the light faded . . ."

She shook her head, explaining, "He shouldn't have gone that way. If Doctor Carrel had only been there! He's saved such cases. He need not have died."

And always, one asks, "Why don't the tears come into this woman's eyes?" One prayed that they would and that the stiff, stern figure would relax a little. The gesture with which she raised her tea cup was angular, somnambulistic. Her son stared at her with a round, pallid, and expressionless face.

"You may have another cake, little one," the widow said.

He munched it without words until someone asked: "And what are you going to do when you grow up, young man?"

"I am going to be a soldier, like papa," said the boy, his voice as expressionless as his face.

The widow made a swift movement, and said, "You see? And I have had nothing to do with it—nothing at all. It is in the blood of the orphans. Must we lose them, too? Why do you want to be a soldier, son?"

"I want to kill the Germans, because they killed my poor papa," he replied. And his face twitched into an expression at last. As he continued to sip his tea, great tears rolled down his cheeks and fell into the cup. But the widow didn't cry.

In the great munition factories most of the women wore mourning, too, and the eyes of many were disturbingly like the eyes of the widow. It was not easy at first to watch their slender, dark-clothed figures, their soft and pretty faces, bent over tasks of preparing death and mutilation for men. You wanted to turn straight away from the contemplation of their deft fingers pouring shrapnel bullets into completed casings, or from the easy skill with which they molded and polished ammunition. Then that look and the dryness of their eyes stripped from their labor something of its dreadful incongruity, gave to it a tinge of justifiable revenge. And it was impressed upon the observer more than ever that in the fragile hands of the women lies the power that someday may obliterate war.

It is this grim, matter-of-fact determination of both sexes, of all classes of the French, that arrests one. It is, in a sense, hypnotic. Even from the little boys playing at soldiering in the street it projects itself. For me, it found its culmination in a review I watched one afternoon in the Place des Invalides.

Infantry, cavalry, and several batteries of those canons, the famous 75mm cannon known as the *soixante-quinzes* filled the place, where many times Napoleon reviewed his brilliant corps, with sober color. Eyes wandered from the quiet, helmeted ranks to the dome of Les Invalides, beneath which the great emperor lay. His tomb seemed to brood over the review, and in neighboring faces you read a perception, nearly superstitious, of the soul of the inspired leader who had brought so much glory to France. Then the band burst into the Marseillaise. As the ranks swung over the bridge, the crowd cheered. I have never heard such cheering. It wasn't a matter of volume; it was a curious choked quality that arrested one. It was as if these people tried to give vent to an emotion beyond physical expression and were angry at their failure. Yet, for them, the music seemed to express everything.

Chapter Five

LORRAINE AND THE DEVASTATION

One learns to shrink from the great railway terminals in wartime. On several occasions, I left Paris by rail to visit the front and each time the excitement of the prospect died at the ticket window. I think it is because these stations have witnessed too many departures for battle, too much of the tearing of warm life from warm life and the definitive rupture of romance, too many broken returns, too many shocked greetings.

My first introduction came not long after dawn of a gray morning. The foreign office had asked if I would like to visit Lorraine, suggesting that I take the day train for Nancy, where a staff officer and an automobile would meet me. An elderly English Quaker, who was interested in Red Cross work and the rebuilding of devastated villages, joined me and together we drove through the scarcely awakened streets to the Gare de l'Est. We entered to present our papers and accomplish the formalities that are necessary before one may take a ticket.

With a pronounced reluctance the dun light penetrated the great hall, which had an air of mourning. Soldiers crowded the wide spaces, shivering. Their uniforms were soiled; some retained the white marks of the trenches. The young faces were drawn, unhappy, wondering. For the most part these fellows were *permissionnaires* returning to the trenches after eight days of home and love and hero worship. They had swung their backs on all that knowing, if they were not hit, it would be many months, perhaps a year, before they could experience such blessings again. They were like a band of men of whom a certain number has been chosen for some violent discipline and who are left in doubt as to the actual selections.

The place was saturated with melancholy. Instinctively, we left it. Across the plaza we saw a café whose name was in harmony with the spirit of the station: Café du Depart.

"A cup of coffee?" the elderly Quaker suggested, for neither of us had had any breakfast.

We sat on the terrace among the soldiers watching regretful faces above faded uniforms. Accoutrements littered the pavement between the tables. One or two men spoke to us formally, and we answered formally. Beyond that there was no companionable morning chatter. We all stared at the gray façade of the station. The huge clock mocked us, pacing the minutes too quickly. In the eyes of the soldiers their doubt smoldered. Would they enter through that portal once more? Would they look again upon the familiar and the desirable?

From the summit of the façade the stone figure of a woman gazed back. There would have been no mistaking her even if she hadn't been labeled: it was the figure of Strasbourg. She appeared to be summoning the staring and melancholy soldiers through that portal and on to the East for a violent and necessary redemption.

Our compartment was filled with officers. Even my Quaker companion wore a uniform of the Red Cross. On that long train, I was the only one in civilian clothing. We glided quickly into the district entered by the Germans just before the Battle of the Marne. About the bridgeheads many buildings lay in ruins. We passed the once charming little town of Sermaize-les-Bains, where scarcely a wall showed more than two feet high.

An officer spoke, explaining, "They say it was because the mayor of Sermaize failed to come out and greet the commander of the entering forces. That offended the commander. Wherefore . . . " he said, making a circling gesture with his hand in the direction of the accusing rubbish.

All morning and during a portion of the afternoon, we were carried through the war zone, pausing at towns whose names have become immortal. And in the fields between, we saw many graves marked with crosses, here and there supporting a faded cap. Around the graves the fields were cultivated, yet no mound had been disturbed. The French have come to look upon the random tombs of the men who fell saving Paris as national monuments. They impress one as the most imposing memorials a nation has ever constructed.

During this trip, I received one or two examples of the social justice of compulsory military service as it is practiced in France. My Quaker companion and I were gossiping about Japan during luncheon in the wagon restaurant. Next to the Quaker sat a pleasant, middle-aged man wearing the uniform (made of a sort of overall material) of the transport corps. Suddenly, he turned and spoke in excellent English.

"You are interested in Japan?" he inquired.

We then embarked on a random conversation. Quite naturally it became evident before we were through that the man in overalls owned coal mines in Japan, South America, and Belgium.

"Of course," he said smiling, "the Belgian mines must be looked upon for the present as a bad investment." This man in overalls, who drove soldiers and supplies to the front, was a man of exceptional wealth! "I'm going back after my first permission in more than a year," he added.

"You despise such work?" we asked.

He shrugged his shoulders and explained, "One does one's share, and that is arranged according to the best interests of France. My task has its compensations. For instance, at the commencement of the Battle of Verdun, when things looked rather dark, I helped in that marvel of transport. You must have heard. We moved fifty thousand men in motor trucks from Revigny to the Verdun sector in a twinkling."

A little later we passed Revigny and our companion waved his hand, buttoned his coat of overall material, and left us.

Across the aisle a colonel shared a bottle of wine with a private; they chatted amiably. Yet to be seen was the discipline the French developed, which produced results that more than matched the iron-bound German system.

We halted at Bar-le-Duc, the base for Verdun, where the third great attempt of the Germans to break through was in progress. The booming of the guns came to us across the rolling hills. There was scarcely an entire pane of glass left in the railroad station. The squat, barrack-like, temporary hospitals filled with the martyrs who had entered the turmoil to return shattered along the Voie Sacrée or Sacred Way sent forth an air of suffering and misgiving. For the Germans at that time were in the habit of raiding Bar-le-Duc with their air squadrons. The day after my last visit, indeed, they dropped a shower of bombs, killing and maiming more than thirty civilians.

After Bar-le-Duc we left the main line, taking a tangent to the south to avoid the salient at Saint-Mihiel. At the first station west of Nancy, the *contrôleur* told us we must alight. "The train," he explained, "does not stop in Nancy itself, because of the *bosche* bombardment."

We were greeted on the platform by a stout, hospitable man in the uniform of the *état major*. He drove us into Nancy whose chief beauties, in spite of the bombardment, remained intact. There were ruins, however, in the vicinity of the station, which the Germans had been unable to hit directly. An apartment house in the middle of a block had recently been struck, and all that survived was a heap of rubbish in a yawning hole. More pitiful, more anger-producing, was the rubble and charred beams that marked the site of a children's school. If it was the purpose of the Germans to make the innocent suffer in Nancy, they have achieved an admirable success. We noticed, in particular, the wreck of a house. "That," our captain explained, "was struck by a great shell and afterwards bombed by an airplane."

Strangely, when I was in Champagne sometime later, I met an

officer who after learning I had been in Nancy asked me if I remembered this particular ruin. "It was my home," he said simply. "Fortunately, my family was not there when the shell struck."

Close to this circle of devastation lay the hotel, so far practically untouched, in which we were to spend the night. "Perhaps," our officer grinned at me, "a shell will fall through *monsieur*'s bedroom and furnish America with a *casus belli*." I patiently explained to him that I entered the war zone at my own risk, but his wit intrigued him, and each time he repeated his joke we tried to laugh.

Affairs in Nancy, there was no doubt, progressed much as it did during times of peace. "Why not?" said the inhabitants I talked to. "We go along. We merely hope that the next shell won't fall near us."

On the walls of many houses we saw the cross of Lorraine, painted in red. "Why?" we asked.

"Because," the captain replied, "each one of those marked houses has a cellar. When the bombardment commences, people caught in the street enter the nearest house marked with a cross, and the inhabitants must receive them and give them shelter."

The elderly Quaker shook his head and asked, "Why should Nancy be bombarded in this fashion?"

The captain shrugged his shoulders, answering, "It might be a little pique. You see, just before the Battle of the Marne the Kaiser and the Crown Prince were decked out in all their plumage and waited, mounted on horseback, to make a triumphant entry into the capital of Lorraine. At the last minute they had to change their plans. That was very sad—for them. I think they have never quite forgiven us. Tomorrow, in the devastated districts, I will show you worse things. Wait until you have seen Gerbéviller." As he said this, his eyes showed a disturbing promise.

In our hotel, surrounded by shattered buildings, we dined comfortably that night. Other officers came to our table from time to time with the gossip of the sector. One of them, a charming young fellow, a captain in the machine-gun service, was particularly pleased to find an American, because he had heard a good story that day about one of my countrymen in the Foreign Legion. Over coffee, he told me the story with much joy.

"You know," he said, "that the soldiers have been in the habit of making finger rings out of the aluminum they gather from shells of the *bosches*. They send them to Paris where they are sold and lots, I daresay, have found their way to America."

I told him that as far back as a year ago I had seen such rings in New York.

"Then you will understand," he went on, "how eager the soldiers are to get this material, which in good condition isn't very plentiful. They are quite jealous about it. The other night, it seems, this American in the Foreign Legion was on solitary duty in a listening post between the lines. Those places are never very comfortable, as you may learn for yourself someday. The *bosches* try to locate them with their artillery and, when they do, they simply blow them to pieces. That night they got within range of this post and turned their guns loose. Your poor countryman thought the end of the world had come. His escape was cut off—the short trench, or sap, leading out was obliterated by great shells. There was nothing for him to do except to stay and take his chances, and they were pretty slender. At the end of an hour, nothing whatever was left of the post except a heap of formless earth; yet, through one of the miracles of war, the sentry remained untouched.

"As soon as the fire halted, the poor devil crawled back to the front-line trench and climbed the parapet. He expected to be greeted as a hero, as the savior of France. He pictured a deputation welcoming him at the parapet with the Croix de Guerre, with the Military Medal, with the Legion of Honour. There *was* a deputation at the parapet—of *poilus* [front-line soldiers], crowding around him with anxious and envious faces. They greeted him in an excited fashion. 'You lucky devil!' they cried. 'For the love of Heaven, let us see! How much aluminum did you get from those *bosche* shells?'"

The machine-gun officer, in spite of his appreciation of the incident as humorous, expressed a visible pride in its climax. He sipped his coffee and said, "That legionnaire will be a better soldier for his adventure."

"How," the Quaker asked him thoughtfully, "can anyone hope to defeat soldiers who take death and war with that *blagueur* attitude?"

With his quiet reply, the machine-gun officer offered unconditional assurance: "We do not believe such men can be defeated."

We thought of the guns of Verdun, which we had heard that afternoon roaring from the German lines, expressing their desperation and their anger. For some time we continued to chat, talking of nothing but war, for that is all there is to talk about in Europe these days. A general officer strolled in, nodding pleasantly to the diners.

"We must make an early start tomorrow," our staff officer said, "so shouldn't we go to bed?" He then showed us to our rooms and made sure that we were quite comfortable. He brought a map, explaining the trip he had arranged for the next day.

"Of course," he said, "they might send a shell in here tonight, or an air raid isn't an impossibility."

I hoped he was being humorous again—and yet his eyes were uncomfortably serious.

"If that doesn't happen," he said, "you will see some things that will surprise you." And again his face altered with that disturbing promise. "Among other things," he added softly as he turned to go, "you will visit the ruins of Gerbéviller—of Gerbéviller-La Martyre."

Chapter Six

THE SINISTER INVASION

We started early the next morning, threading a course among the pleasant hills of Lorraine. For brief periods, the idea of war seemed a distasteful imagining. For a reminder, it was necessary to glance at the helmet of our military chauffeur, or we would glimpse a battery of *soixante-quinzes* in a patch of woods. It was a Sunday, and often the artillerymen would be washing their clothing in a swiftly running brook, or were stretched out in thick grass and lost in a book or the re-reading of a letter from home. We might pass a column of infantry, covered with dust, crowding to the side of the road to make way for our *état major* automobile. And here and there, we met lines of the busses that had disappeared from the Paris streets at the commencement of the war. Covered with netting and painted a dull gray, the buses carried fresh meat for distribution from point to point behind the lines.

We swerved into Lunéville, whose outskirts saw vicious house-to-house fighting during the first weeks of the war. In a number of streets, the buildings were scarred so intricately from rifle- and machine-gun fire that it seemed incredible a single soldier could have emerged untouched. Our driver then hurried us into the countryside again.

The staff officer, with that look of disturbing promise visible in his eyes, spoke sadly: "We are entering the devastated district of Lorraine." Almost immediately, we flashed through a village whose simple peasant houses were without roofs or showed jagged breaches where shells had entered. "We got as much of the civilian population out of these towns as we could," the officer said, "but it is hard to move Frenchmen who think they have a right to stay, so plenty of them suffered."

As we went on, the villages displayed harsher scars. In some, only a few walls were left, but we could see rough shelters constructed from the wreckage, and old men and children wandered around with a furtive air, as if in anticipation of another catastrophe. In the midst of all this destruction we came to a village that was quite untouched.

"Why is that?" I asked.

The staff officer shrugged his shoulders, responding, "Who can explain the vagaries of the *bosches*?"

We all wondered if the charming hamlet had been spared because someone lived there who had been of service to the enemy.

"Spies . . ." the Quaker began.

What I learned about the vital work of the spies in Europe I shall relate in another chapter. The subject was forgotten at that moment, for we left the village and crossed a broad, flat plateau, in which innumerable French tricolors waved lazily amidst the grass, like the fronds of a strange and beautiful plant. We saw that beneath the tricolors were mounds of varying size, containing graves ranging from that of a single man to a trench tomb of a hundred bodies. There were mounds on which no flags waved; these were decorated with plain black crosses.

"The German dead?" I asked.

The staff officer nodded, explaining, "As far as possible, we have taken care of their dead as carefully as our own. On that cross you will find a row of numbers. The families of those German soldiers can know where their men are resting." He pointed to a tiny mound with a small black cross set at an angle above it. "An officer," he said.

"There is a German name on that cross. Lieutenant or Captain von so-and-so."

"In view of those ruined villages," the elderly Quaker said, "such charity is admirable. It is very French."

"We are not *bosches*," the officer laughed. "In spite of the crimes of the Germans in Lorraine, we have no quarrel with the sorrowful families of the dead."

"Both sides fell very fast here?" I asked.

"It was a hard battle," he answered. "You might say the *bosches* were turned back as they were at the Ourcq. It is, as you can see, nearly at the other end of the line. Because these men fell, the Kaiser was forbidden to trot into Nancy, and something was repaid of the debt we owe the *bosches* for Gerbéviller. We are getting very close now, before long we will see it."

The name had acquired in my mind—and, I think, in the Quaker's—a symbolism of inexpressible wrong. We shrank a little from the fact. The automobile approached the edge of the plateau too quickly for us. There was, however, in our first glimpse of the dead city, an unexpected relief. It snuggled (barely visible because of the pleasant shrubbery) in the center of a shallow bowl, and the charming little Mortagne River wound through the fields and patches of woods and lingered behind the nearest of the half-visible walls.

Then we understood it was a lack of definition that furnished the pleasant deception. The wall against the trees, for instance, became the torn and eyeless front of a factory. Behind it there was nothing— our hearts sank, for of all the fragments of buildings we could see from that point, the factory was distinctly the largest.

It is the approach to Gerbéviller from the plateau that makes its tragedy insupportable. So far, very few have been permitted to inspect this record of the German invasion, this monument to the Teutonic campaign of terribleness. To those who have driven down, like us, from the plateau, there must have come this thought: "After all, the French have exaggerated. It might have been necessary to bombard the garrison defending the place. And the destruction isn't really as shocking as in Nancy."

Then, as the shrubbery falls away, exposing the skeleton, every visitor must have cried as we did: "But this is incredible! This isn't bombardment. It is systematic and wanton destruction."

"There was no garrison here," the officer said. "When the army retreated at the first shock, only sixty *chasseurs à pied* were left to guard the bridge at the other side of the town. Only a few shells have fallen on Gerbéviller. It is the work of the incendiary, of the man who destroys property as a child knocks down a house of blocks, because it pleases his unconsidered impulse to be cruel—to smash!—to laugh, as he sees things go Smash! Smash! Smash! Sister Julie, if she will, can tell you better than I, because she was here. She lived through each minute of the dreadful three days and, since she is a *religieuse*, what she says will not seem so far beyond belief as the story that I know only by hearsay. But first you should see the château and its chapel."

We entered Gerbéviller and, for a short distance, we were threading through streets flanked by walls scarcely two feet high, like those of Sermaize-les-Bains. They were eloquent with the story of their fall. They seemed to be trying to explain to us that after the conflagration dynamite had been used, that their skeletons had been torn to pieces by stained and vicious hands.

For a long time we saw no one. Then a child appeared, walking at a demure pace, her eyes downcast as she picked a path among the ruins. We paused in a weed-choked plaza. To the right a wall rose for two thirds of its original height and through its empty windows we saw the trees of a broad and luxuriant park. The rear wall and most of the side walls had been levelled. There was only enough left to tell us that here once stood one of the most beautiful Renaissance châteaux in France.

The officer then nodded towards the opposite side of the plaza: "The chapel," he said. We gazed with mounting anger at this jewel, which had been shattered with repeated and difficult blows. Through the breaches of the façade a desecrated altar was visible, roofed only by the sky.

"There are no shell holes," the Quaker said, with a flash of temper spreading across his placid face. "I am a Quaker, as you know," he went on simply, "but in this place I want to tell you that I have two sons who are Quakers also, but they are both officers in the British Army."

The staff officer smiled, and said, "Perhaps it is just as well that you, yourself, are beyond the military age."

"It spares my conscience," the Quaker agreed.

"What regiments did this?" I asked.

"Bavarians," the officer answered. "We had always thought, too, that they were rather kindlier than the Prussians. In the grounds of the château there is a grotto where they detached the ceiling mosaics piece by piece. That is what hurts so in Gerbéviller—the careful, the systematic, devastation. It is difficult to understand how men could go to such minute pains to destroy."

We re-entered the automobile and went on through the ghastly streets of Gerbéviller. Before long the car stopped; a heap of stones blocked our way. "I can go no farther, sir," the soldier-chauffeur stated. We alighted, made our way around the rubble, and continued on foot.

"It is worse than Pompeii," the Quaker mused. "That ancient city is more habitable, would be far simpler to restore."

Ahead was a wooden shack, constructed against a piece of ruined wall, prompting the staff officer to say, "The old and the new . . . but that is about all that has been done towards the restoration of the city. It is so hopeless. So few of the inhabitants have clung to their homes, but after the war something will be done for them."

"The Germans made a more thorough job here than in Louvain," the Quaker commented.

"Nothing could have been more thorough," the staff officer answered. "Where there were originally 475 dwellings, just twenty have emerged from the ruins comparatively intact, and that is due to Sister Julie. They are all clustered about the Hospice of St. Charles, of which she is the superior."

We quickened our pace, for we were anxious to meet and talk with this remarkable woman, who had saved what little is left of the city. We knew that General Castelnau, after the defeat and the flight of the Germans, had mentioned her in army orders. To decorate her with the Cross of the Legion of Honour, President Poincare had come himself to Lorraine and to the hospice. In Nancy the night before we had heard Sister Julie mentioned with a sort of reverence.

At the head of a narrow, sloping street we saw several comparatively complete buildings. We entered one through an archway surmounted by a cross. A sad-faced sister ushered us into a parlor with walls of freshly splashed plaster. We didn't need to be told that many bullets had torn through those walls.

Sister Julie entered. She impressed us as a short and stout woman, rather beyond middle age. From her pleasant and sympathetic face, dark eyes snapped. On her habit—of a *religieuse*— shone the Cross of the Legion. From time to time, as she talked, she fingered the medal. She greeted us warmly, but at first she seemed a trifle reluctant to speak of that unbearable occupation of her city by the Germans. As she went on, however, her gestures assumed a rapid and varied intensity. At times horror slumbered in her eyes, at other moments anger awakened them.

"There wasn't much bombardment," she began, verifying what the staff officer had said. "The town was little hurt by that. Only sixty Chasseurs à pied held the bridge across the Mortagne. But, alas, they were too magnificent. The Germans were so angry at their superb stand, declaring that the old men of the town must have helped in the defense. They came in at nightfall—Bavarian troops who had fought hard and marched hard. It seemed that they were tired, and their general thought they should have a little relaxation. He issued orders that in Gerbéviller they were to do what they pleased."

She shook with disgust and pointed from the window, explaining, "They amused themselves. No bombardment could have been so complete. They used explosives, oil, all the flammable material they could get their hands on. When a house was burning, they clustered about the cellar entrance to welcome the women and old men who had to leave their refuge or roast. The men were bound and made to watch the welcome of their women. One finds it difficult to speak of such horrors. Then many of the men—old fellows, for the youngsters were all at the war—were tied in groups of five and, while they questioned with eyes like the eyes of an animal one has accused unjustly, they were shot down. For many hours we heard the firing, and we muttered prayers for departing souls, while we tended to the wounded. One girl, rather than face such things, hid in the Mortagne with the water up to her neck. She was in the river all one day. It killed her, but she was more content to die that way."

We remained silent before the sad conviction of this woman of the church, who spoke of what she had seen with her own eyes.

"In the night they came here," she continued. "Their work of destruction had progressed far. I had many desperately wounded men, some German, and a few gray old fellows who had sought refuge at the hospice. The Bavarians came and fired and told us we must leave in order that the hospice be destroyed like the rest of the town. The officer in charge had a pistol in one hand and a sword in the other. I pleaded with him, saying, 'The thought of your mother will not let you commit this crime. The building is full of the wounded and the dying, and some old men who are incapable of bearing arms, and I have Germans.'

'Point them out to me!' said the officer.

"And they entered and went to the cots where the wounded Frenchmen lay. I tried to keep my eyes closed that I might not witness this crime, for they tore the red bandages from the wounds and the blood flowed again, staining the beds. When I cried out, they sneered that it was necessary for them to search for weapons beneath the bandages. Rifles and bayonets beneath bandages! I grasped that officer's arm.

"'Do no more evil to these poor little ones,' I told him. 'Burn no more. See! I care for your wounded, as I care for our own.' I pointed out to him the violent, scarlet sky above Gerbéviller and said, 'Save this little corner for sickness and death.'

"And he went. But later, when the French returned, some of those men came back. We saw our ruddy executioners, our firebrands, pallid and torn and asking for help. So we took them in until the little hospice was like a shambles. The blood! It ran from their resting places on the floor. It ran so thick in the corridor that I arranged a mop as a sort of dam to turn it into the street. Angry at retreating, those that were unhurt tumbled over the walls of the houses they had burned. That is why we are not like a city that has been bombarded. That is why so many houses are only heaps of bruised stones."

She arose and spread her arms. On her dun uniform of a *religieuse* the Cross of the Legion glittered. "Is it any wonder," she said, "that all the world will forever speak of our beloved dead city as Gerbéviller-La Martyre?"

We left Sister Julie and Gerbéviller. We went out of Lorraine with a sense of flight, before a sinister invasion perilous to the entire world, of unusual and ruthless creatures, suddenly unmasked by the tearing claws of war.

Chapter Seven

THE PERSISTENT BOMBARDMENT

On my return, the familiar beauties of Paris acquired a new and precious meaning. It was possible to more accurately estimate the value of the epic moment when General von Kluck's flank was turned and the sinister invasion broken, almost within sight of the fortifications. So, I got a military permit and visited the region where General Manoury's taxicab army flung itself on the extreme German right.

The flags waving over the graves were thicker than in Lorraine. They were like a strange and colorful grain. And scattered irregularly behind the pierced walls of the graveyards, in each little village, were the sepulchers of soldiers, buried where they had fallen. Behind an ugly breach in a cemetery wall was the tomb of an officer, set at an angle.

"The captain, you see," one of the natives told me, "was leaning against the wall watching the effect of his men's fire on the enemy when the shell fell just there. We came out in the evening and buried him."

He took me to a flowering tree not far away and pointed out a polished round hole in the trunk. "That," he said, "was made by a shell nearly spent. It struck its nose in and exploded its entire charge backwards. It killed two lieutenants who were standing in consultation just where you are. Here are their graves, at your feet."

The inhabitants will relate a thousand such intimate details of the Battle of the Marne. They understand it in no other language. It is, in fact, impossible for the layman to gaze across the field, sewn with tricolors, and interpret the miracle in any broader terms. But of the most intimate and desirable detail of all, there was no one who could speak surely about it. I looked at a quiet and picturesque farm where von Kluck had had his headquarters. I wondered what dramatic event had happened there, perhaps during the course of a moment or two, that had urged him to give the command to swing in across Paris. Had he run ahead of his supplies? Had an order been misinterpreted? Was a fit of petulance responsible? Had he lunched too well? It was there that the German structure of forty years' growth had tumbled, but no one could tell what had happened at the pretty farm during that decisive moment. The closeness of the thing was impressive. As I stared, I could hear the dull booming of guns up ahead, in the vicinity of Soissons, and only a few miles behind me lay Paris.

Already, a number of monuments (by direction of General Gallieni before his death) have been raised on the field of the Marne, yet it isn't the mecca for Frenchmen one would expect. The authorities have seen to that—they make a visit almost as difficult as the entrance to a front-line trench. There are military reasons for this, of which it is better not to speak. They will probably keep the Marne closed to the ordinary visitor until the end of the war. I found it necessary to show my pass there more frequently than in the actual zone of operations. Civilians who visit are always accompanied by a staff officer.

The staff officer I had during my visit to the Champagne front was a charming young fellow—small, constantly smiling, inclined, as far as one might be, to take war as a part of the day's work. He had been severely wounded in one of the early battles. That seemed to be the only portion of his own experiences that he thoroughly resented. "It keeps me in a staff job," he mourned.

I asked him what his sensations had been on first hearing the shells. He laughed, and said, "When the first shell whistled whoo-ee-ee, I commanded my men to present arms. That amused them, and was good. Then I told them to lie down."

This officer met a party of us in Épernay and drove us first of all to Rheims. The desecration of the cathedral there is by no means a thing of the past. The bombardment continues according to the fancy of the German gunners. We drove in past miles of shell screens, constructed between the road and the enemy, of sheets of cheese cloth or masses of dead foliage. A soldier was our chauffeur, an orderly sat at his side. Above their heads helmets and a rifle were suspended.

Out of the gray and rainy morning came the rumbling of guns. The houses of the suburbs were marked with shell fragments. One or two men and women glided silently past us, clinging to the shelter of walls. We swerved into a vast open space. At first I didn't realize we had arrived. Then I left the car and, holding my breath, unconscious of the rain, stood gazing upward.

The cathedral of Rheims proves how absurdly conservative photography is. A picture of the twin towers and the rose window doesn't give you a sense of the unbelievable tragedy, nor does it create an instinct to speak not at all or in whispers. That is because the horror of Rheims Cathedral, from the front, is a matter of detail. The left hand tower rises in the shade of ashes. The semblance of figures on the façade, featureless and stripped, nevertheless have something human about them. They are like victims of the ancient trial by fire. Instinctively, one glances at the brave, little, bronze figure on horseback that has miraculously survived each bombardment. More than ever, Joan of Arc belongs here—her pose with flag uplifted is one of inspired command. She seems about to lead the wraiths of the cathedral to a stern reckoning.

We entered the desolate structure. When I removed my hat, a staff officer shrugged his shoulders and said, "That is not necessary. So many men have been killed in here that the edifice is no longer consecrated." His comment expressed, perhaps, more than its intention. For there is a depressive feeling within the cathedral, the source of which is certainly more remote than the emptiness and the battered walls and pillars. The emptiness reaches you first of all. The aisles are vast, the open spaces appear endless. Pigeons, flying between the tracery of the eyeless windows and about the roof, accentuate the sense of distance. And it is out of this emptiness that the feeling of depression steals. With me were officers and soldiers hardened to the filth and corruption of war. Some of us had seen devastation more complete and no less excusable than this. Yet no one failed to respond to that sense of suffering that seemed to have survived its physical source. It is, of course, impossible to say how far our knowledge of what happened here gave birth to such thoughts. It is merely significant that we all experienced them. One visualized rows of bandaged and groaning men, stretched on the straw or crawling about with awkward, incoherent motions like mutilated insects. The vaulting seemed to retain the echoes of cries and curses. Openings showed where the Germans had sent incendiary shells to burn their own wounded. Such anguish leaves something behind it.

We went about softly—almost on tip-toe. Through the emptiness we experienced a sense of obstacles. We walked carefully so as not to stumble over the shadows that remained.

Out front again, we had a moment to appreciate the shattered surroundings of the cathedral. The miracle of the preservation of the statue of Joan of Arc was more impressive. Within the entire range of our vision, it was the only object that had not been violently disturbed. No wonder there were flowers at its base and flags at the pedestal. No wonder the inhabitants had devised a prayer, printed it, and placed it on the iron railing at the front. We read it with a thrill:

"Joan of Arc—Pray for us—Bring to France the victory," it said.

We turned away and were taught by our staff officer that the ruins in Rheims isn't limited to the vicinity of the cathedral. We wandered with him through the gardens of the archbishop's palace, staring at the ghosts of that structure, which are nearly as famous as the cathedral itself. Roses were in bloom along the hedges. Their growth in such a place seemed to symbolize a mockery of the Prussian spirit of conquest, a reminder of the indestructibility of the soul of beauty. We continued to wander, sadly, through the best residential district of the city. The few houses that were still inhabitable were marked with numerals. "The number of people the cellars will hold," the officer explained.

While the greater proportion of the population had left or had been killed, those that remained were quietly illustrative of the extraordinary determination of the French. Two women, whom we had met in that mass of rubbish of homes, remain in my mind. We had been compelled to leave the automobile, and for many blocks we hadn't seen a habitable structure. As we climbed around a corner over a hill of rubbish, I heard a feminine cry of surprise. Ahead was a house that by comparison had suffered slightly: the glass had been replaced by boards; the front door could not be closed; countless pieces of shell had scarred the exterior. A young woman leaned from the upper story. The surprise in her face at seeing civilians here matched our amazement at the sight of her graceful figure in such surroundings. We stopped and chatted with her.

"You live here?" I asked.

"But certainly. Why not?" she replied.

"You have a great deal of courage," I told her.

She shrugged her shoulders, and said, "It is my home, is it not? Enough is left of it, so I stay at home."

"And the shells?" I inquired.

She laughed, exclaiming: "The shells! They follow one anyway, and there isn't much to bring them here now."

Farther on, in a less damaged quarter, a little old woman wearing the universal black came up and spoke to the staff officer. A basket was slung over her arm; evidently she was going marketing. "Pardon me, Monsieur le Capitain," she said, "I am a little confused. The hour of the bombardment remains the same? The Rue de la [. . .] is safe at this hour?"

We smiled, but the captain, who was accustomed to such queries, replied seriously: "The hour is unchanged, but I wouldn't advise it *madame*. The Rue de la . . . is likely to be unpleasant at any time."

She shrugged her shoulders—that invariable gesture that has

acquired a quality of renunciation. "It makes no difference," she responded. "Another route will do as well. One must order one's life according to the clock of the shells."

And she wandered away, her basket resting comfortably in the crook of her elbow.

Chapter Eight

THE AMAZING GARDEN

It was in Champagne that I accomplished, for the first time, the much-desired experience of entering the front-line trenches. Such an excursion isn't without its discomforts. We started on a dull afternoon, clothed for rain and mud, of which we had been warned we would find plenty. The officers and soldiers with us were ominously silent; we drove swiftly and began to hear cannon. When it was necessary to sound the automobile horn, the driver was cautious, and the discreet response gave us a feeling of danger. Already we wondered how individuals, not unlike ourselves, ordered their lives amid such dangers and discomforts.

A famous novelist was with me. He spoke no French, and he was considered of such importance that a member of the Chamber of Deputies, who knew his language, had been assigned to accompany him.

While the voice of the cannon grew angrier, we entered a deserted and shell-torn village. Barbed wire filled the gardens and was stretched across the streets, so that we had to zigzag a course through. The shattered walls were pierced for rifle and machine guns.

"It won't do to go any farther with the cars," the staff officer said. "The entrance to the communication trench isn't far."

My curiosity increased. I wanted to know exactly what the entrance to a communication trench was like. I fancied that the pictures again would be wrong, and so they were.

We were walking, I remember, along a sidewalk in the shelter of some ruined walls. The sidewalk had a stone curb. Then I understood. The curb line ran level straight ahead, but a portion of the sidewalk, perhaps two feet wide, next to the curb, sloped gently downwards. In a moment we were walking shoulder high in an excavation such as one observes about unruly gas mains. Abruptly, we were in the communication line.

The next thing to be experienced was being under fire for the first time. The trench stretched diagonally across level fields. It was higher than one's head, so it was impossible to see anything except the white mud (through which we slipped) and the grass overhanging the edges. The guns were a great deal louder. The officer raised himself cautiously above the bank, and I followed his example. There was a railroad embankment ahead, some queer whitish furrows in the distance. One heard curious little gusts of wind.

"When will we be under fire?" I asked.

The officer grinned, responding, "Don't get up too high. We have been under fire ever since we left the automobiles. Listen!"

One of the gusts of wind had a sharper sound.

"Shells," he said.

I experienced a sensation of nakedness. I was glad when he said: "We'd better get down."

We walked on through what appeared to be endless lines of trenches, with a glimpse at a turning and a bit of brick wall near which *poilus* improvised a meal. In all directions, lines branched from the communication line we followed; each was labeled. It was like a hidden city in which the inhabitants carried an air of constant expectancy. Covered with mud, these creatures slipped by us from time to time.

"How are things in the frontline?" our officer would ask.

"Fairly quiet," was the almost invariable reply.

"It is the rain," the officer explained to us.

Yet it wasn't quiet—the roar of the guns seemed continuously closer. No minute passed without a number of detonations, and the gusts of wind had a more menacing volubility. At every turning, we found a machine-gun emplacement. Directly in front of it was suspended, at approximately the height of a man, a great globe twined with barbed wire, ready to be lowered in the event of an enemy invasion of the trench.

"While they are getting rid of that," our officer explained, "the machine gun attends to their little affair."

We came to trenches marked: "*boyau de la deuxieme ligne.*" The *poilus* we met didn't speak above a whisper. We were aware of an empty road winding along the surface of the earth. A flight of steps led upward that was nearly barred by a huge sign that forbade pedestrians to use the road under the severest penalties.

"You mean to say," I asked, "that soldiers have to be threatened from that exposed place?"

"The communication trench, as you can see," our officer answered, "is very warm. The men prefer comfort and the German fire. We were losing too many through such foolishness. Even now it is difficult on a warm day to keep them in the communication lines."

We passed frequent, broad flights of steps. "The units leave that way for an attack or a sortie," our officer explained casually. We glanced at these stairways of death with a vague discomfort, unable to quite comprehend them, and hurried on. We paused before a narrower flight.

"We are just behind the first line," our guide explained. "Now I am going to show you something."

We followed him up the steps into the most amazing garden any of us, I think, had ever seen. It was hidden on one side by a half-destroyed building, on two other sides by brick walls, pierced for defense; on the fourth side, it was hidden by a low structure that from a distance looked as if it might have something to do with the scientific raising of chickens. We entered through the archway of the half-destroyed building. Everyone spoke in whispers. Cabbages, arti-chokes, *haricots*—such vegetables as a Frenchman enjoys—stretched in neat rows.

"Sometimes they get a trifle too much ploughing," the officer laughed softly. "The Germans, I should think, are not neat farmers, but here they do their work unasked."

We soon discovered that we were not there to see the garden, but rather its owner and his home. We approached a building that was like a chicken house. It was less than one storey high and the white earth of the country had been firmly packed over its roof. We went down a flight of steps into a corridor, half subterranean and lined with concrete, from which four doors opened into four long, narrow cells roofed with steel arches painted white. This, we were told, was the headquarters of that sector. The room to the right was occupied by telephone operators. Next was the commandant's apartment, furnished with a cot bed, a bureau, a wash-hand stand, and a chair or two. There were homely touches, including photographs of a woman and two children. Even in these lifeless pictures the faces seemed watchful, apprehensive.

The room next door, occupied by the majors, was much the same, but in the cell at the end of the passage there was a variation. No one had to tell us for what purpose this shelter was used. The sickly ether odor welcomed us; a crucifix was suspended above a bed improvised from three stained mattresses piled one atop the other. A brown blanket, also stained with black and white splotches, covered it.

"*Poste de secours*," the officer said. "A first aid post, directly at the front, yet thoroughly protected."

The light entered reluctantly; the melancholy of the crucifix oppressed us. As we climbed to the surface again, a small procession crossed the peaceful garden. Through the stooping, slow-paced files of figures we saw a still form on a stretcher, covered with a stained blanket.

We turned gladly and followed our guide through the archway and down another flight of steps deep beneath the surface. We emerged into a tunnel-like room, crowded with switchboards before which soldier-operators sat, smoking and calling into the transmitters. The wires strayed across the ceiling like the web of a gigantic spider. We were told that from this protected cave one could communicate with any portion of the front, or with the *état major*. From it radiated black passages designed to furnish shelter for hundreds of men. We were permitted only a minute to explore these with a candle, for other plans had been made for us.

"I am going to show you an artillery observation post," the officer said, "if you are not afraid. You will please not speak above a whisper or make any unnecessary noise."

We followed him down one of the dark passages. The only light was an occasional flash from the officer's lamp. He paused at the base of a perpendicular ladder, which rose beyond the roof through a narrow shaft until it was lost in the darkness.

"Here we are," the officer said. "You can go up . . . if you are not afraid."

Now that we were actually at the front, that chilling statement had become habitual with him. It was possible to do this or that—if we were not afraid. Such a formula must have its ritual answer, and through the darkness we murmured our delight at seeing the artillery post. While I waited my turn at the ladder, a patrol stumbled nearby, flashed his light on a telephone instrument against the wall, then went close and took down the receiver. I heard him reporting to headquarters.

"Very quiet. Oh, four or five casualties. Sending them back. No, no. Nothing at all. Everything is very peaceful," he spoke into the receiver. He then snapped off his light, hung up the receiver, and stumbled away, continuing his routine.

It was my turn. I commenced climbing the ladder, while the water dripped with a perpetual animosity. The succession of rungs seemed endless. Certainly we would emerge at some high point with a prospect magnificent and extended. But such a post, it occurred to me, must to an extent be exposed. I tried to calculate how high I was already. Then far above light gleamed, as an officer had opened a trap door. With muttered warnings to avoid a misstep, he helped us through into what might have been a little shelter, roughly constructed and extremely low, arranged on the summit of some lofty monument. Openings on each side were curtained by dark canvas flaps. The officer closed the trap door, unfastened the flap in front, and raised it.

"Look," he whispered. "Our trenches and the *bosches*."

But the first thing we saw was grass and didn't understand. Then it came to us: after that climb we were at ground level. The officer smiled, and said, "But there is a little ridge here and one can see very well. It is necessary to enter that way, so the enemy will not be suspicious."

For a long time, we stared across the slowly waving grass at the routine of war. Not many yards ahead of us was a deep, wide fosse. At intervals, forms in blue overcoats holding rifles extended across the parapet like statues. A hundred yards beyond them white mounds straggled a parallel course. The interval was a jungle of weeds and barbed wire. A few skeleton trees in the distance stretched their branches in gestures of protest. Poppies, scarlet and significant against the white soil and dun vegetation, drooped everywhere, even in the jungle of No Man's Land. There are so many poppies this year in the war zone! They are like great drops of blood.

A perpetual sighing (like the wind) overhead was accented now and then by tearing screams. The officer looked about uneasily, and said, "They feel all over the landscape with their shells for these observation posts," he said. Then, indicating a row of sentinels in the trench just ahead, he continued, "I am going to take you now to the very frontline." Glancing at us curiously with an enigmatic face, he said, "And perhaps—if you are not afraid—even beyond."

Chapter Nine

BETWEEN THE LINES

We descended the ladder, wondering what the officer had meant. It had not occurred to me that I could go beyond the frontline, nor was I quite sure I wished such a privilege to develop. We slipped from a covered communication line into a chalky wet space between the parapet and a shell-gouged railroad embankment. In the lee of the embankment, blue-clothed soldiers shivered, seeking what shelter there was. Our little party broke the monotony for them. They straightened and, smiling, spoke to each other in low voices, inaudible to us. They were like a party of men playing a game of hide-and-seek, exuding a breathless excitement at the imminence of discovery.

A line captain consulted with our staff officer. The staff man, with the desire to be hospitable showing in his round and pleasant face, came forward. "The captain," he said, "wants to do something for you."

We were appreciative and curious.

"He says," the staff man went on, "that it is a very quiet day because of the constant rain."

Coming in, as I have said, I had noticed that no moment went by without shell explosions. As we talked, we could hear the whining of shells overhead, and at intervals a number would shriek too close for comfort. We saw heads duck automatically. On such a quiet day we didn't want the captain to put himself out too much to do something for us. We asked what his plan was.

"He suggests," the staff man said, "that it might be possible to take you to a listening post in No Man's Land—if you are not afraid. You are not afraid?"

To that formula, of which we had grown well-worn, we gave the customary reply. Moreover, it was an opportunity permitted to few civilians. So, in a solemn file, we followed the staff man and the line captain past a dugout labeled, after the fashion of a summer cottage, "Villa de Venus." We climbed a flight of steps to the platform against the parapet where the sentinels stood.

"Of course," the staff man said, "if you go, we can't promise there won't be a shell or a hand grenade."

In response, we made gestures of indifference and then looked at each other suspiciously. There were no signals of retreat. Even the famous novelist, who had large and dreamy eyes, was willing to go. More than once I had questioned if he fully understood the conditions amid which he walked. He wore a long, black cloak buttoned to the throat. It had been warm work coming through the communication line and now, at the top of the steps, he unbuttoned his cloak, throwing the flaps over his shoulders. A group of soldiers nearby scattered, laughing silently. Our conductors started, gave the familiar renunciatory shrug, then continued on with an air of hesitation.

The flaps of the famous novelist's cloak were lined with vivid scarlet. He trudged ahead with the hospitable captain and, as we passed, sentinels snickered behind their hands and edged away.

"Why don't you tell him to take it off?" I asked the staff man.

"He's too distinguished," the officer replied. "I'll guarantee the captain will make him walk low through the sap." We watched the captain motion to the novelist, then stoop and disappear. As we came up we saw the opening of a narrow sap that led at right angles from the main trench into No Man's Land. Ahead, the scarlet cloak led the way; we followed at a discreet distance.

Soldiers have written and talked a good deal about listening posts and yet, like nearly everything else at the front, the actual thing was unlike one's preconceived notion. The shallow, unfinished appearance of the sap advertised it as a temporary structure that could be abandoned at any time the German fire should make it wise to do so. As I crouched in the post, strands of the overlapping barbed wire caught at my hat and the weeds, evidently encouraged to mask the narrow ditch, brushed against my face. The cut merged into a small square pocket where a solitary figure rested motionless and somber. His rifle barrel protruded through the grass, a box of cartridges lay on a dirt shelf to his left, and, convenient to his right hand on another shelf, was a wicker basket such as those old women used for their knitting. It was filled with corrugated black objects, the shape and the size of pears—they were hand grenades.

This further proof that we were actually between the lines and within hand-grenade throwing distance of the Germans warned us to take our places one by one in the pocket, with our guide and the sentinel, as stealthily as if we were afraid of awaking a light sleeper. And we looked at the man closely, for we knew we were seeing one of the riskiest and most unpleasant details of trench work. Here a man watches alone, listening for enemy miners, alert for the first sign of activity from the opposite trench not many yards away. As everyone knows, it isn't simple to be brave when one is alone. At the front, you acquire thorough admiration for the men who assume the strain and the solitude of such assignments.

Our guide was still inclined to hospitality. He produced a map of the enemy trenches made from air photographs. Each trench was labeled. There was, I remember, the "Boyau Unter den Linden," the "Boyau Bethmann-Hollweg," the "Boyau Bismarck," and many others that illustrated the play of French humor. I was instructed to peer through openings in the grass and the wire in the nearby mounds of white, wet earth that marked the German trenches.

"That communication trench coming up is the Boyau Unter den Linden. Can you see it?"

asked our guide.

Thoughtlessly, I answered, "I am not quite sure. No, I don't see it."

The hospitable captain made a gesture of disappointment, a peculiar clicking sound with his tongue. "You should see," he said. "It is very interesting. What can I do? Ah, yes. There is another listening post a little nearer the *bosches* to which it might be possible to penetrate. You would see better there. You are not afraid?"

I followed him back to the main trench and crouched along another sap to a pocket where the occupant clearly disapproved of our presence. Through the grass and wire the confusion of trenches appeared much the same, but when the captain asked me if I could now see the Boyau Unter den Linden, I replied without hesitation: "Perfectly. It is surprisingly distinct." Nor did I keep him in suspense about the other objects he pointed out. I recognized all the *boyaux* with a miraculous ease. So eventually we stole back to healthier regions, both of us very pleased.

We all thanked our host and commenced the return journey. That was halted almost at the start, while we studied a picture that seemed better than anything I had seen at the French front that symbolized the waste and the distortion of war. In the background, there was the main street of a ruined village almost directly behind the first-line trenches. The street made a slight arc between walls that, for the most part, gave only a sketchy illusion of habitation. Many of them were unsupported, offering views through eyeless windows of emptiness and desolation. Here and there, a building maintained a semblance of completeness. Its doors might have gone, its windows may have disappeared save for jagged pieces of glass, and its roof may have been pierced by shells, but by very contrast it was serviceable. From one such survival, the odor of ether slipped with a sickly stealth. It was a first aid post whose attendants worked under risks nearly as great as those of the men in the frontline. The cold and brutal agony it housed reflected itself in the scarred brick wall and the tile roof, from which the rain dripped with a suggestion of inexhaustible mourning. It was good to turn to another structure, the cook house, from which a savory scent joyously emerged.

At the end of the curving street a tower arose. Even above the debris of the town it presented an abhorrent spectacle because it was the skeleton of a church. Like a mutilated sentry, it seemed engaged in the pitiful occupation of guarding what was no longer worth the trouble. Shells shrieked overhead and, through the heavy air, the gross petulance of the guns continued uninterrupted.

Poilus strolled against that background, a little wraith-like in their damp, blue uniforms. They carried out of the cook house tin pails from which fragrant steam arose, or beneath their arms they hugged great, round, loaves of bread. As they went, they laughed or talked quietly. One by one they disappeared behind the shattered walls or into burrows beneath the earth.

The commander of that sector stood in the middle of the street with a number of his officers. He glanced at the picture that must have become all too familiar to him. "There was hand-to-hand fighting in each of these houses," he said, "but it was worth it, for it brought one more village back to France." He pointed to the devastation and sighed. "The last village," he said.

"And how," an officer asked, "would they like villages like this in America? Is it possible there is a country that isn't full of villages like this? In such a country they can't understand. They can't understand."

The clouds grew a little thicker; the light faded. It seemed as if the whole world must be like this. These men appeared to know, in the past or the future, no mode of life beyond this. Stern-faced, physically contented, unafraid, they had an air of guaranteeing the redemption of those familiar fields ahead, which reluctantly sheltered the invader beneath a sullen sky.

The officer was right, even now it is hard to understand such things in America.

Chapter Ten

WITH THE BRITISH IN FLANDERS

I received the coveted invitation to visit the British front the morning after my return to Paris from Champagne. The provost marshal started me off adventurously enough. I was to report to the landing officer at one of the great seaport bases the next day, at one o'clock—daylight saving time. That variation of an hour confused everything.

"You can only make it by the military train at 11:40 tonight," he said. "You'll have twelve, nice, sleepless hours for a journey that ought to take four or five. But then war is never convenient. Goodbye, good luck, and cling to your headquarters pass."

At half past eleven, the façade of the Gare du Nord with its staring yellow clock was sufficiently forbidding. There were no hurrying crowds, no babel of voices, no porters. A *gendarme*, unreservedly surprised at the presence of a civilian, trundled my bag through. The great shed, inadequately illuminated, had an unfamiliar air. A single train of low and antique carriages stretched to the north until it was lost in a darkness relieved only by red and green signal lamps that were close to the ground, vague in a slight mist, like will-o'-the-wisps.

No one reached the quay without a catechism from the soldiers and *gendarmes* at the barriers. A khaki-clad figure stood with the others—the first Tommy—the extreme rear-guard of the British lines. He grinned, struggling with what he conceived to be the American idiom, and said, "Give my regards to the boys..."

The train, crowded with *poilus* and officers, threatened to be insufferably stuffy. Therefore, until the last moment I paced up and down the murky platform, listening to subdued voices chanting popular army airs, oppressed by the wailing notes of an accordion. Through an open window I had a glimpse of the player. His eyes were upraised, his face was dull with mental pain, and his hands on the accordion swayed apart and came together with slow, caressing gestures. His companions, in dirty blue overcoats, sat facing each other on parallel benches beneath a dim light. They swayed unconsciously in rhythm with the music, muttering inaudible snatches of words. Eyes and ears were challenged by a sense of despair nearly voluptuous.

I paced on, made very sad, very lonely by this haggard playing, by the instinctive response drawn from its hearers. A squad of solemn and weary soldiers tramped down the platform. Bent beneath full knapsacks they shuffled along, clinging to the butts of their rifles with an air of reaching out for help. Suddenly, with tired motions they swung into a ragged platoon formation and waited dumbly for the command to break ranks.

A thick and unreal atmosphere invaded the melancholy shed. These fatigued and over-burdened figures; the crouched forms in the dusk of the third-class carriages; the persistent lament of the accordion; and the vapors curling about the few lamps, like dying moons high in the roof, all welded themselves into a conception of the exotic —of more than that—of the barbaric, of a helpless and primitive fatalism. This could not be Paris. These stooped and soiled figures sent forth for killing, many of them for death, could not be educated reasoning men. Then, close by, an officer breathed the word "Verdun," and the unreality dissipated. The picture assumed harder, surer lines. It had grown cold in the shed.

There were four officers in my compartment. Two others climbed from the platform and lounged in the meager corridor. It was one of these who had spoken of Verdun. He had, we discovered, been there.

He sketched his incoherent recollections of its deadly turmoil. He broke off and glanced up with an abrupt reluctance. "Without doubt you recall so-and-so?" he asked.

The other officer responded with a nod, explaining, "You may have heard. A piece of one of those high explosive shells—a great fragment, all ragged . . ." There was no dismay at the intelligence, scarcely any surprise.

From the darkness beyond the shed, the locomotive whistle shrieked. That sound alone was appropriate, for it was comparable with the sudden grief of a woman. The train crawled into the obscurity, writhing through the yards like a gigantic reptile. The two officers moved away. In the close, dim carriage we curled ourselves in corners and tried to sleep. But it was difficult not to watch these uniformed figures, outstretched in awkward attitudes that mimicked the appearance of human refuse on a battlefield. Moreover, the train constantly halted. At each station, a stocky little fellow would open his eyes, spring up, crash the window down, and demand at the top of his lungs if he had reached his destination.

Finally, an elderly officer stirred and asked with an accent of pity: "Don't you know, my friend, that you've still twenty miles to go? On this train that should permit you several days' complete rest. Sleep well."

After that, our compartment was quieter and we dozed. One by one, the officers gathered their baggage and left. The last clambered sleepily out in a grudging dawn at the first large English base. After that, it was clear enough that we were behind the British lines. British faces, British khaki, British methods filled the frames of the windows.

At country stations hospital trains lay on sidings, ready to receive from temporary hospitals and ambulances their grim and scarlet freight. Their drab sides were relieved only by red crosses painted in white squares. But in each car, clusters of field flowers added splashes of color. The wide, plate-glass windows were open to the air. Orderlies in white jackets moved about the beds, slung where the seats had once been.

An airplane, a swallow-like speck, appeared to the right, flying in our direction. It came up rapidly until its lines were silhouetted against the sky in the east. The track curved and the war plane glided gracefully after us. I was on my feet about to reach for two sandwiches I had stuffed in my raincoat before leaving Paris. They ceased to interest me. Officers stood in the corridor, gazing tensely from the window. Those who boast they can identify war planes are invariably uncertain at such a moment.

"Is it a *bosche*?" one officer asks.

"If it is, he's sure to drop his card on us," said another.

"This train isn't such an easy hit. Hello!" the first officer responded.

Conversation became general in the carriages. Someone laughed. Without warning the airplane had swooped downwards and disappeared behind the trees. My dry sandwiches drew glances of envy. Before they could be eaten, the railroad line swung towards the sea, and with the first sparkling of water came the sheen of innumerable tents. This coast, remembered as a mecca of holiday makers, had become a vast encampment for Kitchener's volunteers—the men soon destined to be brought up for the "great squeeze" [Battle of the Somme].

To the right, in a field that rolled broadly towards green and treeless hills, several companies of infantry seemed involved in some incomprehensible game. A hundred yards in front of them stood a series of posts between which cumbersome sacks were placed at approximately the height of a man. The arrangement suggested the tackling dummies one sees on a football field during early fall practice. Then I commenced to understand, for other sacks, equally fat, were sprawled on the ground. The soldiers themselves illustrated the rest. Released by the flashing of an officer's cane, they dashed precipitately forward and assaulted the contrivance with their bayonets. Some lowered their points and pinioned the prone sacks, while others chose to target the sacks representing standing men. Steel gleamed, ripped through canvas, emerged on the other side, and was withdrawn with quick, twisting motions. All the while the sea rolled in with an exceptional placidity beneath a smiling sun, and a clean wind blew across the dunes and the fields. But it was clear that these new soldiers saw nothing, felt nothing, beyond the sacks, inert and pig-like, on which they rehearsed with a frantic obstinacy the killing of men.

Farther on practice trenches scarred the sands or were in the process of construction. A minute efficiency appeared to have been brought to the training in attack and defense of these men who recently had stood behind counters, or bent over desks, or perhaps tilled peacefully such fields as these.

The train drew up before the station of a fairly large town— formerly a legendary summer resort. Two youthful and attractive Red Cross nurses entered our compartment. A sub officer—fresh-faced, slender, typical—had come to see them off. They smiled back at him with an attempt at brightness. He didn't quite hide a slight nervousness and the sadly prophetic expression in his eyes. One of the girls spoke impulsively to him, saying, "I am sorry you are going up to the front."

The sub officer glanced away, tapping at his shoe with his walking stick, replying, "Stupid, isn't it? And just when I'm beginning to know and like the people here." Certainly he meant that; it wasn't the familiar English emotional screen. He immediately followed this comment with: "I wonder what will get me up there?"

It was symptomatic of a vital evolution in the Englishman who has experienced this war. I have seen many examples since. Such a shift of psychology seemed more important to the Allied cause than the rehearsal of a bayonet charge I had recently witnessed. Nor was there any attempt on the part of the young nurses to shirk the hard facts.

"At any rate, you can choose your own hospital," one of them suggested.

The officer's petulant striking at his boot continued. "Wish I was sure of that," he said, "but I fancy they send you where it's most convenient."

I looked at him again and saw that he was straight and unafraid in spite of the prophetic dullness of his eyes. So much youth, so many possibilities tossed among the chances of a war in which death is simple and kind! It was impossible not to forecast, not to question if he was to be the destination, some days hence, of a bullet or a shell fragment or a gas attack, or a flash of the improved liquid flame. To walk into that sort of thing for an indefinite period with your eyes open! No wonder they've largely given over shirking the hard facts in France.

Something lingering, wistful, nearly sentimental, colored the farewell of one of the women. There was, it appeared, romance here. Some concession from her was to be expected yet, when the train started and the sub officer dropped to the platform after clinging perilously to the step until the last possible moment, she turned to the window with a sigh.

"Poor fellow!" she sighed. "I wonder. Oh, dear! I wonder."

There were no tears, no comfort, from the other woman, no further allusion to the young man—only an anxious discussion as to whether they would be in time for the English boat. It seemed rather cruel. Then I remembered the hard facts: for many months these women had worked in hospitals sheltering wounds unbearable merely to see; they had watched young men go forth not to return; and they had helped others back to a mutilated, useless existence. The romance in Flanders isn't the old romance. It is there, nevertheless, and it is greater than the old romance because it is definable only in terms of undisciplined truth.

Such fugitive experiences are always impressive in the war zone. I, too, carried from that sunlit station a sharp regret. The momentary glimpse of this young soldier had left a sense of acquaintanceship. It seemed incredible there should be no renewal, no knowledge by and by of the resolution of his future that then had appeared so brief and futile.

Those poor girls didn't catch their boat for England. We puffed into the noisy, dusty seaport base an hour late. An excitable porter scooped up my bag and piled it on a truck with their luggage. Before I could stop him, he was careering drunkenly along the docks at their flying heels. Fie, the military landing officer who rescued us, was sympathetic with the nurses and he promised me an hour for a bath and a noon breakfast before the arrival of the transport with the rest of the party heading to the front.

Later, while we waited for the boat, he chatted amiably. "I'm one Englishman," he smiled, "who knows you don't hunt Indians or shoot buffalos on Fifth Avenue. Several years ago somebody tried to show me all of New York in three days. I'm still convalescent."

He indicated two gray cars rolling down the quay driven by young men in khaki. An officer sprang from the tonneau of one and hurried forward. He was introduced as a staff captain from headquarters who would be my cicerone for the next few days. This Captain Williams was sympathetic about my presence there at such an hour; it illustrated for him that interminable journey by night. Explaining he would have waited for me, he at once made me feel welcome and at home. The English don't ask many to see what they are accomplishing in Flanders, but when you are there they reserve little, and they never give you a feeling of intrusion.

Two transports came in today. As they made fast to the quay, I saw that the decks were cluttered with life preservers. For on these transports everyone is compelled to wear a lifebelt from port to port. The men commenced to troop off—Tommies, subalterns, and generals. It seemed fantastic that so many human beings were crowded onto these little boats. There were no smiles on the sunburned faces. Men coming to Flanders for the first time, or returning after a leave, don't smile easily, but when a boat goes forth for the chalk cliffs of England even the menace of submarines can't kill a breathless gaiety.

Captain Williams collected our party, including a man from the foreign office and two Japanese—one straight and slender with a face of a Samurai, the other short and round with a gentle, nervous manner of speech. During luncheon in the maritime station, Williams outlined his plans for us. That afternoon we were to see interesting but not dangerous places. Later, we might learn the vital mechanism of army service and ordnance. If we wanted to, he would take us to the front-line trenches. We could visit Arras, possibly

Notre-Dame-de-Lorette, and other notorious points of the fighting line that for the present must remain anonymous.

That program was carried through and we saw, I fancy, a little more of war than anyone intended. Therefore, the peace of our first afternoon was heightened in retrospect, for we were not to know this peace again during the trip.

Chapter Eleven

HOSPITALS AND HEADQUARTERS

First of all, we drove to a temporary hospital on the cliffs. The adjective "temporary" prepared us for a comfortless and hastily thrown together affair, but instead we found another monument to that admirable efficiency that the English, since the commencement of the war, have developed at the cost of a multitude of traditional fetishes. Grass plots and flower beds flourished there, and throughout was a network of macadam roads put in by the Royal Engineers. Only one or two of the revered marquee tents survived; for no matter how the satirist of British tradition may sneer, experience dictates everything in Kitchener's army and long, narrow, wooden buildings of one storey proved themselves more serviceable, more adaptable to cleanliness, and, curiously, less expensive than the tents that served for field hospitals for so many wars.

A colonel of the medical corps greeted us, offering to direct our exploration. "Each one of these huts," he said, "is a ward." The name drew a laugh of surprise from us. "Anything," he laughed back, "that we put up of wood in the war zone is christened a 'hut.' Don't know how it started, but it's easy to say and everybody knows what it means."

He opened a door to the long building, which was filled with a pallid green light from the curtained windows and stretched a long way in an interminable vista of suffering. Above the beds, set in a double row at right angles to the walls, were odd contrivances of wood reminiscent of cotton looms. They gave the ward an appearance of a factory whose activity has suddenly been arrested. Then gradually, from the mesh of posts and beams, drawn faces detached themselves and stumps of limbs protruded. These faces watched us curiously while the surgeon led us down the aisle, pointing out the elaborate system of weights and pulleys arranged on wooden frames, to take the strain from injured legs and arms. Some poor devils lay on their backs with both legs and both arms in the slings.

"Several of these frames have been used before," the surgeon said with a little pride. "Others—this one, for instance—have been built here since the beginning of the war." He braced his hand against the wood, leaned over the patient beneath, and inquired, "You tell us what you think of it, Jock."

The soldier grinned, having evidently progressed well, and forecasted a sound escape. He moved his bandaged limbs to show us how beautifully the machinery responded. "And it doesna hurt much," he said, "and a man can move about a little and go twist like on his side. Watch, sirs." Then he did it—a trick as difficult, doubtless, as a contortionist's masterpiece—and conquered with heaven knows what agony, secreted behind the features suddenly stripped of their grin. Certainly, one should be grateful for that much—when one has suffered for eight months, it must be pleasant to move a little and to twist onto one's side.

But across the aisle was slung one of those tragic stumps, and the face beyond it was sunken and feverish, and the eyes could not conceal a despairing restlessness. The surgeon spoke to the man gently, asking him how it went.

"A good deal of fever," the mutilated fellow answered dully, "but all right, I guess."

It became clear that he didn't care, that for him the future held no energetic lure. The horrible stump of scarcely healed flesh quivered in the sling. His eyes closed. We didn't want to see the man's grief, so we hurried on. We had to call upon a bleak cynicism, equal to the surgeon's, to recall that the most likely end for the youth of Europe is a room like this, or else a common grave, or a resting place un-blanketed even by the friendly earth.

In another ward, we saw in the end cot, above the bed clothes, a young, square face. "A prisoner," the surgeon explained with a smile. "We were afraid we were going to lose him, but he's coming right enough now, and he likes it here. He's a great favorite with the nurses." The German did, indeed, have an air of contentment, but he glanced at the Tommies in the neighboring beds, at the pleasant quiet nurses, at the surgeon who had pulled him through, and his expression held a great question, as if he wanted to ask why he had been commanded to strafe such friendly and lovable people.

We drove across the plateau to a convalescent camp, where the commandant, an elderly gray-haired man in a colonel's uniform, welcomed us into his official family. He was really like that—a paternal type, a father with a gigantic brood of children—and the grounds of his camp were his front yard and his fields. Immediately he boasted a little as the heads of thrifty households do. With pride he reckoned what he expected to get for his crop of hay—much more than last year—so just that much more for the government, for even here efficiency was a deity. This efficiency was visible in his brood: working at their own trades, remaking shoes, converting jam and butter tins into pails and sprinklers and gasoline funnels, seeing that no small piece of rough material went to waste.

The colonel made us gasp at the sight of the extraordinary process of feeding the British soldiers. (Assuredly, this must have been a painful scandal to the Germans.) It was only a little before tea time and, in the dining hut, long deal tables were neatly arranged with plates, cups, and saucers, with huge loaves of bread between and bowls of jam, butter, and cheese. On serving tables arose pyramids of egg cups. With his air of a thoughtful parent, the colonel indicated these and said, "Any boy that wants it can have a boiled egg with his tea. And look here, if you like." He took us into a kitchen as wide as a barn and as clean as a dairy. Pails of tea cooled there, and sharpening our own appetites were splendid rashers of bacon brought in from the storehouse for tomorrow's breakfast and legs of lamb beyond counting for tomorrow's dinner. I've learned since that there was nothing exceptional here. Tommy fights on such food unless his supplies are cut off by an unexpected bombardment or, in the event of an attack, he is caught for the night ahead of his transport.

The colonel grinned, as he explained, "Now and then they complain if they don't get just the type of cheese or jam they're accustomed to. But that sort of thing's looked after."

We followed him breathlessly to a hollow in the grounds, where a hut stood that was reminiscent of the Y. M. C. A. shacks I had seen in Panama during the construction of the canal. The colonel told me there was a chain of these behind the front, furnished according to a familiar pattern with a store at one end, a billiard table at the other, and often a miniature stage for concerts and amateur theatricals.

"So," he said, "if a boy gets hungry or doesn't like what we give him up there, he drops in and buys some chocolate or a cup of tea or coffee and maybe a handful of biscuits." Somebody ventured the opinion that over eating can be as deadly as bullets. The colonel remained placid, telling us, "When you work as hard as these boys do, you get awfully healthy, and you need lots of food. Besides, when you're going into battle you don't worry much about your liver."

Here—and in this respect the camp may be taken as conformable with the ordinary cantonment—the Y. M. C. A. had no monopoly of recreation work. There were two other huts, one furnished by the government, the other endowed by individuals. Near the latter was a garden where two young women, gloved and wearing rough straw hats, toiled with rake and hoe. We paused and, following the colonel's lead, chatted for some moments about their potatoes and beans and cabbages. As we walked on the colonel laughed a little, explaining, "You know that very handsome young girl with the rake is Lady so-and-so. The war is changing things rather, don't you think?"

That brought no dispute from us. As I have indicated, this truth is everywhere impressed upon one—the war *is* changing things rather. Lady so-and-so has forgotten the interval that formerly gaped between her and Tommy so-and-so. The hard facts have really leveled that. The presence of death, its constant threat—for even here the Lady and the Tommy are equally subject to an airplane bomb or an unlovely Zeppelin attack—make one's recollection of such social rifts a little abashed. And that's the best thing that can be said for this war, the finest thing that can survive it. The individual has learned largely to seek his own level, holding within easy reach a universal and attainable goal.

In this very camp a soldier pointed out a working example. The three recreation huts sift the men into an instinctive classification. "In one," the soldier said, "you can toss your fags on the floor, lift your feet on the tables, and shout your blooming head off, if you please. In the second, maybe ash trays don't grow but the floor's the place for feet and shouting's not tolerated. The third, over there, is a regular little club where you behave like a gentleman and read the papers and magazines and improve your mind." He glanced at his neatly brushed uniform, and added, "I like that place, and it's funny. Most of the men after they've been here a while drift up that way. Anybody likes to be respectable if he gets the chance."

Our party entered the officers' mess for tea and sat down, blessing the Army Service Corps for all it had placed before us. In the confusion names had been lost, but in addition to the medical officers there were two men whose khaki carried the black facings of the church. The chaplain next to me—tall, slender, a little gray-haired—had spent a good deal of time in America. We discovered common friends and asked each other's names. As his is a *nom de plume*, perhaps the censor will let it through.

"If you would know of me at all," he said modestly, "it would be as G. A. Birmingham."

The thought of this writer's rollicking Irish stories and plays made his presence here seem an injustice, and proves again that the so-called British apathy is a lie. It was one more example of how every social and intellectual class is feeding this monster of war. While we talked, someone produced Harry Lauder on the gramophone, a hymn or two, and a waltz. Williams closed the entertainment with the announcement that we had a forty-mile drive to General Headquarters ahead of us.

One goes rapidly in these military cars, for there is no speed limit outside of villages where transport parks, cavalry, or *billets* make it necessary. In each of the cars sturdy men, whose khaki carried a black armband with M. P. in red, stepped out and regulated our passage with the assurance of a London bobby at Oxford Circus. Where the traffic was congested and diverse, staff officers and Tommies, truck drivers and airmen, bowed with an equal meekness to the mandates of these calm, stern creatures. Yet, the military policemen have never once seemed in key with the disorder of war. It is hard to appreciate that such clockwork detail makes that vast disorder possible at all.

For long stretches the drive seemed as if it was just a pleasure jaunt. But a black and yellow board at a crossroads, pointing the route to a Belgian field hospital, was a momentary reminder. The long road, lined with poplars or lime trees, bisected a highly cultivated countryside. Our entrance into one of the two general-headquarters towns, which have replaced Saint-Omer since the extension of the British line, had nothing to offer of the panoply of war. Instead, a brook rippled beneath an ancient bridge; gray stone houses, half hidden among trees, terraced a steep hillside; and a Gothic church tower raised its sharp silhouette against a sky already sprinkled with gold.

"This is one half of the heart of the British Army," said Captain Williams, his words sounding like a joke said in bad taste. Yet the other headquarters town a little farther on was equally rural, quite as picturesque. In the sleepy buildings officers worked at desks, disturbed by the roar of cannon only when an unusually heavy bombardment conspired with a favorable wind. You pictured Sir Douglas Haig, even farther removed in an isolated château, seated in a somnolent library, the cradle of every detail of routine and death. Peace at headquarters and horror at the front, but not an ounce of glamour left in war anywhere!

Our own home shared the restfulness of the headquarters

villages. We came upon it for the first time on the edge of this golden sunset. Far at the end of an avenue of huge and symmetrical trees stood the red and white façade of a château. Two time-stained gate houses were outposts. A clock stared from the top storey, justifying Williams's hurry.

Two dogs ran around the corner, greeting us excitedly, and military servants took our bags. I was led into a comfortable room and stared from its broad windows at a great park bounded by evergreens and elms. I saw a sundial in the center and magpies flying with a gentle rustling of wings among the trees. It was difficult to remember the real reason for this visit. And it was always like that at the château. To be sure, staff officers came to dine with us each night, and we talked continually of war, for to discuss shop isn't bad taste in the British Army. But the impulse to all this chatter seemed far removed from the dining room and the quiet movements of the servants; one might talk just so in London or New York. From the first it didn't seem possible that less than forty miles away lay that open wound in the body of civilization that we had come to probe.

It was brought nearer as we started for bed the night of our arrival, when Williams appeared with an armful of khaki-colored bags, slung from straps. He handed one to each of us. "These are gas masks," he said seriously. "On no account forget them tomorrow."

With an assumed indifference, we asked what kind of gas the Germans were using, to which Williams replied, "An improved variety." He lit a cigarette and between puffs, added, "If you get the alarm, hold your breath until you put your masks on, because three whiffs of this new stuff is certain death, and it isn't a pretty way to go either."

Even with such a prompting, utter weariness won't let you dream of war.

Chapter Twelve

UNDER FIRE IN A FLAT LAND

After an early breakfast we started for a point of the line already sufficiently historical, but not to be mentioned here. We glanced regretfully back at the château as the cars scurried away and up the avenue. Captain Williams was with me and, following his advice, I examined the workings of my gas mask. It was designed to cover the head completely and to be buttoned into one's coat collar. Goggles were fastened to the brown cloth and beneath them was a wooden tube (with an elastic band) that was to be placed between the lips for breathing out. Through the chemical-soaked meshes of the cloth itself, sufficient air filtered for breathing in. It was an unlovely, uncomfortable, and odorous contrivance. We were careful to keep ours slung over our shoulders—as Williams carried his—and every officer and man near the fighting line carried one. The necessity for such a precaution revolted your sense of decency, aroused a sort of anger.

We hurried through the dew-soaked morning, still a trifle misty, and came upon significant pointers measuring our progress towards the front. Beside the gray lines of old churches, modern automobile trucks were drawn up—out of place, grotesque, nearly laughable. We passed many such trucks on the road, forging ahead beneath giant loads with a noisy stubbornness. In one village, side by side with a crowded, loquacious native market, stood a traveling-motor repair shop. Inside a huge truck's machinery whirred and grimy men busied themselves making whole the parts of many smaller trucks that were clustered around like a lot of patient animals. We had had trouble with our ignition, so we paused and asked for a new spark plug. The uniformed mechanic waited only to know the model of our car; a moment later he was back with what we wanted.

We dashed on towards the trenches with a breathless haste. We had to reduce our speed to pass a long line of lancers, trotting beneath the trees, raising the dust higher than their waving pennants. The sight led someone to exclaim, "Then they'll use cavalry again!"

Everywhere was evidence of the approach of a "great squeeze." The appearance of the villages altered. In each one, now khaki-clad forms swarmed and bronzed faces looked at us interestedly. Beside the entrance of each house and yard a sign had been painted: "Billets for fifteen men," or "Officers' billets," or "Stabling for ten horses." Restful legends for troops fresh from the trenches. We didn't have to be told that these men were not from the class in training. The lines on their faces, their air of confidence and pride, marked them as veterans. We were getting very close.

It is a curious fact that always on approaching the frontline you experience a sense of reluctance mixed with a desire to accomplish just that from which you shrink. It is possible at one moment to resent each turning of the wheels and the next to wonder at your good fortune in traveling in such a direction at all. But long before you reach your goal and you are aware of that strain that makes it wise to send men back to *billets*, and all that day the strain grows and colors the days ahead less alluringly.

Beneath a bland sun that had routed the mist, our nearness became evident when we swung into a road along a poplar-bordered canal. A sullen roar, exactly like the distant explosion of a giant cracker on the Fourth of July, disturbed the peace of water and shrubbery. For a moment it deadened the bird song. It seemed natural that we should sweep past barges painted white and stained with huge red crosses. It rendered quite superfluous Williams's explanation that the wounded, who suffer too harshly for ambulance or train transport, are carried in these craft smoothly to the sea and the hospital ship set for England. Its repetition, its constant recurrence now sketched a morbid picture, blurred with smoke—a sort of hell to which men go before they die.

We entered a large village and drew up before the headquarters of a division general. After the car engines stopped, the cannon chorus grew throatier, as if warning us back in a titanic fury. Captain Williams got out. "I'm going in to report," he said, "and to find out, if I can, what the Huns are up to. I don't want to get you fellows strafed if I can help it."

We sat in the cars listening to the ugly roar while we studied this nerve center of the fighting system. The headquarters was a large brick château set across a wide and pleasant yard. On the high verandah a group of officers lounged and smoked, and with puzzled faces appeared to listen, too. Sentries paced swiftly up and down before the steps. From a wooden shack at one side a brass horn, like an automobile signal, seemed perpetually ready to scream. Any doubt as to its purpose was resolved by a large sign on a house across the street: "Gas Post."

Our driver, exposing a friendly intelligence, explained, "Men caught in the street by an alarm go to one of these posts and await instructions. We don't take any chances with gas. A few days ago there was a high wind, and people in villages ten miles back of the line were slightly affected."

Williams, looking rather sober, came out and a bright young officer from the headquarters followed him. They climbed in and we twisted out of the village on to a road that crossed open fields. One guessed that it was in view of the German artillery, but we hurried along it towards a hamlet above which a shattered church tower was like a storm-swept beacon. The roar of great guns, no longer muffled by trees and houses, was thrillingly louder.

"What does it mean?" someone asked. Williams didn't answer

and the division officer, whose face was also a trifle perplexed, said, "Just a little hymn of hate." Suddenly he pointed and exclaimed: "I say! The Huns have got a sausage up. That means business." Above the tree-divided fields, seemingly quite close, an observation balloon (the shape of a sausage, indeed) floated at an angle. Two or three airplanes, with the appearance of gigantic butterflies, drifted lazily about in the sunlight.

We experienced a shutting off of all the wider future. We were merely grateful to get off that naked road and among the trees of the village. When the car engines were stopped again at brigade headquarters, the roar of the guns was perpetual and close, and interrupted now and then by heavier explosions. Clearly there was something ahead more exciting than Champagne or Lorraine had offered.

A brigade officer—a charming fellow with red hair and freckles—came out, shook hands, and announced that he was to be our guide for the trenches. He shared the seriousness of this event. "I see you have your gas masks," he said to me, "but you'll want helmets."

He waited as if for a reply and it seemed necessary to say something: "Yes, thanks. It would add a little to the romance," I told him. No matter what impression you make on other people at the front you have no illusions about yourself.

At the officer's command, an orderly brought a cluster of round, flat, steel hats. "They're good for protection against small shell fragments," our guide offered. Then grinning, he said, "They wouldn't stop a forty-two, you know. You've been to the French front? What do you think of their helmets? Both types are good, I guess."

The helmets served. Under fire any trivial topic, once started, is worn threadbare.

It seemed strange that this town, which the Germans must have known was a feeding place for the trenches, wasn't under constant bombardment. As we drove off, the brigade officer shifted from the subject of steel hats. "They've a town like this just beyond their lines," he explained. "If they throw a shell in here we retaliate, and vice versa. So for the most part it's hands off. Since they knocked the church tower about they've been pretty good but, of course, it's likely to come at any moment."

That contingency ceased to interest us, for already we were among the fields again, not immune like the town and on this side, nearer the enemy, ruined farmhouses and ragged trees scarred the landscape. Suddenly, the officer bent towards the driver and whispered. With a startled locking of the wheels the car stopped, then turned around, while the driver with jerky motions signaled the other car back. All at once there was a noticeable tenseness about the uniformed men with us. For some distance we scurried back the way we had come. We took a turn around a smashed farmhouse in the direction of the trenches. Beyond such signs of wreckage, beyond the rising clamor of the guns, there was something about that flat country basking in the sun that meant danger. We were in the heart of a vast army yet, except for ourselves, there was no human being to be seen. It was an interminable uproar in an empty place!—the ground seemed to writhe beneath it. The devastated landscape had an earthquake-like appearance, at which the bland sun mocked.

I shouted, asking why we had made that startled turn, why we had chosen this new road.

"Because," the brigade officer answered, "the Huns are strafing the road I had planned to take. I thought when we started their sausage looked a little close. This seemed safer."

But was it? It was obvious that the observers in the balloon, if they looked our way, could see us crossing the level fields. But our dash was brief. We drew up at a crossroads marked by the usual blasted house. An officer and a soldier sprang from behind the ruin, their gas masks striking against their hips as they hurried towards us. Beneath his steel helmet, the officer's face was troubled and disapproving. He hit at an automobile tire with his cane. "Get those cars away from here," he commanded shortly. "This crossroads is a nice place for shells this morning."

Several craters nearby were sufficient testimony, so we clambered out and, at Williams's direction, threw our hats in the cars, put on the steel helmets, and made sure that our gas masks were safe. We followed our guide around the ruin, while the cars with an air of flight dashed away. The brigade officer led me down a lane that offered scarcely more cover than the road. The others followed in a straggling line. My guide glanced back, nodding approvingly. "We're a less tempting target that way," he said.

I looked ahead. Fully a mile away, at the end of a lane, arose another ruined wall—the nearest shelter from the eyes in that distorted balloon. It assumed the remoteness and the desirability of an explorer's goal; the confusing roar of gun fire indicated its distance. Overhead, shells commenced to scream and, as we walked on, that evil sound came oftener and grew louder, until it too was near and perpetual.

Sometimes it was only a querulous whine. Sometimes it was like the hurtling of a great sky-rocket. Now and then, because of caliber and proximity, it reminded one of a racing automobile with all its exhausts open, streaking past within a few feet, yet unseen because of some obstruction. We looked up expecting to see the source of that hideous sound. Each steel scream—from its whining commencement to its crashing climax to a series of receding high-pitched sounds— was a matter of seconds. Something must be outlined up there against the sun. But always there was nothing and we walked on, wondering how men could dwell perpetually in such a racket. We were taught immediately that there are other irritants for a soldier's nerves infinitely harder to bear.

Rat-a-tat-tat.

It cut, apparently close at hand, under the curtain roar of cannon fire. Rat-a-tat-tat for long periods, a momentary cessation, then a recommencement. It suggested a woodpecker, gigantic and restless. It is the gloomiest and the most abominable sound of this war—a perpetual reminder that machine guns can spray more death and wounds than shell fire. You can't be sure of the source or direction of machine-gun fire—it may be after a number of targets, including yourself. The red-headed brigade officer, experienced in such estimates, walked a little faster and hesitated, before saying, "I daresay they've seen a couple of our men coming up with a water cart."

One feels a swift sympathy for those men and a desire to know if the soldiers for whom they had started would have to wait for water. But sharp fire begets selfishness, and just then shells began to drop in the field to our right. The sound of a number of screams did not diminish. They ended instead in fat, puffy explosions; in the cloudless sky, clouds—snow white and beautiful—were born.

"Shrapnel!" the officer muttered. "What are they after?"

From the rear came Williams's voice: "What do the Huns think they're strafing out here?"

And above the roar another anxious query from the officer, "Can they see us from their sausage?"

Before anyone could answer, four roars at intervals of less than a second heralded four formidable detonations and, not far off in the field, four sable curtains of smoke belched (apparently from the grass) and were drawn by the wind into ugly and impenetrable curtains. The idea of an earthquake-shattered landscape was strengthened, for about these sudden eruptions was the monstrous fortuitousness of nature.

A map that the officer had commenced to unfold was for the moment forgotten. Strangely, it was possible to express curiosity, as if these things passed on a cinema screen.

"I suppose they're high explosives," mused the officer, and the ruddy head nodded in response. Four more shells hurtled into the field, but only three volcanoes joined the black pall against the sky, leading the officer to shout, "Hello! A dud!"

The cause of his satisfaction, the meaning of that word, were apparent. Somewhere in the field lay a shell, from the supposedly perfect German ammunition factories, which had failed to explode. Meanwhile, other shells came too close for a civilian's comfort. We glanced at each jet curtain of smoke and studied the innumerable craters in the road. Doubtless, we all wondered if another crater would be formed too near at hand.

Our minds were reluctant to grasp or hold details, making it difficult to recall incidents that occurred only a few minutes ago. In short, it had become necessary to drive the memory to its task. From officers and men, I have learned that this closing of the mind to everything except the immediate future is nearly universal. For this they express a rather pitiful gratitude.

So we walked on, and nothing came too close.

We reached the goal of the shattered wall and caught our breath for a moment behind it. A straight highway receded between torn trees. On a split signboard a name was decipherable, familiar to anyone who has motored through Belgium and northern France. There were shell craters in every direction. The machine guns had resumed their hateful petulance. We knew that the communication trench must be near. No one asked, though, as it was easier for the moment not to talk.

The brigade officer folded his map and thrust it in his pocket. He led us around the wall and into a screen of bushes from which a narrow passage sloped downwards. We descended only a little way and found that the walls had been artificially raised.

"That's the worst of trench digging in this blessed bog," the officer said. "Go down two feet and you strike water. Trench walls have to be raised like these. They're a lot easier knocked over by shell fire, too."

We had no criticism to offer of the communication line. To be sure the close sides admitted none of the pleasant breeze. Those steel helmets were demanding the price of their protection—they bound one's temples and constant perspiration rolled from beneath the brim into one's eyes. But I had never dreamed what a friendly place a communication trench could be. It was good to touch the yellow walls, supported by rattan work, to know that a shell would have to make a direct hit to limit our progress now. Here and there, as a matter of fact, there were breaches in the walls, but for a little while the crying in the sky was mournful rather than angry and the explosions were muffled and farther away.

We circled a number of the usual traverses and machine-gun emplacements, but the trench was surprisingly short. It scarcely gave us time to smile at Tommy's imagination expressed on neat signboards at the junctions. These had, it appeared, the official stamp, for our guide spoke of such thoroughfares as Oxford Street, Kingsway, and the Strand, as if he had been conducting us through the peaceful racket of London. The Strand went straight to our destination, and we emerged from it into a wide plaza, enclosed on the opposite side by a parapet of interlaced logs and sand bags. A few silent figures, with rifles through loopholes, braced themselves there. We walked with an air of stealth. When we spoke our voices were lower! We were in the frontline.

Chapter Thirteen

THE DAY'S WORK OF LIFE AND DEATH AT THE FRONT

Frequent traverses, of the same construction as the parapet, stretched at right angles to protect the men as far as possible from shell and grenade fragments and from the enfilading fire of machine guns. We were to learn the wisdom of that precaution before long.

A trench officer strolled around the end of a traverse. He wore a uniform of the same quality as his men's, for the hard facts have been realized here, too, and officers no longer expose themselves or engage in bursts of foolhardy bravery. The German sniper has a little difficulty now in distinguishing officers from men. The officer, with his round helmet and oriental appearance, came up and greeted us gratefully. We evidently broke the monotony of his watch. In his eyes was something of the universal strain, but he spoke easily, asking the question that had spoiled our walk and troubled us all: "What were the Huns strafing back there?"

The fact that we couldn't tell him illustrates the vagueness that surrounds everything for the individual in this war. Out here men even die with a certain vagueness.

"How are things with you?" Williams asked him.

"Fairly quiet," the newcomer answered, "just now." He glanced quickly around as if expectant of something.

We walked on with him, subdued by the gun roar and the constant sight of those armed figures, who were braced against the parapet, peering through loopholes, and quite motionless yet expectant, too. Openings to dugouts made black patches against the reddish earth at the base of the parapet. The men at the parapet were sentinels, but the larger part of the command must lurk in these holes. Upon entering one, I saw three forms, quite the color of the earth on which they lay, crowded into this tiny cave. Their log-like sleep suggested the cultivation of a log-like mental attitude, or the type of fatigue that was beyond the debate of one's nerves.

"What about the rats?" someone asked the trench officer as I emerged. "See any rats down there? At home they say the rats are so bad they actually eat the soldiers' faces."

"I can only speak for my own men," the trench officer said, spreading his hands. "Most of them, when the rats begin to eat them alive, wake up and say, 'Shoo.'"

There has, perhaps, been as much written about the vermin as bullets. For a moment the subject clung—probably because it kept us from looking too far ahead. It is impossible to exaggerate the bullets, but we began to suspect that imagination had played with the other, for these men were fairly clean. While their uniforms were marked with last night's mud and whitened with this morning's dust, they required no more radical antidote than a brisk brushing. Trenches are dirty and uncomfortable, but I didn't see here the disorder of body and clothing as is observable among any gang of laborers engaged in excavation work.

"Conditions," the trench officer said, "are naturally better than during the winter and early spring, but experience as well as the weather has got something to do with it."

"What about the activities of certain unpleasant small life?" someone asked.

He paused. Across the plaza we saw a few groups under non-commissioned officers, twining those deadly globes of barbed wire, invented by the French for the blocking of communication trenches. Others worked with trowel and cement at machine-gun emplacements. Some made repairs on the parapet, where an ugly lack of uniformity recorded the recent entrance of a shell.

"Those chaps don't look particularly fidgety, do they?" he asked. "If our little companions have largely left us it's because shorter periods in the trenches, compulsory baths, and a complete change of clothing once a week have made us less enticing for them, and a lot fonder of ourselves."

A harder burst of firing directed his glance towards the parapet. We crowded at his heels in the direction of a periscope. "Their sausage is keeping them busy this morning," he said over his shoulder. "By the way, any of you fellows heard news of . . . ?"

The freckled face of the brigade officer darkened. Williams wanted to know what about him?

"Went up in one of our balloons yesterday," the trench officer answered. "A lucky shrapnel shot cut the cord and we could see him from here drifting over the trenches while the Huns shot their heads off."

"I heard this morning," the brigade officer said, "that somebody had seen him cut loose his parachute."

"Not much chance that way," Williams mused. "The anti-air guns would get him for sure. He'd have dropped in their lines anyway."

"Nice chap, he was," the brigade officer muttered. "We've been hoping for news all morning."

Putting his eye to the periscope, the trench officer said: "I wondered." After a time, he looked up and continued, "Perhaps you'd like to see the Hun trenches. If you raised your head above the parapet you'd make good practice for one of their snipers. Try this."

In the glass at the base of the periscope appeared a forest of posts rising from a jungle of grass and barbed wire. Beyond, very close at hand, lines of yellow dirt and sand bags zigzagged across the landscape, curving towards us to the right and left. A trifle puzzled, I glanced back at the British trench walls and saw that to either side they fell away before these sudden swoops of the enemy's lines. We were, it appeared, in the apex of a small triangle and subject consequently to attack from three sides. Phrases I had skimmed in the official reports flashed back with a new eloquence. I understood quite thoroughly, now, the meaning of, "We straightened a small salient today."

"That's our line," the trench officer said, grinning. "Great salients and small ones. Little fellows like this breed local trouble. Only comfort is, it's as bad for the Huns as it is for us." He drew from his pocket a narrow cylinder, not unlike a small telescope. "It's a hand periscope," he explained, "rather useful thing—magnifies a bit. Want to try it? Put the end over the parapet and squint in the eye hole. That's the notion."

I gave it a try and the ugly yellow ridges seemed closer, the waving grass more distinct and larger. There was no use looking too carefully because of the sinister souvenirs of night attacks and patrol work the grass in No Man's Land nearly always harbors. But the ridges fascinated me. They were like furrows ploughed by a drunken giant. They offered no evidence of the multitude of men they sheltered; and yet, if it hadn't been for the gun roar, we might have called across to them without raising our voices much. We could picture a routine within their hollows similar to our own. But at any moment a trivial variation over there might send death stalking close to us.

"How far are they?" I asked.

"Something less than a hundred yards, I should say, from here to their frontline," he replied, shifting his weight from one foot to the other. "You know, they're not bad at potting periscopes."

At that distance, they could recognize this mahogany cylinder for an officer's periscope. Just then a machine gun jibed at the heavier roar. Rat-a-tat-tat—spraying death as a garden hose sprays water. I glanced up at the top of the periscope to see if it trembled.

"I say, that thing was a Christmas present. Move it about a bit," said the trench officer. He seemed relieved to have it back again. The machine gun subsided. "Might give them some of that back," he said, pointing to a group squatting on heels about a sergeant.

"The hornets seem stirred up enough this morning," one of the others offered.

We joined the group and found in the center of it a machine gun whose mysteries the sergeant explained with the deportment of an old-fashioned schoolmaster. He was glad to have fresh scholars. He opened and closed the breach, inserted a belt of cartridges, and commenced to run it through.

The trench officer stooped down and ordered: "Throw that safety block back!"

The sergeant obeyed with an aggrieved air, while mutely we thanked the officer for preventing drawing any unusual attention to that particular traverse. In place of a practical demonstration, then, the sergeant pressed with both thumbs on a steel plate. The cartridges swirled through, flashed into the breach, and out through the escapement.

"As long as I keeps pressin' down on this plate," he said, "she keeps spittin' and somebody don't like it. The water in the jacket boils when she's spittin' hard. You have to watch out for that."

Evidently we showed a little distaste for the brutal perfection of the thing. The sergeant was a trifle offended, I think, at our haste to leave his class. Around the next traverse we ran into another scholarly group. This time a flimsy tripod stood on the trench floor. One of the Japanese, who had so far observed without saying much, was now aroused to ask a question: What was this used for?

"It will interest you," the officer said. "It's one of the things with which we make ourselves most scandalously miserable in the trenches." Behind his banter was a wistful seriousness that you understood as he went on, "It's for throwing rifle grenades." He picked up a black, pear-shaped object that differed from the ordinary hand grenade in just one particular: a long, slender, steel rod protruded from one end. The hand grenade, he explained, was satisfactory enough when the trenches were within throwing distance, or for a swift dash across No Man's Land and a retreat through the night, but there were many hours of daylight in a place like this when it wasn't wise to let the other fellow feel too much at ease. He passed the grenade around, cautioning us not to release the safety pin.

"The usual pattern," he said, with a reminiscent frown. "When you draw the pin the spring flies back and fires the fuse. If you don't throw it then there's general hades. Maybe you've heard. A couple days ago in a bombing school a new man was standing by the instruction officer who was showing him how to release the spring and throw. The soldier had drawn the pin and, as new men do now and then, had gotten a sudden touch of panic. The instructor shouted at him: 'Throw that thing away, man! For God's sake, throw it away!'"

"Poor devil!" he continued. "You see in his anxiety about the other he'd quite forgotten he'd drawn the pin in his own grenade." He ended with an exclamatory gesture.

Stroking the corrugated surface of the grenade, Williams said, "Not so large, yet one of these things will do in a score of men."

The trench officer took it from him and slipped the end of the rod through the apex of the tripod. A soldier, whose bent attitude was suggestive of worship of the toy-like affair, placed a blank cartridge in a tube at the base. The officer lowered the rod against the cartridge. The soldier stooped closer, manipulating a graded quadrant.

"Range is correct, sir, to drop it straight into their trench," said the soldier.

Williams started to speak, but the brigade officer laughed and said, "No, thank you. Our friends over there are jumpy this morning. They'd send a few back in our direction."

What happened then had the blind irony of chance. It was, indeed, that slight variation of which I had spoken of before. From a point not far ahead came a sharp crack, barely audible and lost at once in the general uproar. Williams seemed inclined to hold us back, but we went on after a few minutes. As we turned the corner of a traverse we saw a quiet form outstretched. Someone had already flung a blanket over his face and shoulders. Five minutes ago that form must have been alert at sentinel duty on the parapet. Now, someone had taken his place and he lay there, exactly the color of clay except for his boots. Those boots were too black, too heavy, the stillest things you have ever seen. Feet held in such a way ought to twitch occasionally. The multitude of studs on the soles were designed to keep that man, who would never do anything again, from slipping.

We knew why he lay there. A grenade had come in from just such a machine as we had been inspecting. He lay there in order that the other fellow shouldn't feel too much at ease. And how many more lay like him the length of the trenches that morning with studded boots outstretched in a sickening stolidity?

We walked neither slower nor faster. We didn't vary our talk about the catapult we had just seen, about the further clever tricks of trench warfare designed to keep the other fellow from feeling too much at ease. I remember Williams mentioned the whiz bang—too jocular a name for a shell that drops in and discharges multiple explosions—and the trench mortar, a huge and awkward ball that tumbles on the opposite parapet, where it either kills directly or buries men alive because of the blasting explosive it carries. Two thousand casualties, they told me, in this division since December, while the enemy opposite had suffered probably a good deal more, and all from this process of keeping the other fellow from feeling too much at ease.

"I can remember," Williams said as we walked along, "when the sight of a dead man stirred me up most unhappily. Now I don't pay much attention. You can't. Understand? You simply can't."

You can't and keep on at war, is what he meant. That explained, too, probably, the astonishing ease with which one learns to like or dislike men at the front. You can form a thorough-going friendship in a day. That's because a man realizes his opportunities may be limited.

Other officers greeted us and walked a little with our party, chatting above the noise of guns. In London I had seen soldiers leave Charing Cross railway station with the trench stains still on their uniforms. They had seemed a little mythical. Out here, at their daily tasks, they were quite human, as if the whole world were like this, as if it had never been cleaner or kinder, as if it could never change. So we strolled on, with an expectancy lurking in everyone's eyes—an uncertainness, as to whether beyond each traverse some sudden and monstrous surprise wasn't waiting for us. I was glad to see a new man smile as he pointed to the entrance of an officer's dugout.

"Like a peep at the palace?" he asked us, with a pride behind his smile that was perplexing. We followed him down half a dozen steps into a small chamber of uncommon neatness. The walls were boarded and adorned with racy pictures torn from a French weekly. There was also a cot bed, a deal table, and a stove. Expecting it was the office of at least a general, we searched for him in the dusk of the corners. Two young subalterns, however, alone greeted us, and we recalled that generals don't go to the trenches if their staffs can keep them out. Someone in our party congratulated the subalterns on their stove. One of the youths patted it as if it had been a pet, and said: "It is a comfort on a cold morning, and it's often quite cold even this time of year." He, too, let slip a little of that prideful air. We chorused a demand for its source.

The man who had brought us in waved his hand, explaining: "You see, when he was on this front, this was the home for several nights of the Prince of Wales." In a mournful tone, one of us expressed a hope that in those days the racy pictures of scantily draped femininity had not decorated the walls. One of the subalterns, with a meek air, accepted the responsibility for the pictures.

We went out, smiling, but more convinced than before of the dynamic democracy of this struggle, for there was nothing palace-like about that dugout. It was not, as we define such things, even comfortable. It was, we found, almost next door to a kitchen. I ventured in there on my hands and knees because of the meager-size opening. A soldier, bent double as I was in the shallow, smothering chamber, grinned a welcome. He brushed the perspiration from his face and lifted the covers from three camp kettles beneath which coals glowed. Bully beef steamed appetizingly. Low shelves were filled with bread and jams and tins such as I had seen at the convalescent camp. The cook waited, apparently, for some congratulatory comment from me.

"This looks pretty good. And it smells good," I told him.

"I hear mighty little grumbling," he responded, with a grin broadening on his wet face.

Imagine, the usual culinary pride in a place like this! If we could have carried it from the firing line, that meal wouldn't have offended any of us. As I backed out, I heard the brigade officer's cheery voice.

"Maybe you'd like to see one of the few men out here who doesn't worry much about his dinner," he said quietly.

We nodded, a trifle mystified. Cautioning us not to raise our voices, he led us into a protruding section of the trench and beckoned a corporal who was clumsily sewing a rent in his uniform. We waited in front of a dirty, brown, canvas curtain that veiled a portion of the face of the parapet and was perhaps six feet wide and three feet high.

"It's a sniper's post," Williams whispered.

The corporal knew what we wanted; without words he slowly lifted the dirty canvas, disclosing a nest in the parapet cased with steel plates. A stout young soldier crouched in the heat and the darkness of that place. He swung around as if grateful for the light and the air. His face was wetter than the cook's but he turned back, replacing his eye at a small loophole in the front wall.

"Wait a minute, Owen," the corporal muttered. In response that round, young face studied us again. "What's your bag this week?" the corporal continued.

The sniper's lips opened, showing teeth; the grin colored his tone. "My bag?" he said. "Ten periscopes and five Huns." Death is such an impersonal thing nowadays—his pleasure seemed scarcely more out of keeping than if he had spoken of rabbits.

"Pass out the boy that did it," the corporal said.

The grin failed and the rifle was offered to us reluctantly. While we glanced through the telescopic sights the sniper remained crouched, as if ready to spring upon us if we took any liberties with his treasure. He didn't relax until he had his gun in his hands again. Then he dragged it in front of him and turned away. He was exactly like a child whose favorite toy seems threatened by the incomprehensible curiosity of a grown-up. He uncovered a hole large enough for the sighting of a rifle.

"Not so fast," the corporal warned, and to us he apologized. "The Huns are pretty sharp at this game, too. With the curtain up they might put a lucky shot through that hole into one of you." He dropped the dirty canvas and rubbed his hands. He was as proud of Owen as Owen had been of his rifle. Why not? Five Huns! I have heard a good deal of argument as to the value of this sniping, and that did seem a good bag for one man. As a rule, however, some of the French argue such tactics make the Germans too wary. It is more profitable, they think, to encourage carelessness, to foster a sense of security until men gather in gossiping groups. Then a shell from *a soixante-quinzes* at close range bags more in a second than a week of sniping will drop. The Germans too, I understand, are divided as to which method produces the better result. Either way it is also designed to keep the other fellow from feeling too much at his ease.

Chapter Fourteen

THE APPALLING MINES

We walked on discussing this and forgetting the most gargantuan and terrible practice of all. A serious-faced subaltern, standing with his elbow braced against the corner of a traverse, reminded us. From a distance he had an unusual appearance. As we came up we saw it was because of the degraded state of his uniform—worse than any private's we had seen. It wasn't the familiar yellow mud that stained the brown cloth, that dried on a soldier's cheeks and hands. This man was nearly blue from head to foot.

"Where is there blue mud around here?" we asked.

Something of the subaltern's haggard expression was reflected in Williams's eyes, as he replied: "Blue mud?" he repeated. "There!"

We could see now, behind the stained man, a heap of bluish, shiny soil from which water still oozed, running blue and shallow across the floor of the trench. Blue mud! Blue water! Williams introduced the subaltern to us, and he made a wry face and tried to rub the muck from his fingers before shaking hands. He glanced doubtfully at Williams who drew him aside, speaking quietly. "If you wish," the subaltern said, nodding. And, with a stealth greater than we had exercised at the sniper's post, we followed him along a narrow gully that had walls heaped with the blue stuff and a floor like a stream. "Walk carefully," he told us.

Because of the slimy footing, it was really difficult to remain upright. We constantly caught our balance against the yielding soil. Therefore, we didn't see at first the grotesque and uncouth figure that crawled from an opening similar to the entrance of a dugout. We paused, startled by this figure—a prehistoric-like creature leaving his lair and sizing you up for defense or attack. From head to foot he was blue and dripping; the mud was in his ears and thick through his matted hair. Before he could rise the officer spoke to him, and he remained squatted in the opening.

"How deep are you?" the officer inquired.

Scarcely expecting intelligible words to issue from such a creature, he answered with rough good nature: "I'll climb down with a candle so you can see." In that narrow hole, there was room for only one man at a time, and it was necessary to enter as the other had started to emerge, on hands and knees. "Don't slip," the blue man grunted.

In a moment your eyes grew a little accustomed to the light. A wooden platform, burdened with pipes, overhung an apparently bottomless pit.

"The pipes are for carrying off the water," the creature said. "We have to pump almost from the first spadeful and it's pump, pump, pump every foot we go down and, when we get down, every foot we go out." He struck a match and applied it to a stump of a candle. He swung over the brink, fumbling with his feet for ladder rungs. I heard him scramble down, holding the candle in one hand. His face was no longer visible, his candle was a mere speck, and when he called up his voice was muffled and far away: "We strike out from here," he said.

Yet no sound of tools came up. In almost complete silence, that sap was creeping towards the German trenches a hundred yards away. This uncouth creature and many like him were daily accomplishing a task that made ordinary ore mining look like pure recreation. He came puffing up the ladder. The sun and the outside air were pleasant to him.

"How fast do you go?" I asked the subaltern.

It varied, he answered, sometimes two yards a day, sometimes more: "It depends on the soil and the size of the sap. Usually there is room only to work and pass the excavated soil back in baskets." The baskets of soil, we saw, were raised to the surface and used to strengthen old parapets or to construct new ones.

We looked at this officer, who was scarcely more than a boy, with unqualified admiration. The fact that all along the line, from the sea to the Vosges, other men were performing identical tasks made no difference. He reminded us that the Germans were pushing similar saps in our direction, and that one might explode beneath our feet at any moment. A rather depressing thought, but he encouraged us with a smile that cracked the mud on his cheeks.

"I think we have a better system of listening than the Huns," he assured us. "We like to think we can detect their saps here before they get too close."

His easy talk called up a whole gallery of unhappy pictures—men crouched in listening posts, or creeping towards the German trenches at night, from time to time pausing to lie with an ear to the ground, in constant fear of a star shell that might point them out to a sniper or a waiting machine-gun crew. But more compelling was the recollection of that crouched, filthy creature. I imagined him stretched in the narrow tunnel, digging away as stealthily as possible the soil in front of him, quite at the mercy of the German listeners, perhaps breaking through into a rival sap-head (a listening post) and fighting murderously in a narrow hole. When a mischance occurs during this mine work, a burial isn't often necessary or possible.

As we walked on after that, we thought about what might be going on beneath our feet. Certainly mining alone is enough to keep the other fellow from feeling too much at his ease. Imagine trying to protect yourself, day after day, from all the enemy's noisy devices of death, knowing each moment that mines are creeping towards you, wondering each moment if your particular section has been chosen, anticipating each moment the crumbling of the earth beneath your feet, and then a roar, a disintegration as important for you as the end of the world.

One must visit the frontline to put life into the dry-as-dust phrases of the official reports. Now lines such as "we exploded a mine and consolidated the crater" carries more horror than the blackest tragedy ever written.

We were glad to continue on and follow the brigade officer up a path marked "Sniper's Avenue," which proved to be a communication trench that led us out of the reach of mines. I wondered if all of us had counted the hours spent on the frontline. We had, I know, glanced at our watches more frequently than one does at home. I wondered if every soldier who is condemned to the trenches for days also counts the hours, the minutes, until he can walk along a communication trench away from the things that keep him from feeling too much at ease.

At a turning where the wall had been broken down a little by a shell, we were greeted by two sharp reports like the snapping of a whip. We had an uncomfortable feeling of having been shot at, but surely the noise had been too close.

"Those were probably our snipers," the foreign office man said.

The brigade officer shook his head. "Huns, I think," he answered shortly, with his freckled face losing its good humor. Though the puzzle concerned us all, he would say nothing more. We then climbed up, a little reluctantly, from the communication trench to a shell-torn road. But Williams looked over his shoulder and said, "They've pulled their sausage down."

The brigade officer glanced at his wristwatch, saying in a matter-of-fact tone: "About time they knocked off for luncheon." When he read the surprise and distaste on our faces, he laughed and continued, "Friend *bosche* is methodical if anything. He usually has his hour for a comfy feed."

It was evident that the fire from the other side had diminished. In desperation, some of us took the insufferably hot helmets from our heads. Trusting our guide's perfect faith in the German schedule, we followed him across a field and were disturbed by nothing more than an occasional shriek from the sky.

"I told the driver," the brigade officer said to Williams, "to have the cars at Snipers' House."

If ever a name suggested a dramatic incident of stealthy warfare that one did but, in common with most of the soldiers' christening of landmarks, its origin was clouded; nor, when we had come to it, did it offer any evidence of its own. It was the familiar roofless quadrangle of shell-shattered walls. Whatever its romantic past, it was now a prosaic rendezvous for members of the transport service. Nearby, a narrow tramway descended to a communication trench and ambled to the frontline. We scurried from the shelter of Snipers' House along the devastated roads to brigade headquarters.

"With their sausage down," the brigade officer said by way of farewell, "you ought to find the road to division headquarters comfortable enough."

We did, but we took it in a rush. Once there, the general welcomed us for luncheon in his chateau. There's no question that he drew, in everyone's memory, a firm and impressive portrait: tall, powerful, yet with an easy manner of movement and speech. It was only his iron-gray hair that hinted at his real age—about sixty, someone confided. Although he had retired from active service some years before, he had enlisted this entire division, trained it, and commanded it during six months at the front. He explained he was sorry that a corps conference had prevented him from seeing us that morning.

That quiet hour granted us by the German routine was happily out of key with the rest of the day. Those staff members who weren't on duty sat with us around an oval table, skillfully laid and served.

"Any news about...?" someone asked.

The general shook his head, but a captain said: "His balloon fell in our lines. It was riddled."

"Splendid chap," someone added softly.

Luncheon commenced. There was tactful talk of America, our position in the submarine controversy, our political conventions, the possibility of our entering the war. There was—as always at such gatherings—an undercurrent of wonder, never quite reaching the surface, that America should have found it to their best interests to have remained aloof.

I gathered, not particularly from this conversation, rather everywhere in England and France, that a belief had grown since the beginning of the war of America's lack of homogeneity. We were, it was suspected, incapable of direct and concerted action. In those days, the men who were actually treading the exhausting mill frequently placed upon us—whether justly, who can tell?—the taint of many races, the incoherence of too vast a variety of creeds and desires and antipathies.

The general called my attention to the officer on my other side. He wore the facings of a major. He was small and of a scholarly type, so it was unlikely that any extraordinary experiences lurked behind those quiet eyes. A moment later we learned it was a miracle he could sit with us at all: he had landed with the first expeditionary force under General French, had fought at Mons, had survived the nightmare retreat that had ended with the officers' corps cut to pieces. He spoke of it quietly, yet with no false hesitation, no careless clouding of the facts. With the rest he had learned out here to face facts for what they were worth. He wasn't surprised at our interest. He wasn't bored by our questions.

"Individually we didn't know much except that we were going back, turning and fighting Huns without end," he began to explain, "and slipping out of the net when it got too tight. The men were mad . . . through and through mad, because it's harder to fight and die on the run than any other way. At night, black and fagged out as we were, we lost rest asking when we were going to turn. After an eternity one evening the word came. The French commander had visited ours. The next morning the stand was to be made, the great battle fought. Tired as we were, we didn't sleep much that night for the relief and the joy of it. And when day dawned the word came to fall back again—we went with heads down, sullen and ashamed. It lasted for two days more. You can't know. Then the definite stand was made and the push to the Marne and beyond. It was what we had craved, because we were like people caught in a fog."

Another inevitable question was asked: "How, with the German artillery on the hills and the bridges down, did you ever cross the Aisne at Soissons?"

The major smiled (his scholarly face was very pleasant when he smiled) and replied, "I rather fancy they set a trap for us there they never had the strength to spring. Probably we were intended to cross to the other side where they expected to fall on us and finish us off. It's obvious, isn't it, when the men crossed in small boats or walked across stringers of which the Huns must have had the exact range?

"I paddled over," he went on, "with a squad in a row boat. You know, the tiny tub had Titanic painted across its bow. Really gave me a start. It seemed an omen . . . a properly bad one. But, thank heavens, the omen didn't work. That Titanic made a safe crossing . . . didn't get a shell near enough to make us jump."

He poured thick cream over a fruit compote, eating the mixture with a visible appreciation. Later, he smoked a cigarette with the same air of a sybarite. Clearly, like so many out here, he had learned to draw from each moment its maximum gift.

After luncheon the general led us to a rear veranda overlooking a formal garden with shrubbery, in which portable huts nestled for the housing of his staff. But we were chiefly interested in the facsimile, placed on a square table, of the entire countryside occupied by his division. Each hill was there, each road, each house, each line of sheltering trees, every slightest branch of the German trench system. Even with modern air scouting, such minute knowledge of the enemy's position drew exclamations of admiration from us. The general showed us how it was obtained, summoning one of his staff who brought handfuls of airplane photographs that he fitted together end to end, side by side, diagonally, with the minute difficulty of a jigsaw puzzle, until it was possible to see a complete photograph of the war-scarred countryside. When the officer hesitated too long or put into his puzzle a piece that didn't fit, the general rebuked him gently with the manner of an employer in a business house or a factory. Men are killed and money is made with precisely the same discipline.

"Of course," the general said, "the Huns know just as much about us as we do about them." He had that hospitable willingness like all the officers I had met to answer questions. He even promised to take us later in the afternoon to inspect some of his hidden artillery. After viewing the facsimile, the general took us into a wooden shed, furnished with rows of benches, telling us of his trip to Paris to purchase this cinema outfit.

"Every night they come here in hordes," he cried. "The men pay a penny and the officers a franc. You know, if the war lasts long enough, I wouldn't be surprised if we got back the price of the affair." His enthusiasm led him to close the doors and run a reel through the machine. It happened to be a review of this division by the Queen before its departure for the front. The long rows swung by and the officers began to recognize faces and to talk. There were some we remembered—the general's, for instance.

"There goes poor so-and-so," said one, "the Huns did him in with a trench mortar a month ago."

"Hello! There's Jerry . . . home, minus a leg," said another.

"The men like this thing because they see old friends," commented one officer, "who they won't see any other way now, walking along with them."

It was an abominably depressing performance. Something in the mechanism stuttered, the light flashed out, the screen went dark. Our active showman was full of apologies as he ran stumbling about the stage.

The film continued. For a time, the sergeant hurried the men through conventional evolutions. Then a new manual, born of this war, followed. The sergeant snarled out the commands as if he hated them, as if the words had to overcome a revolt in his throat: "Put on . . . . gas . . . masks!" he cried. The men then sprang into clumsy attitudes: resting their rifles in the crooks of their left arms, they tore open the bags at their right hips, snatched off their caps, and drew the masks over their heads, buttoning the ends into their collars. With a straggling haste, they took up their rifles and returned to attention.

My first impulse was to laugh. The brown faces were featureless save for round, staring goggles. They retained no individuality, no human semblance. These hideous figures might have been visitors from a far planet, or monstrosities escaped from this earth, too violently disturbed. As they walked through squad formations, the voices of the file leaders were choked and tongue-tied. "Halt! Takeoff. . . masks!" shouted the sergeant. The last word had the quality of a shriek, angry and threatening. You glanced at your own mask, responding to the sullen temper with which it had always filled you.

"They're quick," the instructor boasted. "Some of the men need scarcely half a minute. It's wise to be quick at that game. Want to see the gas house?" he asked us.

Chapter Fifteen

GAS SCHOOL AND THE ARTILLERY

The instructor led us to a small, unpainted shack in the center of the field. The joints of its doors were constructed so it could be hermetically sealed. A single cylinder stood in the corner.

"What the deuce is this for?" someone asked.

The youthful instructor, who ought to have been at a different sort of class himself, smiled and explained, "It's a splendid institution. I put every man through this at least once. Go in with him, shut the door, and turn on the gas. He knows he's getting it thicker than he ever could in the trenches. When he comes out he's got confidence in his mask. He doesn't go around mooning and scared to death about the next gas attack. It teaches him to know the difference, too, between gas and phosphorous bombs and smoke pots."

We confessed our own need of preparation.

"This new gas," he said, "41, is terribly hard to see. If it shows at all, it is like a slight mist. It's the other way around with phosphorus and smoke pots. Sergeant, bring up some of those bombs," he instructed.

Again, we settled ourselves into the attitudes of spectators at a game. The sergeant came up with a basket filled with fat candles and tins of the size and appearance of tomato cans. The officer picked one up, touched his cigarette to a fuse at the end, and tossed it on the grass a few yards away. "Don't move," he grinned, seeing our startled expressions. "Only enough explosive to set it off well."

The tin puffed like a faulty firecracker and out of it sprang an unbelievable volume of pure white smoke that formed perfect and beautiful curling patterns as it blotted out the lower end of the field. The sergeant threw one or two more and placed candles nearby, from which vast clouds of smoke, sooty or orange-colored, hissed wickedly. A thick, velvety curtain banded with yellows and whites and blacks was drawn across the field. In its fringes, the form of the sergeant was lost now and again.

"The merry villagers," Williams said, "will picture the Huns at their doors."

Indeed, we heard one or two shouts and, as we walked through the drift of smoke, we saw French children squatting on the fence and pointing and laughing and admiring.

"French children aren't easily alarmed," the instructor grinned. "I'll wager they can tell you the caliber of each one of those guns you hear firing over there. They know just what I've been up to. It's as good for them as stealing a peep at a cricket match."

I held my breath as I walked through the vapor and I asked if the fumes weren't dangerous. He shook his head and explained, "They sometimes use a smoke curtain to veil a gas attack, and at home I daresay cinema devotees fancy this stuff is gas. It is useful to veil any kind of an attack. Whenever it appears over the trenches it keeps the other fellow guessing."

We shook his stained hand and returned to the cars to keep our rendezvous with the general. At his headquarters, the general's limousine was waiting in front. He came out, climbed in, and the cars wound out of the village. With a sense of shock, we recognized that road. The shattered beacon of the church tower was straight ahead. We hadn't realized that in order to visit the batteries we had to return to the brigade headquarters village. And there was a noticeable change: instead of the one balloon we had seen that morning, two observation balloons of the enemy were suspended in the sky like monstrous planets visible by day. Our drivers responded as if to a signal and the cars jumped ahead along the naked road. The lull of a moment was lost in a sudden rush of sound. Perhaps we had been seen from the balloons and a range signaled? Above the roaring of guns we heard shells shriek; overhead, puffs of smoke were born. The roar became continuous, other puffs appeared.

"Look at that!" the driver of our car cried.

The other cars were far ahead, so we sprang after them. The wind shrilled past us. We tore through a black curtain that had followed a heavy explosion. Jet black sheets of smoke waved close at hand. There was nothing to do except to get every ounce of speed out of the cars. There was no point in leaning forward. The cars were like great beetles scurrying from a foot that tried to crush them.

In a moment, we were skidding to the right among the trees of the brigade village. As we reduced speed, I saw a number of French civilians run from an *estaminet* [a small café] towards the boundaries of the trees, where they stood gaping at the rolling black smoke.

"Why aren't they hunting a cellar?" I asked.

The driver snickered. "Those old Frenchmen! You see they live here. The village isn't bombarded much. Some of those shells came pretty close. They don't want a cellar. They want to see why the Huns are strafing so near their front doors. And say, they don't want to miss anything anyway. But they'll be mad to have their appetizer disturbed."

One felt rather sorry for the Germans, because all along they've thought they could scare the French. That's one of their excuses for being horrible.

The incident was a prophylactic for our own apprehension. We were grateful enough to drive up, unscathed, to a battery commander's headquarters at the edge of the village. We saw the general standing in the middle of the road, surrounded by anxious officers. Williams drew me aside and laughing nervously, he said, "The general has been asking if you fellows know you've been under heavy shellfire. A piece of one of those high explosive shells, he said . . ."

"I think I know it," I responded meekly. The moment of apprehension over, the immediate future was the vital concern.

A ruddy faced colonel walked from the house, as thoroughly
disapproving at the sight of the general as the staff men were. He
opened with the question that had become altogether too familiar to
us: "What are the Huns strafing over there?" The general couldn't
answer this, but someone said, "They seem to be after a lot of things."

"At any rate," the general then proposed, "these fellows have been
strafed so hard today I want you to take them out to a battery and give
them the pleasure of seeing some strafing back."

"Run your cars down the road and back of that shed," the colonel
suggested.

"I'll have to be getting home," said the general.

The discontent of the staff officers increased, probably at the
thought of his returning on that road, but the general smiled, saying
goodbye easily. We saw him leave with a real regret. We listened
anxiously for a fresh burst of firing from that direction until we knew
he had had time to reach his headquarters.

The colonel got his walking stick and led us around the house.
"You don't mind crossing a field?" he asked.

Outwardly we expressed that our route was a matter of indifference, but I think we had all had enough of fields. In the open country
the twin balloons were like the eyes of an angry god. Certainly it was
just one mile (as the colonel had mentioned) to a farm that showed
amazingly few scars. Within a stone's throw of it, the battery nestled
in a scanty grove of trees—a row of log and sand bag redoubts that
appeared to offer no real protection from scouts in airplanes. But
every battery I saw, every huge gun brought up for a bombardment,
seemed dangerously unprotected. Actually, a few twigs, scattered bits
of green, make an impenetrable veil against the prying airmen.

We opened a wooden door and descended into one of the
redoubts. Half a dozen men, scrupulously clean unlike the trench
Tommies, sprang to attention in a circle about the breech of a
howitzer. The gun was as clean as its grooms—wickedly beautiful
and capable. The colonel muttered orders to a sergeant who nodded

to the artillerymen. One lifted a projectile from a compartment in the wall, others inserted the charge behind it, and a corporal closed the breech. The sergeant entered a cubicle at one side where a desk squatted beneath a telephone instrument. He bent over a piece of paper pinned to the wall and from it rattled off a series of numbers like a football signal. In response, the neat men elevated the gun's great nose with an impudent ease.

The sergeant glanced up, and demanded: "All ready? Lower your screen."

A soldier released a cord and a shrubbery screen before the mouth of the gun fell away with a slight rustling.

The colonel glanced at us, and said, "Maybe you'd better put your fingers in your ears."

I noticed that everyone in the small chamber had his mouth open, as if gaping at an unforeseen phenomenon. For the first time the sergeant's voice lost its monotony. It made us jump. "Fire!" he shouted.

The sleek barrel sprang outward, then staggered back upon itself as the cylinders took up the recoil. The men's mouths snapped shut as they flung back the breech and prepared the gun for another charge. Our ears still sang. The air in the redoubt seemed thin and of an odd odor scarcely like burnt powder.

"Where did that one go?" asked the foreign office man, with his voice no longer vibrant.

The colonel smiled, responding, "The range was for a headquarters, so it's safe to say we stirred up a colonel at least."

"Maybe spoiled his tea," the foreign office man said.

"Do the Huns take tea?" someone asked.

Quickly we tried to trace the result of that shell—its possible immediate destruction, its effect, perhaps, on a faraway household where women and children and old men would weep and put on mourning. The absurdity of such an exercise struck each of us. Certainly the men who had sent shells in our direction that day hadn't troubled to forecast. They were getting back what they had offered to this army. The sense of a personal grievance is a powerful backer for patriotism in keeping men at war—that, and the impossibility, as in this case, of seeing the result of your shelling. I wondered what these neat, gentlemanly figures would do, what they would say, if they could witness the death and the maiming and the tears sent forth from their clean and remote hands. Close-in fighting, it was clear, had nothing in common with artillery work. A temporary insanity of self-protection and retaliation lets a man look on what he has done without nausea and stark horror. In the faces of many soldiers you see an eventual understanding, an effort to stifle recollection.

Chapter Sixteen

THE BASE

Whatever the custom of the Germans, tea wasn't neglected at our base. After we had visited the other guns, we walked—still tingling from the noise—to a hut in which rows of young men sat at a table between lines of cots, laughing and chattering amid a rattle of cups and spoons. A heavily banked bomb-proof shelter was convenient to the entrance.

"And it wouldn't take long to get there," the colonel said grimly.

Upon observing two men and a woman standing in the yard of the farm, I said, "They're not quite so near. It's like living on a powder magazine."

The colonel nodded and said, "They're probably doomed. Sooner or later the Huns will get them. What can we do? We can't move them. Truly the French are a wonderful people."

The latter is the most persistent phrase of this war. The woman waved her hand gaily at us, wishing us a safe walk, as we started back across the field. We paused at the gate of the colonel's cottage, waiting for the others to come up. A subaltern rounded the hedge.

"Where are the others?" the colonel asked irritably.

"Taking it in a long line, sir," the subaltern said. "It seemed safer that way."

The colonel led me into his dining room and, while we waited for the others, ordered tea. Across the wall his range charts were spread, as well as his *tir de barrage* plan—like an architect's blueprint. About the plan, the colonel said, "It makes an absolute curtain of shells on their trenches." Then, he demanded, "Where's that tea?"

A private with a startled expression left the room, returning with a huge, blue-patterned teapot. The others straggled in. We sat down and drank and ate biscuits and listened to the gun roar, which, even with the approach of night, scarcely diminished. Suddenly, the colonel laughed and fumbled in his desk to find a clipping from one of London's most revered newspapers.

"Seen this, Williams?" he said.

Williams scanned the clipping and passed it on. It was a letter from an officer to his father, reciting a strange ornithological experience in this neighborhood. During several nights this young man had, he declared, heard shells whistling over his *billet*. They had, however, been preceded by no sound of guns. Investigation of the ghostly incident proved that the shell whistling had come from chickens in the yard. These clever birds had, after many months, learned to imitate precisely the distant passing of shells.

The colonel finished his tea and lit a cigar. "We've devised," he said, "a letter which I fancy the editor will have to print or else acknowledge he's been made a fool of." He found the letter, put on his glasses, and read it with an air of satisfaction.

The army in this section, it regretted, was seriously affected by loss of sleep. The crickets had acquired a most annoying practice of imitating machine guns. Constantly they disturbed rest by firing an apparent salvo in a man's ear. The squirrels made a noise like approaching whiz bangs. Worst of all, a big bullfrog in a pool near his headquarters had caught the raucous trick of the gas alarm. "'It's a rare night when he doesn't sit on his bank and call us forth with our masks on," concluded the colonel. "'So far he has resisted our best snipers.'"

For a moment, in that little room, our laughter was louder than the gun mutter. Williams left us to telephone somebody, probably about going back across that naked road. After a long delay word reached him and he told us we could leave. Through the twilight, we took the road on the run and sped rapidly out of range. When we could no longer see the twin balloons, we felt comparatively safe.

The countryside had a peaceful appearance. As we approached headquarters, the sky was gray except for an ugly, dull, red splotch in the west. It was like an old blood stain, like a wound in something already dead. That night, the peacefulness of the chateau was unnatural. Out of habit, we raised our voices. The silence taunted us.

The next morning, we drove into one of the great bases and it was there that we heard the news. But bigger than the news itself was the manner in which the officers received it. No clearer example of the shift in British psychology could be asked for.

A man from the commandant's staff had joined us. We stood in the yard of an ordnance depot. Williams and this man were whispering. All at once Williams's face shared the expression of the other's— something I had appraised at first as a natural surliness. Quickly Williams beckoned me.

"We've had a nasty smack in the eye off the coast of Denmark," one of them said.

It was our first word of the great naval battle, of the garbled report that indicated a sweeping German victory. It was the same report that the army in the field got, and the army took it as these men took it, with a sullen anger and a fear that it might lengthen the war. If anything, it strengthened the determination on the young faces. It made us feel what a hopeless task it is to try to discourage this growing British Army. But the most arresting element was this new willingness to face the hard facts, to polish nothing for themselves or for the stranger within their gates.

"Sixteen of our ships gone and only one Hun!" the staff man groaned. "It won't sweep us off the seas, by gad, but it's tough."

Someone questioned whether the heavy fire we had experienced the day before hadn't been the German fashion of expressing joy. If that was so, such a celebration wouldn't wear itself out all at once. It made the trip we had arranged for Arras less than inviting. The day's inspection lost its interest; we went about grumbling.

"When can we get a paper?" we asked everyone we met. "Transport isn't in yet," was the usual reply. Then we commenced asking everybody what time the transport would be in.

Only once that day did the old attitude creep through and it was properly squelched. We were lunching in the maritime station with the staff when a very nice, elderly officer said pleasantly: "In my opinion we lost those ships winning a great victory."

"Sixteen to one!" a man scoffed, turning to me. "Did not one of your politicians win a great victory on those figures?"

"Well," the elderly officer persisted, "we drove them back to their base."

A quiet chorus of protest arose. The hard facts were stated to him plainly. He subsided, but his elderly face was a trifle bewildered. Probably he hadn't been here long, probably he had never been in the trenches. Perhaps he was wondering, too, about the fruits of this new attitude that must certainly grow in economics and politics after the war. He joined our restlessness when someone entered and said the transport had been sighted.

The official statements in the first papers we saw were cheering, but by no means all the truth. They made it possible for the officers to glance over the list of birthday honors that were printed that day. They sent us with some interest through the great hangars where provisions and munitions were passed in a constant stream from transport to train. They gave us breath to exclaim at this minute efficiency, which had been developed in two years from almost nothing.

This all expressed itself most strikingly in a great factory building once owned by a German. There, endless sacks of flour were lifted to the upper floor on chain elevators and great soft mattresses of dough flopped down steep slides into the hands of a regiment of bakers, white-clothed, covered with flour, with the appearance of clowns only half made up. At the entrance to each room, a sergeant would remind us that these comic figures were soldiers, regularly enlisted, and he would sing out: "Bakers! 'Shun!" And the long, ridiculous lines would stiffen. Only the staff officer's careless "Carry on" would send them back to their labor of turning out more than two hundred thousand pounds of bread before night. Efficiency also stared at us from posters that carried minute instructions to be followed in case of an air attack. But all about the most peaceful and industrial occupations fell the tattered garments of war. In a shoe shop thousands of pairs of stumpy, studded black boots busied an army of workers. Rows of shoes dripped oil after their bath to soften the leather.

"You see," the officer in charge explained, "these are all old shoes in process of remaking. Dead men's shoes."

The odor of oil and wet leather was sickening. From the first glimpse you knew what those rows of dripping, studded, stolid boots reminded you of—boots too still, on the feet of dead men.

"You see, we don't waste anything," the officer said prosaically.

Even among the little children in the Belgian *orphelinat*, where we had tea that afternoon, the war dominated. It lurked in the black uniforms, in the young faces where that eternal question was more pitiful than ever, in the heap of hay at the end of the yard that the babies with a perfect seriousness modelled into the semblance of trenches and redoubts.

After dinner that night we heard Williams telephoning in his little room. Afterwards he joined us, laughing with satisfaction. "Word's come in from General [. . .] headquarters that [. . .] has shown up. His parachute was shot so full of holes that it's a wonder it didn't drop him, but the wind carried him inside our lines and he wasn't touched. War's full of miracles. Blessed good thing, too. [He's] a corking good fellow."

We had never met this man, but it somehow cheered us a lot to hear of him spoken in the present tense. However, the prospect of the trip to Arras the next day drove this fellow almost immediately into the background. There seemed to be a feeling of some doubt. There was a good deal of talk about the city's proximity to the German trenches, about the necessity of walking close to the house walls because the Germans could see down the streets and had the range of each corner. It is not surprising that we wondered just what Williams meant when he said, "It promises to be a pretty interesting program." Another in the group did little to encourage us, by adding, "Oh, you're almost certain to get some shells."

Chapter Seventeen

THE MAD ACTIVITY OF A DEAD CITY

The next morning was dull and depressing and as cold as early winter. As one does out there, we studied the direction of the wind first of all and inspected our gas masks. We wondered if with less sun the German cannons might bark less viciously, but as we drove on, huddled in our coats, the clouds promised to break.

Williams left us for a moment at a division headquarters. No officers lounged there. The streets were nearly empty of uniforms. When Williams came out, he looked as if he had heard something unexpected. "The Huns are strafing the main road," he said to the driver. "Go the other way."

Outside the village a Canadian Highlander stopped us and examined our passes. He seemed very particular and appeared to be wondering what the deuce we wanted inside the lines that morning. Just beyond we left the main road and twisted through country lanes, while out of the morbid, threatening morning was born the hateful gun mutter. The foreign office man and I clutched at the trivial. We talked of automobiles and fishing and hunting, but always we were conscious of the sinister and growing chorus. A big gun crouched at the roadside; it would have been good to hear it shout back. Somber and undisturbed, a Hindu orderly sat on his horse in a field.

"Like a graven image," the foreign office man said.

The increasing roar discouraged talking. We tore past and entered the outskirts of a town. The streets were deserted. Holes gaped in the house walls, doors were pockmarked, windows mostly gone. A popping noise from the front of our car—not unlike the explosion of a shrapnel shell and under the circumstances about as discouraging —indicated that a tire had gone. The driver sent a startled glance at Williams.

"Annoying!" the foreign office man said.

"Where are we?" I asked.

"Outskirts of Arras," Williams snapped in response. Then he sprang out. At such a moment he was sheer efficiency, for most assuredly he didn't want us to get strafed. "Pile out," he ordered, "and stand close to the wall."

"No, no," he then cried out to the Japanese in the other car, "not you." He directed them to remain in the car while their driver backed them between two house walls. The two chauffeurs commenced to change the blown tire with frantic haste. A military policeman appeared from some hiding place and walked briskly up.

"It's a bad place for that, sir, this morning," he said.

"Things seem pretty warm in here this morning," Williams replied.

As the military policeman waved his stick, he said, "Just had a piece of shell through my window, sir. Listen for yourself."

The foreign office man and I lit cigarettes, hiding our misgivings with a vast indifference. "No comfort smoking in the cars in this wind," he said to me.

Moving about restlessly and close to the wall, Williams asked the policeman: "What's the best way in?"

The policeman, pointing down the deserted street half blocked by rubbish here and there, replied, "Five blocks straight. Turn to your right at a busted lamppost marked *roo dulla hop-pittle*."

One would ask the route so on an ordinary motor trip. The military policeman had done his duty; after warning us he didn't linger. The drivers sprang erect, the jack rattled down. I've never seen a wheel changed so quickly. Racing drivers couldn't be more agile. At a nod from Williams we got in again. We threaded through a dead city, crowded with a noise that gave the lie to its apparent dissolution. The quality of the unnatural ride increased—it seemed incredible, a hallucination, that nevertheless possessed a momentary and terrible reality.

We faced ruins that gaped back at us. At a turning, the façade of a hospital had suffered rather more than anything in its vicinity. Its breached and riddled walls had an air of surprise and indignation. Farther on, a bed on the third floor of a house, whose front was gone, hung over an abyss. The bed clothes were tumbled, pictures were awry and still clinging to the walls. A bottle of wine remained upright on a shelf.

"That couldn't have happened long ago," the foreign office man said.

"Every time I come in," Williams answered, "some ruin has been ruined a little more. Not a very prosperous looking town now, is it?"

I had seen Messina, Italy, after the earthquake. That disaster was scarcely comparable with this manmade one and, in Messina, there had been many women weeping over ruins that were sepulchers. This was sadder because for a long time there was no one—an emptiness pervaded everything. It was more shocking than the reverberations of many guns.

We entered a street that was once, I suppose, the pride of Arras. The grass plot in the middle lined with trees reminded me of Park Avenue in New York. We drew up: on our side was a high garden wall, on the other, beyond the grass and the trees and the roadway, was an old French barracks torn to pieces.

"I'm going to take one of the cars and drive to the provost marshal's," Williams said. "I want to find out what we'd better do now we're here. While I'm gone don't move from under the trees. It's the safest place for you."

He was off. One of the Japanese asked if it was dangerous. The driver of the other car, who had joined us by the fence, laughed above the cracking roar. He stooped and picked up great, jagged pieces of shell casing from the grass. He offered them for the Japanese man's inspection, while the foreign office man mused, "Sounds like a gigantic fireworks exhibition."

The sun now and then struggled from behind the clouds, but always the atmosphere was dull, and abnormal, and frightening. A sifting dust colored it. "Maybe the end of the world would look something like this," someone said.

Williams dashed back, a strained and hurried figure in the middle of the rear seat. He had grown confidential with me and now told me that the provost had had a shell through his building. And, he said, "We might as well walk about. It's as safe as hanging around here."

"What time do the Germans lunch today?" I asked. He looked at his watch, for that information was evidently of real concern to him.

"I take back what I said about fireworks," the foreign office man cut in. "This isn't the least like fireworks."

Nor was it, for there were detonations louder than the reports of cannon from the neighboring streets and scattered crashes like the crumbling of walls where shells had exploded. There was something wanton about this bombardment of a dead city.

Breathing with distaste the strange, repellant atmosphere, we hurried across a market place with empty shelters of corrugated iron half tumbled down. Two officers came swinging by, sticks in their hands, and helmets low on their foreheads. They didn't talk and moved with smooth haste. The striking of their feet against the paving was inaudible because of the turmoil. They were like figures seen in a dream.

All the houses were skeletons from which the flesh had been rabidly torn. We glanced down a narrow street, arrested by the sight of two women emerging from a cellar beneath a heap of ruins. One of them carried two chickens, nicely browned, while the other had a tin of fried potatoes. A group of military policemen leaning against the opposite wall moved languidly forward and took the appetizing food. They smiled and the women smiled, but as far as we could tell no one spoke. The entire transaction had an air of good-natured stealth.

"Women in Arras!" we cried with surprise.

Williams nodded and said, "A few have stayed. It's orders during a bombardment for everyone to remain in the cellars."

"The cooks ought to have decorations," someone said.

"They wouldn't think so," Williams answered. "The French are hard to scare and they love their homes. Last time I was here I saw a French soldier and I asked him what in the world he was doing. Said as calmly as you please that he was home on his first permission since the beginning of the war. Fancy that! Taking your vacation from hades in the same climate. You bet the *bosches* couldn't interfere with his coming home, even if there was only a cellar left. But . . ."

And Williams laughed, pointed, and said, "He didn't come on the *chemin de fer*."

Across a broad, semicircular plaza arose the wrecked railway station that Williams had pointed to. Following his lead, we sidled around the curve and slipped in through a doorway. Grass had grown through the shattered floor boards, rain had come in and mildewed the splintered benches and ticket booths. In a door-less closet a girl's summer cloak hung. There was a card attached to one of the buttons, which Williams fingered, but over the course of two years the writing had become undecipherable.

"Must have been warm that August day she came through here," he mused. "Maybe on the last train, fleeing from the Huns. Couldn't have known they were so close or she wouldn't have left her coat. Hope she didn't get strafed if she came back for it."

Like the cathedral at Rheims, the hall was filled with somber and unthinkable memories. We picked up some tickets scattered nearby on the floor. "Arras to Douai, par Vitry-en-Artois," they read.

"A short trip," I began, "straight across the trenches. When you English take it . . ."

"The war," Williams broke in, "will be getting on Kaiser Bill's nerves, don't you think?"

Something was clearly on Williams's nerves. He hurried us through and gave us only a moment to glance at the broken girders and the twisted rails in the train shed. Among the splinters of the platforms, where crowds had once thronged eagerly, the long grass waved with a slow melancholy.

"It's not very far," he reminded us, "to the Hun trenches, and they have a nasty habit of dropping whiz bangs in here. There's no bomb proof. Let's go."

We had scarcely reached the shelter of streets lined with looted shops when a soldier came running up and spoke to Williams. He turned to me, sharing another of those confidences that made me wonder why I had ever come to see war.

"What I was afraid of," he explained. "The Huns are strafing the station—dropping whiz bangs in from the trenches."

Probably the German observers had seen us leave. It was the luck of war that they hadn't caught us going in. Now, we climbed a small mountain of stones and beams at the end of the street and emerged into La Petite Place, a short time ago one of the finest examples of Spanish architecture in Europe. Opposite us, the Hôtel de Ville had raised a few sections of interior walls and the stump of its tower— white, formless, ghostly.

"I was in Arras a few weeks before the war began," Williams said. "Had to change trains, and was just too short of time to run down and see this place. Isn't much to look at now, is it?" Of the old Spanish houses on the market square, several were completely down, others retained just enough form to expose the brutality of their wounds. With a sense of sheer gratitude, we followed Williams down stone steps into the cellar of one of them. The bombardment was a trifle muffled here. An elderly French woman and her pretty daughter greeted us.

"You're not afraid to stay?" I asked them.

The girl tossed her head and the woman laughed, indicating a cook stove, a table, a bed, a rough counter, half a dozen chairs. "They've driven us downstairs," said the elderly woman, "but why should we be driven from our home and our business? We are quite comfortable, and we do a little trade with soldiers. Monsieur has seen Arras during the bombardment. Perhaps he would like to see what it was like before. An *album artistique* might interest monsieur."

Smiling at my bewilderment, she fetched a tastefully made up blue book with silk cords and tassels. It was impossible not to buy the thing—it was a collection of photographs, many of them taken at grave risk, and sold under a risk nearly as great to the Tommies to send home to their families.

"And you've been doing this . . . living like this since the beginning of the war?" I asked her.

"But certainly," she replied. "Through that door I saw the first bombardment of the Petite Place. I saw the shells bring the great tower of the Hôtel de Ville crashing down. That was cruel. It was the glory of Arras. When it fell I thought of the judgment day."

"You mean you didn't barricade that door?" I wondered. "Why?

"Because the shells came from behind us," said the woman. "If they exploded too close the fragments were likely to fly on towards the center of the square. Besides . . ." With an air of secrecy, she then opened a door to a flight of stairs leading downwards. "You see there is another cellar. Come," she invited. After lighting a candle, she led the way down for many steps. The vaulting was ancient. We found ourselves in a labyrinth, corridors leading in all directions. The walls were of a soft limestone. The stone, one guessed, that had been used for the Hôtel de Ville and many other buildings had been quarried here. But there were fresh breaks in the walls and sometimes the corridors were partly blocked.

"The shock of the shells brings pieces tumbling down," the woman said. "That's why we find the upper cellar more comfortable after all. Wouldn't we be more comfortable there, now?" We agreed, and as we went up she told us how Arras was honey-combed with these cellars. With real regret we left her for the strange light and the racket outside the cellar. We reached the vicinity of the cathedral over a hill of rubbish.

"Palladian," the foreign office man said.

Indeed, the stark remnants of the cathedral were more impressive than the untouched building—a bad example of the late Renaissance —would have been. Its size must have been enormous.

"Usually it's all right to go in," Williams said, "but I wouldn't

advise it today. Do as you please, but if one of those walls should fall .
. ." We didn't argue the point, for we had learned to believe in
Williams's judgment. He glanced continuously at his watch as we
went on. We knew he was trusting the luncheon hour to give us an
opportunity to slip out of Arras in comparative safety. By the time we
had returned to the market place, in fact, the roar had receded and
the explosions of shells were less frequent. The drivers seemed glad
to see us.

And so we left, dodging new holes and obstructions, casting quick
glances at the driftwood of that morning's strafing—torn shell
screens, split trees, a twisted bicycle, scattered heaps of stones. We
thanked heaven for the German appetite. We prayed it would persist
for some minutes longer.

Chapter Eighteen

WHERE MEN ARE LIKE ANTS

On the last afternoon I spent in Flanders we went on a picnic. It was a most extraordinary picnic, intended to give us a panoramic view of war, as it is fought nine-tenths of the time under modern conditions. It took us to a point of the line that saw some of the hardest fighting of the Champagne and Artois offensive. The French, who had manned it then, had progressed in spite of overwhelming odds and frightful casualties. Now in the hands of the British, it was still one of the knottiest problems of the entire front. We would understand why, but first we had our picnic.

After we had left the cars in the shelter of a village, Williams chose a spot behind a steep hill. At his direction one of the chauffeurs carried the baskets up a grassy bank and deposited them beneath a grove of trees. Trampled box hedges straggled here and there. It was a very pretty spot and we congratulated Williams for hitting on it.

"Yes," he answered, "it's just the thing, because the Hun airmen can't see us and disturb our luncheon." He distributed sandwiches and, lamenting the absence of a corkscrew, he knocked the neck from a water bottle with some skill. "Isn't much healthier around here than it was in Arras," he continued. "Have some of this cold ham? This was a kitchen garden once. There was murderous fighting here less than a year ago."

As we ate Williams's foresight was justified, for we heard the whirring of airplanes and, from beyond the hill, the booming of guns. After luncheon we lounged in the grass, smoking. We wondered when Williams had lighted another cigarette why he delayed leaving.

"Of course it's pleasant here . . ." the foreign office man began.

Williams glanced at his watch. "I'm waiting," he said, "to see if the Huns are going to give us a strafing. They amuse themselves by dropping shells on this empty hill every now and then."

Although the firing became general, no shells, as far as we could tell, exploded near us. So, bent like a party of scouts, we went through a fringe of bushes and around a ruined tower that already had the sentimental interest of a medieval survival. We walked through a house, which had no doors or roof, into an overgrown backyard. Williams stooped, kicking through the long grass at something. We went closer and saw him staring at a faded German uniform coat with sinister tears and stains about the back. An object, long and white, lay nearby. In our own minds, we hesitated to give it a name.

"All that's left of some poor devil," Williams said, moving on. "I told you there had been hard fighting here."

As far as possible we kept out of the grass after that. Grass and weeds grow too quickly in the war zone; they permit too much to remain. We came to a barn with gaping holes in its sides and roof. Beyond it, half-destroyed buildings clustered around a square with a monastic appearance. Between them and a ragged wall yawned an open space, perhaps ten yards across.

"Take that in a hurry," Williams commanded. "The Huns can see us there."

We dashed across, circled the end of the wall, and went into a small enclosure that was all that remained of an outhouse. Wire netting had been stretched across an eyeless window in the front wall through which the panorama of war was visible below us. Names rang in our ears that connote almost as much horror as Verdun. Not far from us stretched a brick wall, pierced for rifle and machine-gun fire. Just beyond was a ruined farm notorious for some of the worst hand-to-hand fighting of the war.

"From behind that wall, and from the farm, after they had got it," Williams said, "the French went forth to capture that network of trenches off there, just behind our present frontline."

You stared, not because of the familiar name of those trenches, but because it seemed impossible to you that men could have crossed the several hundred yards of open ground between the wall and the network. Even with artillery preparation such an attempt seemed suicidal. But, as Williams told us, men had fallen all around here. To the left, we could see deserted dugouts that were captured in September. At some distance, a spur of land thrust out on a broad plateau that was absolutely bare. Before the war it had been thickly wooded.

The present British and German trenches made yellow scars along a low ridge. The German line was a little above the British one; both passed through a ruined village. As we watched, the bombardment became more violent. We could see the effect of every hit— shell after shell of high explosives sent black clouds springing from the yellow earth. The lines were so close it seemed inevitable that mistakes would happen or that an imperfect fuse would shower death on the gunners' own men. But the accuracy of the fire was appalling: each shell appeared to fall directly in the sorrel ditches. When the spreading smudge had cleared away, we would detect breaches, but the only men we saw were one or two soldiers who ran swiftly along the brown road towards the communication line.

"It's nearly always like that here," Williams said. "Fancy being under one of those planes, those Black Marias!"

Our fancy, however, was directed to a danger more immediate. We looked up at a whirring overhead and saw a war plane flying high in our direction. As if born of the air, five more appeared, sailed over the trenches, swerved back above us, and circled away again. They were too high to make identity certain, so we crouched as close as we could to the wall while we speculated.

Suddenly, the anti-aircraft guns engaged and around each

machine we saw shrapnel burst, but it was too high for us to hear the fat explosions. As long as we remained there after that, there was always a circle of little, puffy clouds around each airplane. The shells came from both sides, so we knew both Germans and British had taken to the air today. Someone suggested that it mightn't be a bad plan to go home, but the spectacle was fascinating. The rest of us begged for a few minutes more. We wanted, I think, to see one of these airmen show some sign of fear. As long as we watched they persisted in their scouting, contemptuous of the pretty white clouds that appeared as if from nothing all about them.

"It's a marvel they're not hit," the foreign office man cried.

"So it is," Williams answered. "It takes young men for that work, young men in whom recklessness is born."

For a long time, we remained there, glancing from the scouts to the trenches where black geysers spouted with an increasing frequency, forgetting for a time the possibility of a slight elevation of a single gun that might send a geyser spouting in the midst of our little group.

"Good God!" somebody burst out. "I can't believe there are men where those shells are falling. This thing makes men seem like ants."

We left at last, reluctant to leave this spectacle of death, in which the victims remained always hidden. Driving along the base of a hill we passed a large cemetery, where wooden crosses stretched in neat rows. The gun roar gave the scene an exceptionally sinister appearance. Even in their long rest these soldiers, we realized, were far from immune to German shells.

"The trench toll," Williams muttered. "Sad enough place! Every time I come here that cemetery's larger."

And just across the road the living busied themselves so the other fellow's cemetery wouldn't fail to grow. Some were practicing at a rifle range, a rattling blacksmith shop lurked under a hill, and men fidgeted about two observation balloons partly hidden by trees—the gross, corpulent things ready to take the air. And always, the guns reminded us that this care for the living and the dead was exercised under heavy fire.

Farther on we gazed with amazement at a football game that swept swiftly through its changing phases in a rough field. The shouts of the players failed to reach us because of the pervading roar. They were like pupils in a deaf and dumb asylum from whose open, eager mouths comes only a shocking silence. But there was no question that they were having a good time, cheering clever plays, and jeering at bad ones. Within their easy view, close to the road, lay a dead man. His stolid, studded boots seemed striving to advance towards them.

"The stretcher bearers are coming for him," Williams whispered.

We swung into the long road again, increasing our speed. If we could only get over that next hill without a shell . . . When we drove up to the chateau it was raining; great drops fell from the eaves like tears. Later after dinner, when I was talking to Williams, I challenged the reliability of that new, frank British attitude.

"I'm looking," I said, "for someone to tell me he doesn't mind shell fire."

Williams snorted, saying, "When you find him you can call him a liar, and the worst of it is you never get used to it. Each time's a little worse than the last."

It was pleasant to look back that night and to forecast nothing on the morrow more exciting than the inspection of passes by military policemen, Scotland Yard detectives, and French soldiers.

I wondered that I had had the effrontery to buy a return ticket. Doubtless, I thought, Paris would seem like a strange city in a peaceful and sorrowful world.

Chapter Nineteen

THE GRIM GAME OF INTELLIGENCE

After such experiences, the quiet of Paris, however, did not make it seem as remote from battle as I had expected. I looked upon the men in uniform with a new sympathy, a broader comprehension, and we talked of nothing but the war.

It was about that time, I remember, that a German spy was caught under dramatic circumstances and shot with a deserved dispatch. It is impossible to write about that case but it reminded me that when I had sailed for Europe, I had planned to find out something about these men and women—not so much their ciphers and signals and mathematical routine, rather the kind of people they are, and the type of drama they play continuously behind the lines. So, I reviewed my own contact with them and the stories I had heard of their daring.

In the first place, in Europe spying has officially ceased to exist. One speaks instead of "intelligence," yet it doesn't make much difference under what label a man faces a firing squad or feels the noose tighten about his neck. For, as a matter of fact, there are more spies than ever—better spies, spies with a nearly superhuman lack of fear.

There is, of course, a good deal that can't be publicly told, but it isn't all tragedy as you'll learn from the curious case of the near-sighted London clerk. Nor do these men perpetually work in the shadow of death. You may not know that an Entente intelligence officer assigned to New York informed London of the approaching Irish excursion of Sir Roger Casement, but you must have guessed the presence of spies of both sides in America. You may have suspected that, often in a legitimate way, they are not uninterested in you. Have you ever smiled at a German waiter's bored expression during an after-dinner discussion of the war? Since hostilities commenced, have you tried to visit England or France? In the latter case, you may be sure that both sides know enough about you and your sympathies to exalt your own importance and to justify your admiration of the system.

After docking on the other side, for instance, as I told you in an early chapter, the passengers are virtually imprisoned in the dining room until the chief alien officer has had his fling. He appears to possess a dossier for each person. In my own case he asked me to fill in blanks on a form, largely repeating the information on my passport. He attached this to the passport. On the London train I asked other passengers if they had been similarly decorated. Enthusiastically they denied it. It seemed definite, since I was a correspondent, that a check had been placed upon my movements. The American Embassy offered that doleful interpretation. When I applied at Bow Street for an identity book, the clerk admitted that the slip was a code for the police. So I went to an acquaintance in the intelligence department and threw myself on his mercy.

"What in the name of heaven," I demanded, "is this soiled piece of paper?"

He smiled, and said, "They gave you your identity book at Bow Street, didn't they? You know it might be a recommendation on information from America."

I explained patiently that I had sailed on two days' notice. His smile didn't alter, and from all that happened afterwards, I know he was right. It isn't simple to elude a system that works so quickly and that's the reason the Germans early in the war ceased getting many spies to England or France through New York. They turned, as a consequence, to Spanish America. That menace too, a distinguished officer of the intelligence corps told me, was well under control. A few days before, he said, a clever attempt to get a man through had been defeated, partly by accident, for the fellow captured had had a genius for makeup. He had looked like a Latin and he had talked like one. On the long journey from South America, he had hoodwinked the crew and all the passengers, except for one woman who had known him for years and who had penetrated his disguise. Still she had been friendly, and he had unlimited confidence in his masquerade.

When the boat reached England, he was one of the first hailed before the alien officer. He went jauntily because he knew his passport was in perfect order. The alien officer found it so, but he glanced suspiciously at the man and told him to stand aside for a few minutes. That was only due to his compliance with recent orders to be careful with Spanish-Americans. As a matter of fact, he suspected nothing out of the ordinary. But the fellow hadn't forecasted anything like that happening and, in a panic, he scribbled a note requesting the woman who had known him to not speak to him in any language except Spanish.

When he slipped it to her, the sharp eyes of the intelligence men saw. They drew the woman outside and got the note from her. They went back and took the man into custody. He laughed at them, showing no fear, declaring his innocence with a tolerant air. They hurried him to London and put him before the official who told me the story.

"I spoke to him in German," the official went on, "and at odd times—suddenly. I couldn't trap him. He said he was a South American merchant on a peaceful commercial enterprise. He didn't know a word of German. I began to doubt, because when I spoke the language his eyelids never moved. It seemed to me he must show some response if he understood. As a last resort I simply shouted out, 'Achtung!'"

The official, smiling a trifle sadly, went on: "His heels clicked together. His chin came up. His hands straightened at his sides. He tried with a convulsive effort to check that mechanical response, but it was too late. I had him and he knew it. He broke down and took his medicine. He was a German reservist. A military command was the one thing to which his whole nature had to respond."

Even if the defense at the ports is overcome, there's an interior net to furnish spies to the executioner. I learned to understand the misgivings of hotel acquaintances, who said their luggage had been gone through, although they found nothing missing. One man complained that the servants were a badly trained lot, for they burst into his room at all hours, retiring with the apology that they had not known he was there. I didn't tell him that his wastebasket refuse and the litter of his writing desk had probably furnished an interesting puzzle for some intelligence officer. Hotel espionage in England and France, however, is a knife that cuts both ways.

It may be indiscreet to call attention to a perfectly obvious fact: the Swiss are a problem for the Entente Allies. Except for natives who have been retained through disability for the army, the male hotel service is largely in the hands of the Swiss. The sons of this neutral nation must have the privileges, the courtesy, and the protection that other neutrals receive, and because of the nature of their employment and its permanence, it is difficult to keep tabs on them. The natives of northern Switzerland often have German names, speak the German language, and subscribe, perhaps, to the German idea. It would take unlimited confidence to pronounce one man a northern Swiss and another a southern German. So while the Entente gets much valuable intelligence from the hotels, it is safe to guess that the Teutons (the Germans) have found the servants useful, too.

I was told that early in the war the top floor of one of London's large hotels had been closed because of suspected signaling of Zeppelins. That night of which I have written, when the Zeppelins were, in fact, trying to get over us, a British floor-valet muttered dark things about the foreign servants, as we gazed at the bursting shrapnel and the searchlights. In his less emotional moments, however, the valet had nothing to say, for it is bad form to audibly doubt neutrals.

But, with all that, the German spy has ceased to be a terrible and unavoidable curse in Europe. Those in authority have probed his methods and chained his activities. He has even become an object of thoughtful criticism. One day this point was under discussion by some of the men who have made that cheerful situation possible.

"German intelligence is universal," one said. "Every German, no matter where he is, feels himself a divinely appointed agent of his government. He sends what he can to the Wilhelmstrasse. He is ambitious to impress the Wilhelmstrasse. Consequently, he sometimes hits false trails and sends the real agents off on wild goose chases. In the long run, it is a weakness to use amateurs in the intelligence game."

About that time, as if to prove that every rule has its exceptions, the case of the near-sighted London clerk came unsolicited to the department. It was valuable intelligence, because it gave solidity to the many rumors about at that time, specifically of Austria's anxiety to make peace. The official who handled the case told me the story with a reminiscent smile.

"It is hard," he said, "to learn just how much is behind these rumors of a nation's desire to make peace. It seemed likely that Austria would be rather better out of it, but you can't place much reliance on newspaper gossip. Then this youth came shambling into my office, white as a sheet, his eyes red beneath huge spectacles, stoop shouldered, trembling as if he had a chill. His flashy clothing looked absurd. Mourning would have become him better. I fancy he expected to be condemned to death. He tried to avoid that by telling all he knew.

"He worked in a city office—clerical work in an insufficient light that explained his eyes and his shoulders and his bad complexion. You know how little sunlight that type gets. You know how destructive to ambition such work is. He plodded along with no bad habits, with no future, an inoffensive, pitiful little chap. Then the great romance came. A visitor was taken through the office one day. The clerk noticed him because he was so big and handsome and prosperous. He was nearly tongue-tied when this impressive figure paused and chatted with him. It developed that the visitor had known the clerk's father. He expressed some interest in the young man, took him to dinner. In many ways he was kind to him. The man declared that he was worried about the clerk. He looked underfed, on the edge of an illness, and something ought to be done about that. After a little thought he slapped his knee. He had just the thing. Business was taking him to a neutral country across the channel for a few days.

"'Suppose you get leave of absence,' he proposed, 'and come with me. I'll pay your expenses because you're your father's son, and because I like you.'

"The young fellow demurred. He couldn't trespass on such generosity.

"'It's all right,' the older man said. 'No charity about it. As a matter of fact, I could use a secretary for a few days. There's sure to be a man or two I won't want to talk to myself, and that sort I can shunt off on you. Meantime, you'll get a vacation that will give you a fresh start and maybe save you a bad illness. Tell 'em at the office your uncle's going to give you a little holiday.'

"The clerk, unable to believe in this sudden stroke of luck, arranged it. His friend gave him a new suit of clothes. His studious expression went well with this new prosperity. They sailed. On the other side there were some aristocratic-appearing men who paced the dock. When the clerk and his host landed, these men came up with bows and words of welcome. For the first time the youth realized what an important person his benefactor was. But the men paid an incomprehensible amount of attention to his insignificant self. They were solicitous about his health. They apologized for the poor comforts he would find in this town. The best available had been prepared for him. He had a vague idea that all this was really meant for the other man. At the first opportunity he asked who these people were.

"'They talked to me so strangely, as if I was a lord or something.'

"'You be nice to them, my boy,' his host said. 'Treat 'em well. Let 'em give you anything they want, and you act as if that was what you were raised to. They're friends of mine, and I'd hate to have 'em offended. If you think they're crazy, keep it to yourself and give 'em their way.'

"There was a private suite at the hotel, a solitary dinner, more grander than the clerk had ever imagined existed outside the covers of a novel. The next morning a servant appeared, announcing that the gentlemen awaited the clerk's arrival in a remote parlor that they had reserved.

"'Go along, sonny,' his host said, yawning. 'I can't be bothered with these people. You go sit down and have a nice chat with 'em, and let 'em get whatever they have on their minds out of the way.'

"'But what do you want me to say to them?' the young man inquired.

"'Say as little as possible, I tell you. You say, I must return for further instructions to England. Yes, that's a nice answer. That can't offend 'em. You say just that, and now and then you might say: Gentlemen, you interest me.'

"The clerk looked at him appealingly, but his host waved him away.

"'Go on. Don't ask so many questions. I hired you to talk to people like this. Do as you're told, and you'll be all right.'

"So the clerk went to the remote parlor, and at his entrance the elderly, aristocratic gentlemen arose, bowing most profoundly. 'Will you sit here, your excellency? You slept well? You were not too uncomfortable in those insufficient rooms? You find that chair to your liking? Suppose we speak informally of that which brings us together.'

"The bewildered clerk leaned his elbows on the table. He wanted to smoke a cigarette, but he thought it might offend the old men. He wanted to say, 'What does bring us together?' Instead, he murmured, 'Gentlemen, you interest me.'

"They smiled at that. They bent closer to him genially. He realized he had made a hit. He determined to use that phrase as often as possible. He had had no idea any one phrase could be so successful. Then his ears tingled, he felt confusion sweep him. He was like a man lost in a deep woods. Someone had said, pleasantly, 'Then perhaps, you will give us your government's terms.'

"For a long time he kept his head bent. He didn't answer.

"'Of course we understand,' he heard a voice drone, 'that this conference is quite informal, and that the terms you mention must, to an extent, be considered tentative. Still it is a beginning, an encouraging one. We must begin somewhere. The tentative terms, please.'

"The drumming in his ears increased. He scarcely heard his own voice murmur, 'Gentlemen, you interest me.'

"This time there was no good-natured response. The others stirred and made no effort to hide their surprise. Clearly something else was demanded of him, so he took courage and completed the recital of his lesson. 'I must return for further instructions to England,' the clerk told them. The others sprang up and paced about the room. They gathered in a corner, evidently consulting. One gray-beard approached him with an air of timidity.

"'What has occurred, your excellency?' he said to the clerk. 'We have heard of no great victory. Yet since you left England something must have occurred. Something must have happened since you arrived last night, when we all spoke of your cheerful attitude.'

"The clerk shook his head. He had only one thought, to escape from that conference about which he knew nothing, yet which was clearly of grave import, concerning matters in which he could have no honest share. Ready to burst into tears, he arose and made for the door, combining his two phrases in a desperate effort to explain his retreat: 'Gentlemen, you interest me, but I must return for further instructions to England.'

"He was aware of consternation and whispered astonishment behind him. He stumbled into the private suite and with a trembling voice demanded some explanation. But his host was more curious as to what had happened at the conference. When he had got all that from the clerk, he rubbed his hands and smiled with satisfaction. 'Just the thing,' he grinned. 'You did well, my boy. You've got 'em guessing. Now you go on home just as you said you would, and we'll arrange another conference a little later.'

"'Who are these people?' the clerk burst out. 'They treat me as if I was the king or David Lloyd George.'

"'Friends of mine,' his host said airily, 'and they're giving me a pleasant experience. I'd hate to have it lapse.'

"So the clerk came back to England, but he couldn't wait to hear from the impressive man. He didn't want a repetition of his glittering holiday. The cold chills were running up and down his back. He came here and told the whole story. Of course we had to get to the bottom of it. The intelligence department persuaded his flashy host to come here. He's locked up now, but I doubt if we keep him. You see, he's a swindler of international reputation. He was a trifle disappointed to be interfered with, but evidently he'd made something out of his game and, really, I don't think such a confidence game was ever attempted before. The Austrian government had been his . . . what do you call it? . . . his sucker. He had actually approached Vienna whispering that Great Britain was readier to talk peace than anyone knew on the outside. The British government, he said, would discuss tentative terms, but it would have to be done informally and secretly. He was the man to arrange matters, to put the thing through—for a consideration.

"Vienna, to all appearances, actually took that bait. Money was no object. If the swindler would bring the British representative to a neutral country, they would send commissioners to confer with him. It became necessary for the swindler to find the British plenipotentiary he had agreed to produce. You know how he got him. Poor little chap! He kept his word. He did come to England for further instructions, and he received them to go back to his desk and forget all about it. I daresay he's there now, bending over figures in a bad light, thinking that a diplomatic career has its drawbacks after all. Meantime the aristocratic commissioners . . . doubtless they are still waiting."

I knew of two secluded rooms in London about which this business of intelligence centered, and of two men, quiet geniuses, who largely controlled it. If for nothing else than contrast, I wanted to see those rooms and those men, for through their inventions England has been pretty completely purged of the spy terror and Germany has been given a spy terror of its own. The thing was arranged. I walked from the smug respectability of the Embankment into the amazing somnolence of Scotland Yard. In the office of a church society one would have found more movement, more irritability, more anxiety. Except for the bobby who strolled away with my card no one was visible.

The man I had come to see sat behind a littered desk. He wore a light alpaca jacket and his necktie was a trifle awry. He had the pleasantest and the sharpest eyes imaginable, which, however, showed something of that strain I was to notice so generally in men's eyes at the front. It was as if, while risking nothing physical himself, he shared the deadly anxiety of his agents at work far from the safety and the quiet of this place. His squarely cut and powerful features suggested a secretive mind. That, at least, was in keeping with one's idea of Scotland Yard. The necessity for it, he let me know, was infinitely graver than ever before in the history of British intelligence.

As I talked with the man with the pleasant, sharp, and tired eyes, I had to remind myself that a secret service net covering Great Britain, France, and large portions of the war zone was amenable to his hand.

Sir Roger Casement had been secretly spirited here after that dramatic dawn in Tralee, Ireland. He had stood there beyond the desk, proud rather than worried. It was impossible not to question how many guilty ones had stood before the desk, reading in those tired, quiet, questioning eyes their condemnation to the extreme penalty. The quiet of the eyes, the quiet of the room, the quiet of the building made such pictures seem incredible. The place offered no appearance of an inquisition, no stagey atmosphere of danger. Now and then a clerk tiptoed in and out, as in any office, leaving more bundles of paper to litter the desk. And yet the room was crowded with shadows. It was full of death.

One discovery I carried out of Scotland Yard: in such places there is none of the common contempt for the spy, none of the customary aversion for the degradation of his penalty. Such men find in the stealthy and anonymous heroism of the secret agent something sublime, the most perfect sacrifice.

The Admiralty isn't far from Scotland Yard. It was there one found the man who within a few months overcame new conditions and largely laid the foundations for England's success in snaring submarines, in policing the channel, in watching the movements of Zeppelins and Germany's high-seas fleet.

One's first impression is quite different here: the moment you cross the threshold of the Admiralty you face an air of secrecy and mystery. There are policemen to be passed and you notice civilians who seem to be on some errand but who watch you with too much interest. When your credentials have been examined, a guide is furnished and you need him, for he takes you down steps and up steps, through interminable dim corridors, extricating you from the demands of guards who appear here and there from the obscurity. He leaves you at last in front of a leather-covered door behind which a great silence broods.

The opening of that door alters everything. Dazzling light floods a large room through windows facing the Horse Guards Parade. A fire burns briskly. There is a solidity about the room and its furnishings that goes with its air of unalterable purpose. Men move about, but immediately one figure catches your attention and holds it. On the padded fender sits a slender, wiry man in a conventional naval uniform. Above his smiling face a broad forehead recedes between fringes of curling hair. Still he looks young and possessed of an abundant vitality. He springs up, smiling in welcome. He lights a cigarette and paces about the room as he talks. His smile never hides the uninterrupted anxiety of his eyes. At first, that makes him seem like the very different figure in the alpaca jacket.

I spoke to him, I remember, of the trawlers I had seen in St. George's Channel and the Irish Sea—hundreds of tossing trawlers, fishing for submersibles and, when necessary, making themselves the bait. I had marveled at the bravery of their sailors. As I watched the smiling active figure, as I saw the smoke curl from his cigarette, I realized that there are harder tasks, that the assumption of responsibility may be a greater sacrifice than the risk of one's life. At any moment, through the leather door, there might slip in a tragic reflection of his system—word, perhaps, of many lives lost through a breakdown somewhere.

Certainly this room was too cheerful. It made it more difficult to picture the details of a story I had recently heard—one of those cases about which little is said, because it involves signaling, the simple word that makes any official tongue-tied. Yet, it is obvious that the German spies have used that form of communication under favorable conditions. At any rate not long before, a trawler's crew had observed a red flash from a distant headland. Those who man these filthy craft are largely of the naval reserve class—men out of comfortable homes and convenient clubs. Consequently, they bring to their work an exceptional intelligence. After seeing the red flash, they didn't dash to the shore in the hope of finding something—that light suggested too many possibilities. Instead, they held their patrol and at the first opportunity reported to the Admiralty.

There have been many rumors of a German submarine base hidden away on the shores of the British Isles. The Admiralty, therefore, ordered this trawler to keep about its routine work while an intelligence man, with the clothing and accent of the vicinity, appeared in the nearest town. He had to work carefully, often slipping out at night and crawling through underbrush and behind the rocks, seeking out the base that the signaling had suggested. He found no indication of a base, no likely cache for supplies. He did, however, report the existence of a cove behind the headland, where there was a beach favorable for landing a small boat. The neighborhood was wild. There was only one house within a radius of several miles, which was occupied by an unkempt old man, who consistently turned back the intelligence officer's efforts, snubbing his attempts to talk. Aside from that there was nothing. Except for the old man, who might be of foreign birth, the people of the neighborhood were loyal beyond question.

The intelligence officer was recalled, but the trawler was kept on that post just at the edge of the radius of the red light. The commander now had a detailed map of the cove and the beach and the headland. He waited. "That man," he told his crew, "can't know his lamp is visible at this distance. Some fine night . . ."

And, on one very dark night, the red winking came across the water. The clouds were so thick that the commander knew he could sail close to the headland unobserved. He felt, in fact, that when he entered the cove his presence was quite unsuspected. (This business of waiting in the dark for the shaping of unknown forces into defeat or victory is the hardest part for the men assigned to intelligence work.) But the red light no longer showed. Although the boat was not many yards from the beach, there was nothing to be seen. There were no sounds beyond the cries of a rising wind. And the minutes lengthened. The commander reached the conclusion that the affair was founded on a delusion, or else some trick of shrubbery through which the wind permitted an innocent light to gleam intermittently. The men lost their caution, murmuring from time to time, so the commander spoke to them sharply. Then, a sudden sound aroused the crew. It was magnified in the black silence, suggesting the scraping of a hard object on sand. After a moment came a guttural laugh, followed by a prolonged hiss for silence.

"Hold that searchlight ready," the commander cautioned. "Not till I give the word. We'll wait a while longer."

A stealthy stroking of oars rewarded him. A small boat was making its way from the beach for the entrance of the cove. It would have to pass close to the trawler.

"Now!" the commander cried.

And the searchlight flashed out, circling the cove with a white eagerness, catching at last at the end of its ray a collapsible boat filled with men. The men stared up at the trawler open-mouthed. One cursed in German, another laughed foolishly in a feminine way. The commander couldn't believe his ears, for a third man began to sing a rollicking chanty. They knew they were caught.

The men were lined up on the deck of the trawler while the commander examined their collapsible. It held nothing except the oars, and there was no indication that it had ever been used to carry supplies. The commander turned to the line of prisoners, noticing that his own men glanced at them with curiosity. As he went closer, questioning, he was met by that absurd laugh and the song recommenced.

"What is this?" the commander asked.

His second in command strolled up to him, replying, "Most of these men, sir, are drunk. Ah, there goes that light again."

The commander turned sharply. The light didn't flash from the headland, it was actually farther down the beach. It went out but its purpose was clear: it had warned away a submarine to which these men belonged, to which they had started to row in their boat.

The commander lit a cigar, relieved to be able to smoke again. Because of the shifting of the light, he knew it might be impossible to implicate the unkempt man on the headland, who by this time must have destroyed all evidence. On the other hand, the intelligence department would be grateful to the commander for he could say definitely now that there was no submarine base in this secluded cove, that it had never sheltered any serious plot. The amazing truth was discerned from the grinning faces of the line of prisoners. It wasn't at all funny though that they should risk so much for no graver purpose than to come to a drinking party ashore. It visualized for the commander the suspense and confinement suffered by these submarine crews. No risk was apparently too great for an escape from that, for a momentary stretching of one's limbs, for a little release from the expectation of being crushed like a beetle beneath the gigantic heel of the British Navy.

Chapter Twenty

TRAGIC SECRETS

The stealthy watchfulness that makes such hauls possible is continually with one in Europe these days. Intelligence has very special phases for the French and for the British in France. In the beginning, spies actually moved through the ranks of both armies. The siege war of the trenches has made that game impractical. Under these conditions the problem of getting information through is increasingly difficult. Code letter writing through neutral countries, while comparatively sure, is a very slow expedient. Often intelligence is demanded in a hurry, must be had at any cost. For a time, carrier pigeons were used with success. They cried, however, their own warning. As a result, there aren't many carrier pigeons in the conquered provinces now, and to be found with one on your property amounts to a condemnation to death. I was surprised to learn that Germany and France both still experiment with them. I was told that an airman not long before had come across the lines at night and dropped a basket full of pigeons in a lonely spot. The conspirator behind the lines was supposed to find the basket, fasten duplicates of his message to each bird, and release them all. The innocent-appearing empty basket would be the only evidence left.

The airplane has revolutionized spying as completely as it has scouting. It's a risky business and even unpopular among the Amy Air Corps—as courageous a body of youngsters as war has ever produced. I have described them to you, sailing through bursting shrapnel, photographing and observing with impudent indifference. In an air battle they will take suicidal chances, but they don't like these quiet rides through the night to lonely places.

It isn't that they are physically afraid. They shrink from the work because it threatens the spy's penalty. The airman, like his passenger, is tried, condemned, and executed as a spy. And these boys—who know less than the quiet, worried men in London, Paris, and Petrograd, or in Vienna and Berlin—have a horror of the spy's work and the spy's death. Still they do it. It amounts to this: among the British and the French the belief in this war is so great that to ask for volunteers for any task is practically to take your pick of the entire army. This particular stratagem, moreover, must be seen through in the face of an enemy intelligence system that from the filmiest hint unravels conspiracy and caps it with black tragedy. One pitiful case comes instantly to mind. It's about the boastful indiscretion of an airman, who didn't want a spy's death, nearly got it, then, through his escape, unwittingly condemned the man who had saved him.

After crossing the lines and safely landing his passenger, the airman rose into the sky with a sigh of relief and started to return. Through one of those accidents no man can guard against, his engine failed and from a great height he dropped swiftly through the night. Failing to right his machine he fell, evidently unobserved, in a field at the edge of a town. But a resident living in a house on the outskirts had heard the noise and was drawn to the field, where he found the unconscious airman. This native was an old man and couldn't lift the airman alone. He returned to the house where he lived with his daughter.

"There's a man out there in the field," he whispered. (They've learned that even the walls have ears in the conquered provinces.) "If we don't hide him," he went on, "the Germans will find him at daylight, and he can't help himself because he's injured. He may die.

Shall we let a friend die or be taken? You must help me carry him here."

"That," the girl whispered back, "may mean death for all of us, will probably mean death . . ."

"A friend!" the old man said.

The girl arose, went to the field with her father, and helped him carry the man to the house where they hid him. They both knew the risks of that journey even in those quiet hours before the dawn. When they had completed it, they glanced at each other and smiled.

"God is with us," the old man said.

And through the weeks that followed they seemed miraculously protected. The presence of the airman was never suspected. They nursed him back to his former ability and started him on his road back. For there is a road back, out of the conquered provinces, out of the hands of the Germans. The murder of Edith Cavell, the British nurse executed by the Germans for helping Allied soldiers escape, didn't close that road back. Innumerable other executions haven't closed it either, because that is something the Germans can't do. In every war such a road persists; it penetrates even the vaunted barrier across the Dutch frontier. So the recovered airman was passed from guide to guide on that road, until finally he slipped from the grasp of the Germans and reported himself ready for duty to his own people.

His exaltation demanded expression. He wanted to shout out his contempt for the German intelligence system that he had so easily mocked. In broad daylight, he flew high over the lines and dropped into the town where he had been concealed, with a taunting letter that stated the exact period he had waited beneath the noses of the Germans for the moment of his escape. Of course he didn't think, and his pride had overcome his judgment. He had underestimated the Teutonic skill. The sequel slipped to him, as more important intelligence slips from beyond the German trenches. That airman has lost his exaltation, and he wonders that his life should have been given back to him. For from the single clue of the note, the German agents found their way to the house on the edge of the town. The gossip of the cafes, shrewd guesses, a painstaking process of elimination were their mileposts—and when they knocked at the door and drew the old man roughly from his house, they were sure. He stared at them trying to shake off their hands, with a great surprise, because it had been so long, because he had forgotten to be afraid.

At that moment, an acquaintance brushed against the daughter in the marketplace. She was directed to a friend's house, where she was told that her father had been taken. So she, too, was placed upon that underground road of sympathy and patriotism. During the dawn of her own escape, the old man was made to stand, blindfolded, against a wall. While he still marveled over the miracle of his success in saving the airman, he was sent abruptly to probe the greater miracle.

In the early days of the war, when there was retreating and advancing, before the neutral zone had narrowed itself to the few sinister yards of No Man's Land, the airplane gathered its intelligence well in advance of the troops. At night, pilots and observers were frequently condemned to strange lodgings, filled with apprehension, where sleep was uneasy. Sometimes they came back with shaken nerves.

I was told of such an experience. An airplane was caught ahead of its division by the sudden approach of a storm at nightfall. The darkness possessed a resistive power; the first rain made it like a soggy, smothering garment. The machine descended in a country still smoking from the devastation of war. To struggle back up to the blackness and the rising wind would have been an invitation to disaster and, with their own eyes, the pilot and the observer had seen the enemy retreat beyond this point.

"At least," the pilot said, "we can't sleep in the fields."

The observer indicated a tiny gleam of light not far ahead, evidently the light of a candle diffused through windows. They walked towards the light and found a small farmhouse; it surprised them because war seemed to have passed it by. They knocked at the door and a French woman with a pleasant, middle-aged face opened it for them. Immediately, both men experienced a sense of something out of the way. There was a queerness, not at all definable, about the pleasant face. It frightened them, made them want to go where its stare could no longer include them. But they couldn't go, the storm had become violent and they were exhausted by a day of labor and perpetual risk. They told the woman they must spend the night in her house. She continued to stare and, at last, she shook her head with a mechanical determination.

"Were there no men?" the pilot asked.

The men were all at the war, she told them, with a voice sharing the quality of her face— pleasant, determined, staring. The pilot and the observer explained that they understood that her house was small, but surely it contained two rooms. And they called her attention to the storm.

"You must see that it is necessary for us to spend the night here," said the pilot.

Again her head moved mechanically, as she replied, "You cannot spend the night here."

"You must tell us why," he responded.

"You cannot spend the night here. I tell you . . . you cannot. It is quite impossible," she said adamantly.

The men glanced at each other and looked again at the woman who stood in the light of the candle just within the doorway. "Queer!" the pilot whispered to the observer, who drew him aside. And while the woman continued to stare at them, without any apparent interest, they consulted about her.

"Looks dangerous," the observer said. "Loyal French are never inhospitable. This woman speaks the language all right, but we're not so far from the frontier. Perhaps she is hiding someone—a German, wounded, more than likely. We know they retreated through here today."

The pilot shivered in the rising storm. "You're probably right," he responded, "and the danger's for the German and not for us. We'll stay in spite of this woman, who doesn't get angry, who doesn't plead, who offers no excuses, who simply forbids us. Naturally we will protect ourselves. We must search the house."

So they went back to the woman and told her that they intended to enter and search her house, before spending the night whether she wanted them or not. "You cannot spend the night here," she repeated, with her mechanical dullness. But she didn't resist when they pushed past her. She only turned to stare after them with her pleasant, determined eyes.

There were, as they had guessed, two bedrooms, opening from opposite sides of the hall. They glanced in the one on the left that was clearly occupied by the woman, as her clothes lay about in some confusion. They opened the door of the other, evidently a spare room, for the bed was larger and it had a canopy and curtains. They passed on to the kitchen. That, too, offered no signs of life. The fire in the stove was out. They glanced back, startled, for the woman was at their heels, moving with the precise awkwardness of an automaton, while her strange eyes stared at them.

"We're getting close," the pilot whispered. "In a moment she'll break down."

He questioned her: "You've had no dinner?"

She shook her head.

"If we light the fire you will prepare our dinner?"

Again she shook her head and said: "You cannot eat in this house."

The pilot made a gesture of impatience. "What is the matter with this house that we can't sleep or eat in it?" he demanded. "We will find out. You are sailing pretty close to the wind. That, I suppose, is the door to the cellar." He opened the door. With revolvers drawn, the two men went down the stone steps, their hearts in their throats, while the woman stood perfectly still in the middle of the room, staring after them.

In the cellar they went carefully but they heard nothing. "Come out!" they demanded, while they held their revolvers ready. They struck matches and searched the corners. Except for themselves the cellar was empty. "Queer! Queer!" they muttered. More afraid than if they had found something, they climbed the steps and looked at the woman who still stood in the center of the floor, staring at them.

"Clearly," the pilot said to the observer, "we are getting a case of nerves. There is no danger here—nothing at all, except this woman who stares and stares and tells us we can't spend the night. I'm tired. I've a biscuit or two and some chocolate. We'll disturb her as little as possible. We'll sleep in the spare room and, if you think it wise, watch, turn, and turn about."

They entered the room and lit candles in a sconce on the bureau. The woman, who had followed them mutely, stood in the doorway. Now she spoke with a mechanical intonation that possessed a certain vagueness: "You can't spend the night here."

This time the men laughed at the reiteration of those words that seemed to possess no meaning. Still, there was something uncomfortable about their laugh, which did not last long. They munched their biscuits and chocolate. The pilot brushed the crumbs from his hand. He lit a caporal cigarette and strolled to the bed to make it ready.

"We'll tumble in here . . ." he said aloud and drew back the faded, red, plush curtains that shook a little, as the candles shook, in the wind from the door. Meanwhile, the woman had come closer. She spread her hands helplessly, as one who is suddenly justified; however, about her gesture was something of despair. The pilot bent over the bed, then he shrank away, and the observer advanced. The woman did not move, her hands remaining extended in that gesture of justification.

For many minutes, the three stared at the young girl outstretched on the bed. There were stains, now nearly black, across her simple clothing and straying to the edge of the coverlet. The young girl's throat had been laid open by a saber, but the stains and the agony hadn't left the pretty face, where they saw a vague and helpless determination, very like the mother's.

"You see," the woman was saying, with a mechanical hoarseness, "you cannot spend the night here."

With awkward and sympathetic gestures, the pilot and the observer slipped past her and quietly left that house. In the turbulence of the storm, they read a welcome. Like hotel espionage, the use of one's own people behind the enemy's line has two sides.

During that visit to Rheims, I heard something there that made me ponder pretty uncomfortably. I knew there must be some explanation of the systematic destruction beyond the fact that the Germans had, for a short time, occupied the town. I remember questioning the cheerful little staff officer. He looked away, and then explained, "The bombardment is directed from within. Some of the *bosches* have remained. They direct the bombardment of their neighbors' homes. There have been many and there still are some *bosche* spies in Rheims. You see great quantities of *bosches* lived and worked here before the war."

Then I remembered that the Germans had always been active in the champagne industry, that many had been employed in the vineyards and the factories of Rheims. Still it seemed beyond belief. "This ruined city was their home," I said. "These houses must have belonged to their friends."

The staff officer nodded. "It is hard to handle them," he said. "They are very clever at reporting damage and offering ranges. It will continue to be so until there is not one of these people left in Rheims. Yesterday two of them were shot." When we heard the very loud sound of guns, the man gestured sadly at the ruins. "Still the bombardment goes on."

I recalled the authoritative statement of the intelligence officer in London that every German, no matter where he lived, believed himself a divinely appointed agent of his government. And I looked at the ruins, wondering.

During my trip to the war zone of Lorraine, I found this give and take of intelligence more pronounced than anywhere else. I have written of the material agony. In addition, I was arrested by a mental distress, born of a situation not unlike that which made our own Civil War so terrible. In these border provinces the population is very much mixed. On the German side, there are many men who through forty years of enemy rule have never lost their true nationality. On the French side, one hears many German names, sees many Teutonic faces. Here naturally was an opportunity that during all these years the Wilhelmstrasse wasn't likely to neglect. Who is to draw the line? Who is to say that this Teutonic type is a loyal Frenchman or a German spy? And on the other side of the trenches the Germans ask themselves precisely the reverse of that question.

It is a dreadful thing to suspect one's neighbors, to search for guilt behind the eyes of those who, before the war, were one's friends. And no spy could expect mercy from these people. The wantonness of the destruction rankles in the border provinces as it never has in any other war, and when you have wandered through the devastated districts, as the Quaker and I did, you understand why. The church in the provinces brought it home. It had no military value as a line of hills rolled between it and the enemy. Yet, it had been blasted by great shells sent from guns many miles away, and the neighboring houses, mere skeletons now, had been blasted with it. Its bronze bells, now distorted and silent, lay in a pool of mud at the entrance. I saw it on a Sunday morning. The officer who accompanied me said: "Now let us look at the real church."

He led me to a house comparatively whole. He opened a door, and within were gathered two or three bent old men, many women, and a host of little children. They sat on rough chairs arranged before an improvised altar with boards draped in white cloths. One had a feeling that the simplicity of their worship concealed a desire for the only justice they could understand—an eye for an eye. They glanced at us with that desire in their faces, and with pride and suspicion. I was glad not to be standing there alone. (I should hate to enter the border provinces at all without iron-bound credentials.) It was, I imagined, pride more than habit that had held these people to the vicinity of their desolate homes. There would have been, their stolid faces seemed to say, a special degradation in seeking comfort, whole houses, and unsoiled churches at the command of Germany's destructive voice. They seemed to be telling me that Germany had had nothing to do with this, that they were making the best of matters after a bad fire or a levelling tempest.

I was glad to have seen that, for it offered a solution I had been seeking ever since my arrival in France. I hadn't been able to understand how the French could develop, largely within two years, their amazingly successful intelligence system. It had seemed miraculous that, at the same time, they should have brought to so little the German system of many decades' growth. In these faces of the provincial French, the answer lay: one saw there an infinite capacity for sacrifice—and that alert watchfulness, that greed for justice. And the entire country is to some extent like that, because there are few who haven't suffered personally from this war with its new and intolerable methods. Every man and woman is a potential trapper of spies.

Moreover, as I looked, it seemed to me that on that simple altar, before which these determined people worshipped, the supernal and France had become inextricably tangled.

Chapter Twenty-One

THE ADVANCE

The gray and crimson tints of all these phases color Europe too morbidly. There is no escape. On my way back to America, I reached Bordeaux, the gay southern city that at first seemed to offer, with a smile, just that evasion that everyone who sees the war with an intimate understanding must narrowly crave. German prisoners working in the fields and on the roads in the outskirts were, to be sure, a reminder, but they appeared to have borrowed something from the warm, bland countryside to which they had been transplanted. Their faces were without anger or regret. They seemed happier than the freemen condemned to the trenches.

In Bordeaux there were fewer uniforms than one sees to the north and less of the eternal military display in shop windows. There was a much-heralded theatrical production that night and, announced for Sunday, an open-air performance of *Samson and Delilah*. But, almost immediately, the black shadow of war showed itself: there was uncertainty as to when the boat would sail; a promise, ever more clearly defined, of an extended delay; and a sense of lurking danger at the mouth of the Garonne River.

And the next morning, when I stepped from my hotel, I heard the throaty music of bugles, and I saw thousands of Senegalese—just landed and about to entrain for the front—march past. Beneath red fezzes, their black and childish faces shone with the heat. As they swung along with a naive pride, one questioned if they foresaw anything of the facts.

America, with its lights and its careless pleasure seeking, attained a visionary quality. Was it possible such a place actually existed? At first one was happy at the prospect of that refuge, but the bugles continued, blaring the truth of this war, and one became ashamed, reading in such a state a vital wrong that sooner or later would have to be paid for.

As the gangway from pier to ship in New York had shown itself to be the threshold of war, so, too, it was apparent, would it prove itself the only exit. For on the boat, sitting in the steamer chair next to mine, occupying a seat at the same table, was a young fellow from Brooklyn, decorated with the Croix de Guerre and the Médaille militaire. He moved about only infrequently because of an artificial leg that he had failed to become accustomed to. I tried to sound out the impulse that had urged him to leave a broker's office to enlist in the Foreign Legion. He could only express it in this way: "I wanted a look in at the last war." One sought to vindicate his anxiety and his optimism; moreover, he was very modest about his medals.

There were other soldiers on the boat who had been decorated—either French-Americans on permission, or poor devils like that boy from the Foreign Legion, cast into the vast and pitiful slag heap of war. And there was a wrinkled Canadian-Belgian in the steerage.

"I am fifty-six," he lamented. "I have been wounded three times, but each time I have gone back to the trenches. Now, because they say my lungs are weakened, they won't let me fight any more. That is absurd. And it was I who destroyed the bridge at Termonde. The fuse had been cut, and the *bosches* were coming across, firing their machine guns from behind shields of mattresses. I crawled along inside a metal cask to the point where the fuse had been cut. And I lighted the broken end. Pouf! You should have heard that! You should have seen that!"

He lifted the medal of St. George that was pinned to his rough tunic. "The King himself," he said proudly, "placed that there, and there are few who have won it."

So through the tiresome voyage there was no escape. Then one afternoon we steamed into New York Harbor, and I saw a city that seemed proud of an incomprehensible ignorance of the meaning of war.

As we drove uptown, the dusk thickened and lights flashed in a strange extravagance. We passed laughing men and women, weaving in and out of restaurants, in and out of theaters and dance halls. It was like a city uninstructed in reality.

After a time, the sense of wrong vanished. One watched these men and women with a quick sympathy, limiting the period of their carelessness. For a question had survived through my months in Europe: "How long before we, too, will be at war?"

As I drove on, the question drifted inevitably into a statement, brutal and unescapable: "We, too, will be at war. It will not be long."

A History of the 305th Field Artillery

by Charles Wadsworth Camp

Contents

TO
THE MEMORY OF
THOSE OFFICERS AND MEN OF THE
305TH FIELD ARTILLERY
WHO REST IN FRANCE

Preface

When the colonel assigned to me the task of writing a history of the regiment, we were billeted at Arc-en-Barrois in the Haute-Marne. Most of the work, therefore, was done at Arc and Malicorne or, practically, under field conditions. One must admire all the more, then, the success of the artists who overcame a lack of proper tools and working space. To Corporal Roos, Private Enroth, Corporal Schmidt, Musician Boyle, Corporal Tucker, Captain Dana, Captain Starbuck, and Private Everts the regiment is indebted for the majority of these lively souvenirs of campaigns and billets.

Tremearne of B and Downs of A were particularly useful in gathering statistics and material. Where statistics lack, or are not complete, it must be assumed that names and figures were either not furnished or could not be obtained.

The historian has thought it of interest to follow his own narrative with an appendix containing contributions by individual officers and men.

The whole, he ventures to hope, will constitute a pleasant record —necessarily imperfect because of its brevity—of a very memorable experience.

— CHARLES WADSWORTH CAMP

Chapter One

THE REGIMENT IS BORN

When it comes to beginnings, regiments are not unlike humans: they aren't pretty objects or self-sufficient; they gaze upon the world with inquiring eyes; and they address it with lusty and surprised lungs. We were very much like that, and our first surprises came with our first days, when the men commissioned from the Second Battery at the First Plattsburg Reserve Officers' Training Camp reported at Camp Upton.

The adjutant's office was in an unpainted, wooden barracks. A line stretched hour after hour snake-like half around it, its head investigating the somber corridor where the adjutant's assistant sat making assignments. Nearby, those who had survived the ordeal stood in groups—ill at ease, wondering.

"What is a casual officer? Something to do with casualties?" one asked.

"They told me at Plattsburg I was in the regimental quota," another said. "That fellow in there says no. I'm in a thing called Military Police, and when I told him I'd never swung a billy in my life, he wanted to know what that had to do with it."

"I'm in the Depot Brigade," a third grinned sheepishly. "Good God! Do we have to run the trains?"

A captain walked from the corridor and came up with a pleased smile.

"What did they hand you?" someone asked him.

In his voice was pride and a vague new responsibility: "I'm assigned to the 305th Field Artillery, National Army."

Several joined in, as in a chorus: "So are we. That's going to be the number of our regiment." Surprise and gloom then deepened on the faces of those shifted, unexpectedly, to unforeseen branches of the service. And thus the regiment was born and baptized, and we heard for the first time the significant number in which officers and men have, to an extent, merged their thoughts, their actions, and their individualities.

Colonel Fred Charles Doyle was the first to report. He came from the Regular Army and received his assignment from Major General Bell on August 28, 1917. For ten days afterward, the officers poured in and commenced to prepare for the men who would arrive in the course of the next few weeks. During those days of its beginnings, without the men it wasn't much of a regiment and yet it possessed from the start ambition, pride of organization, and already—a noticeable factor—an instinct that our regiment was to be bigger, better, and more terrible to the enemy than any other in the Field Artillery.

Yet, we went gropingly at first, asking earnest but absurd questions about equipment and rations, or demanding with concern where we could house even a single section. The welcome Camp Upton gave us was not of arms outstretched and smiling hospitality. We had stepped from New York through a screen of dreary pine wilderness into a habitation both startling and impossible. A division was to be trained here to fight the Hun, but to any observing person it appeared that even if the war should last another decade, Camp Upton would not be useful. It wore an air of having just started and of never wishing to be finished. From the mud a few white pine barracks stretched gaunt frames against a mournful sky. Towards the railroad two huge tents had an appearance of captive balloons, half-inflated. The rest consisted of heaps of lumber of odd shapes and sizes and countless acres of mud, blackened by recent fires—half-cleared land across which was scattered a multitude of grotesque and tattered figures. These workmen went about their tasks with slow, indifferent gestures and their attitudes were suggestive of a supreme faith in the eternity of their jobs.

Some of us gathered on Division Hill the night of our arrival. We gazed from the little that was done to the immensity that remained untouched. "Where are they going to put the 305th?" we asked. Captain Devereux, who had gathered some information, pointed to the northwest and explained, "That's the area assigned to the regiment. We'll live and train there."

For a long time, with skeptical eyes, we continued to stare at that blackened desert. We strolled back to J22, our temporary quarters, depressed and doubtful. In the barn-like upper floor, where we had erected cots, we gathered about a candle lantern and in low tones probed the doubtful future. Colonel Doyle, who was to be the regiment's commander from its birth to its final demobilization, was to us that night no more than a name. He lived somewhere on Long Island and would be in camp the next day. At least we had a colonel, but who would be our lieutenant colonel? We had one major, made at Plattsburg, but what about the other leaders we needed? Lieutenant Derby then pronounced the first of the regiment's innumerable rumors. (It should be said that it's about the only one that ever came true.) He had heard in town that Henry L. Stimson, former Secretary of War, would come to us as lieutenant colonel.

We gossiped about the unexpected shifting about of our friends. Many that we had expected to have with us had been quietly spirited away. Others, whom we had not expected to see after Plattsburg, sat in our circle, assigned to the regiment. From the start we found the army full of odd surprises, which gave us all, for the moment, a sense of instability.

We learned that our commissioned tables of organization, filled out painstakingly the last night at Plattsburg, would have to be radically revised. Nor was that the only unexpected task. We couldn't forget the black wasteland we had seen from Division Hill, and before long the men of the first draft would stream in, and we would have to share in the miracles that would feed, clothe, and house them (giving them the vital initial impression that they would be taken care of in the army). Our doubts increased when we sought our own washing facilities that first night. Who will forget the scouting among piles of lumber, the stumbling over roots and stumps, the escapes from superbly imitated swamps, or the final, triumphant discovery of a single pipe and faucet—surrounded by a mob of soldiers of violent temper. More than a thousand officers had reported to the camp by that time. And, of the twenty-five thousand workmen of the Thompson-Starret Company, some undoubtedly craved that which is next to Godliness. There may have been other pipes at Upton, but for a time that one remained our only discovery and it had a miserable habit of falling languidly over into the mud unless it was supported by a comrade who had the strength and the will to fight off an army. Yet we shaved . . . yet we contrived to look clean. "Horrors of war, No. 1," we labeled our pipe.

So we struggled on, preparing ourselves as best we could for the day when the first enlisted men would arrive. At night we gazed with new interest at the multitude of fires that blazed crimson against the forest, surrounded by ragged groups of workmen who sat for the most part in a sullen and unnatural quiet. Day by day the miracles happened before our eyes, and the wilderness receded, the mushroom city spread. This morning you might walk in a thicket, but tomorrow you would find it cleared land, untidy with the beginnings of buildings. Our faith grew that the 305th would have a home.

Side by side with the above improvements other and more intimate miracles developed. Colonel Doyle established a Regimental Headquarters on a mess table in the mess hall of J1. Whatever stateliness it may have acquired later, headquarters in those days proceeded, as one might say, on hands and knees. Colonel Doyle explained how things should be done and we did our best to do them right. Already, from the pots and pans of J1, "Paper Work" raised an evil head and sneered at us.

Before the table became really untidy with baskets and typewriters and files and reports, other organizations came enviously in and established headquarters on that table too. There was a machine-gun battalion, the ammunition train, and maybe a bakery company or so. Things became rather too confused for an accurate count. We stole away quietly to J20, the upper floor of which had become our sleeping quarters.

That same afternoon Major Wanvig appeared, bearing an oblong signboard under each arm. One he nailed at the entrance of the building, the other he fastened to a post by the road. On a white background in striking black strokes, each sign bore the inscription: "Headquarters, 305th F. A. N. A." We stood about, staring, and inquired aloud, "That's us—the 305th Field Artillery National Army. Are we going to make it big and successful enough?" So far there were at least no visible shirkers and we had already acquired a belligerent disposition to stand fast for the rights of the regiment. That was as it should have been, since we were destined to be among the first of the combat organizations. There was, moreover, need of such a spirit.

Take J20, for example. Once you had a bit of floor space there the whole world conspired to tear it from you or, as more convenient, you from it. Regimental Headquarters had established itself modestly in a corner of the lower dormitory. Officers of high rank in search of sleeping space complained that we were in their way. Brigade Head-quarters sent messengers to measure us broad and long. Commanding officers and adjutants of various organizations, quar-tered in the same building, cast threatening glances in our direction. Low-browed hirelings of the Thompson-Starret Company came, demanding the return of panels of Upson Board and pieces of deformed lumber, with which we had endeavored to barricade ourselves against an eager and conscienceless world. In spite of everything, Regimental Headquarters clung to its corner until, in late October, it moved to its own building in the 305th area. Those few weeks in J20 witnessed our adolescence—when we tramped across the hill we were, indeed, a regiment.

September 6, a day that must be recorded noticeably, saw the first enlisted personnel of the 305th: Frank Dunbaugh. He stood at attention before Colonel Doyle, saluting and announcing, "Private Dunbaugh reports as directed." And behold, we were a regiment—officers and man!

We all, I think, felt a call to take out that pleasant young fellow and give him a dismounted drill, simulated standing gun drill, physical exercise, semaphore, wig-wag, and buzzer. And, show him the beginnings of firing data and scouting with perhaps, in his off moments, a little of grooming and horse-shoeing and the theory, at least, of equitation. But he was a little man and Division Headquarters tore him from us before we could really annoy him. An order had come down: "Private Frank Dunbaugh is relieved from duty with the 305th F. A. N. A., and is attached to Division Headquarters," and so forth.

Paper Work grinned.

For that matter, he had plenty to chuckle over already. Headquarters was aware by now of his portly and increasing figure. General Orders, Special Orders, Memoranda, and Bulletins were suspended in neat wads from the wall. Captain Gammell, the regimental adjutant, threaded his way among them with haughty ease. At his suggestion, indeed, an officer brought from Division Headquarters a bundle the size of a small bale of cotton. We gathered around it, admiring the countless neat forms it contained, all labelled "A. G. O., No. so and so."

"What a system!" everybody gasped.

What a system, indeed! But we couldn't dream of all that those delicate forms portended. Captain Gammell distributed them, Colonel Doyle explained how simple it was to handle them, and we turned again to the apparently more serious business of getting ready. With grim determination the officers, shorn of their enlisted personnel, pounced upon each other. Because there was no reasonable drill ground, we took ourselves to the stumps and the logs of the half-cleared spaces and drilled each other, shouted at each other, abused each other. How, we asked, would new officers and men take this or that?

"If you make a rookie laugh, it's all off," an officer said after an exceptionally piercing cry of command.

"Or," another put in dryly, "if you give him the impression you're going to murder him he won't respond cheerfully enough."

We endeavored, therefore, not to resemble fools or assassins. Sometimes it was difficult.

Each day now, for a time, Colonel Doyle rescued us from our harsh treatment of each other. He took us to the slope of Division Hill where we sat on charred logs and listened to him discourse at length on various methods of computing firing data, or interpreting the Articles of War and Army Regulations, drawing on his long experience in the Regular Army.

Meanwhile, the activity all about us was frequently distracting, unreal, a trifle prophetic. In the rapping of countless hammers, you could imagine the stutter of machine guns. The fall of heavy timbers was suggestive of the crash of rifles of our own caliber. To give a more realistic touch of war, at the base of the hill lay the encampment of the colored troops of the 15th New York National Guard. It should be recalled in passing that these dusky doughboys were a very small oasis of soldiers in a thirsty desert of officers. In salutes and courtesies, they received a maximum of practice.

Lieutenant Colonel Stimson came to us during one of these classes. That was on September 6, and by evening of the next day, the last of the officers sent down from the First Plattsburg Reserve Officers' Training Camp had reported and been assigned or attached to the 305th. Since the majority of them led the regiment into its first battles, a record should be made of their names in this chapter of beginnings. We commenced with the following officers, most of whom had abandoned civil life only three months earlier:

Colonel Fred Charles Doyle, commanding the Regiment; Lieutenant Colonel Stimson, temporarily assigned to the command of the First Battalion; Major Harry F. Wanvig, commanding the Second Battalion; Captain Arthur A. Gammell, regimental adjutant; 2nd Lt. Allen A. Klots, acting adjutant, First Battalion; Captain Douglas Delanoy, adjutant Second Battalion; Captain M. G. B. Whelpley, commanding the Headquarters Company; First Lieutenant Edward Payne, temporarily in command of the Supply Company; Captain Alvin Devereux, commanding Battery A; Captain Gaillard F. Ravenel, commanding Battery B; Captain Noel B. Fox, commanding Battery C; Captain Frederick L. Starbuck, commanding Battery D; Captain Robert T. P. Storer, commanding Battery E; Captain Cornelius Von E. Mitchell, commanding Battery F; First Lieutenants Sigourney B. Olney, George P. Montgomery, William M. Kane, Harvey Pike Jr., Watson Washburn, James L. Derby, Edgar W. Savage, Frank Walters, and Drew McKenna; Second Lieutenants Sheldon E. Hoadley, Thornton C. Thayer, Norman Thirkield, George B. Brooks, Lydig Hoyt, Thomas M. Brassel, Lee D. Brown, Chester Burden, Charles W. Camp, Paul Jones, Oliver A. Church, Roby P. Littlefield, William H. M. Fenn, John R. Mitchell, Warren W. Nissley, Harold S. Willis, Frederick L. Beck, Danforth Montague, Melvin E. Sawin, George P. Schutt, Lloyd Stryker, Lawrence Washington, John A. Thayer, Karrick M. Castle, Harry G. Hotchkiss, George E. Ogilvie, William L. Wilcox, Lewis E. Bomeisler Jr., Darley Randall, and Edward W. Sage.

Almost at once changes were made in this list of our charter members (as one might call them). Officers were assigned away from us, while strangers were brought into our midst. Thirty-five of the charter members accompanied the regiment to France. After the Armistice, there remained only nineteen. The eternal changes of the army system were largely responsible for these losses, as they accounted also for the passing of many enlisted men, but whenever we meet the old friends we think of them as belonging peculiarly to the 305th. Some we can't see again, because the Vesle, the Aisne, or the Argonne holds them forever away—but it is a dreary business to anticipate, and they were very much with us and very much loved at Camp Upton.

So the first week ended and we were on our feet, if still unsteady.

Chapter Two

IT HAS GROWING PAINS

Going into the second week the colonel talked daily with his organization commanders. Such conferences revolved largely around the almost scented forms from the Adjutant General's Office. These forms, we discovered, would have to be decorated with countless, neat statistics when the men arrived. Soldiers, as far as we knew, might go hungry or without equipment but, as far as figures went, they would unquestionably be cared for tenderly. No one would have the slightest doubt as to their most intimate family history, the number of years it had taken them to dribble through public or private institutions of learning, or their degree of proficiency on mandolin, harmonica, or Jew's harp.

The officers at that period filled out forms about themselves in odd moments. The most persistent and suggestive of forms demanded the name of the relative one wished notified in case one should become a casualty. Whenever things in America or France got a little slack, a request for that information would come around. It kept one "on one's toes." We wondered why that bureau never got fed up with paperwork.

Almost at once a sense of imminence crept into these daily conferences. Now the huge bulletins descending from Division Hill were dealing with dates. With admirable detail they described how the first of the draft men would be received. To aid us in this task, noncommissioned officers, it was promised, would be sent from the Regular Army. They appeared one day—a score or so for our regiment.

We looked at them, we looked at their service records, then we looked at each other. We swallowed our first lesson in how to send, on order, one's best men to some other organization. Certainly, in this case, few commanding officers had parted with their jewels, for some of these rough diamonds had been set in chevrons for our needs. There lay their records of battery punishments and courts-martial. We pitied those distant, unknown commanders, for if these were their best, we shrank from picturing their worst. The audacity of the thing caught our imagination and there was, we felt, something to be had from it. They weren't all bad, by any means—some became the most useful of soldiers.

Our Medical Department arrived about the same time—a worried-looking little group that trudged through the dust, dodging piles of lumber. It was led by Lieutenant James B. Parramore, who later became captain and regimental surgeon. Lieutenant Dennis J. Cronin was assigned as First Battalion Surgeon, and Lieutenant Marshall A. Moore as Second Battalion Surgeon.

That very day, Dr. Parramore constructed a table in Regimental Headquarters on which he placed (with proud gestures) a tin of alcohol, a demijohn of castor oil, a few assorted pills, and gallons, literally, of iodine. He then announced himself open for business. Fortunately, business was dull, so the adjutant reached out for Parramore's enlisted personnel, sat them on a bench in the hall, and —behold!—for the first time Regimental Headquarters had orderlies. There was no doubt about it, we were growing.

On September 27, the arrival of our chaplain, John J. Sheridan, was another reminder. And two days later the long dreamed of moment arrived: 535 recruits were assigned to our regiment. These men, of course, did not come directly to us from their local boards. We received them after two weeks' work of reception and assortment, in which all the officers of the division shared. During that phase the once strange term "casual" became a byword, for all the draft men arrived at Camp Upton as casuals.

Officers met the first train loads at Medford on September 15. There are, let it be granted, few days in the history of our country more impressive than that one, which saw the triumph of universal service and the birth of our great National Army. This is evident from a distance, for in the minds of the officers and men who assisted there lingers beyond question, woven with the sublime, a palpable tracery of amazement and mirth. Why? Because, the draft men came in ancient railroad coaches with sides trimmed with placards suggestive of an abnormally swift and terrible march to Berlin via Upton and listed a number of penalties for the Kaiser, very ingeniously thought out.

And then there was the provocative personal adornment. There had been word in the papers that all civilian clothing worn to Upton would have to be cast away. So these young men took no chances: thrust through the windows were men's heads adorned with tattered straw hats; crushed derbies, through which wisps of hair straggled; and top hats, in a few cases so venerable that it was a pity to see them out of their sepulchers. And Palm Beach suits of previous summers were there, and the dinner jacket (an affair of generations), and the suit that had been worn on Sundays long before the owner's maturity. It was an assortment that would have taxed the sanity of a Hester Street dealer in New York City.

We tried to sound the meaning of such a trip to these young citizens. But we could only sense definitive separations from home and comfort and affection; a shrinking from our uniforms, which symbolized a discipline that was terrifying and undesired; and, perhaps, a perplexed apprehension of violence and the end of experienced things somewhere just ahead.

No mind, however, could linger on those thoughts. There were too many races, clamorously asserting themselves; too much had been made of a number of departures; and there still lingered too many souvenirs of feasts.

Out of the shadows slipped an eager voice, "Hay, Tony! Finish off that bottle before these officer guys can grab it." And another, less concerned, said, "Grabba da hell. My gal, she givva me a charm against da evil eye of officers." And some had reached the point where speech ends.

A man in uniform grew disgusted. "So," he grumbled, "that's what we've got to teach to fire a three-inch gun!" But we knew he was wrong, for he had judged them by the highlights. In the really fundamental background of these men, we saw a sober and determined spirit. Even then we felt the presence of some of the best soldier material in the world.

After meeting a few of these erratic train loads, the least confident of shavetails (those newly commissioned officers) could not forecast their garrison tasks with ease of mind, for such recruits weren't simple to control.

When we gathered at night in J20, the gossip of every group revolved around the arriving casuals. Below is a typical conversation.

"How many souses did you have today, Bill?" a soldier asked.

"Two," Bill responded. "One wanted to weep on my shoulder, and the other wanted to give me an uppercut."

"What did you do about it?" another asked.

"Ordered the fighting one to take care of the weeper," said Bill.

"Say! Did he?"

"You bet," Bill answered. "Closed both eyes so the tears couldn't get out and satisfied himself at the same time. I remember he shouted as he swung: 'Hay, Boss! It's a grand war!'"

Those already in uniform, none the less, felt a quick sympathy for the newcomers. Their individualities slipped away from them so easily! At the station they were labeled and assigned to barracks. Then they were herded and marched in long, uncouth lines to the hospital for physical examinations. We formed squads and tried to instruct them in the school of the soldier. Rich and poor, Hebrew and Gentile, short and long, straw-hatted, felt-hatted, or without any hats at all, they faced us eager to learn.

"One, two, three, four. One, two, three, four. Squad halt. Right face. Left face. About face," we commanded these young men. Those who couldn't speak English very well got the commands confused, others had a curious lack of balance, but all had a disposition to laugh at mistakes and accidents, and to discuss and argue about them while in ranks at attention.

At morning and evening roll call, the arguments were the most heated. No linguist existed amongst us who was sufficiently facile to scan that list intelligibly. Sprinkled among remembered English names were pitfalls of Italian, Russian, Spanish, Lithuanian, German, even Chinese names.

"Krag . . . a . . . co . . . poul . . . o . . . wicz, G," struggled one officer, while calling the roll. He looked up, expecting the response his triumph of deciphering the name deserved, but instead he was greeted by a protest, from a fellow with a lower New York City accent.

"Do yuh mean me?" he said with indignation. "That ain't the way tuh say my name. Me own mother wouldn't recernize it."

"Silence! Simply answer, 'here,'" shouted the officer.

In a tone of deep disgust, the New Yorker responded: "Then I ain't here. That's all. I ain't here." An appreciative laugh then rippled down the ranks. Men had to learn "on the job" how to be officers and noncommissioned officers in those days.

Afterwards, the citizen soldiers would get their mess kits and, sitting on burned stumps or Thompson-Starret rubbish, would eat a palatable meal. For the food was coming from somewhere, and the gear to dispose of it.

We had noticed that First Lieutenants Walters, Payne, and Savage were up to something. For long hours they sat in Regimental Head-quarters studying documents and filling out many forms, and then sample clothing and equipment straggled into the barracks. This meant a new phase and we became tailors, hatters, booters. We would begin the night's work by choosing as comfortable a place as possible in the mess hall with a pile of pink qualification cards before us. The queue of awkward and pallid youths would form.

"Name?" we asked, and out would flow names said in various accents. More often than not it would demand painstaking spelling on our part. Then it was education, occupation, average wages, capacity for leadership, ability to entertain, previous military experience—it all went down. But there was one question in which we took a special interest: "For what branch of the service do you wish to express a preference?" Some had already weighed the matter carefully and believed themselves born to the Quartermaster's Corps. The majority, however, had not foreseen that interrogation and the sincerity of their answer was sometimes naive. "Oh, hell! I don't care, just so I lick the Choimans," said one. We concentrated on the finest and shamelessly we proselytized—and out of this impromptu mission came some of the regiment's best.

Those hours of dreary, yawning statistic gathering had their lighter moments. In one instance, a slender young man in the familiar suit of remote beginnings was asked by the officer: "Wages in your last job?" And he responded, "$50,000 a year." That officer, one recalls, rose to the occasion, for the young man was not boasting. "And I understand you wish to express a preference for the Field Artillery?" he inquired.

And wasn't it Second Lieutenant Hoadley who faced a youth just the reverse of this last—that is, flashily tailored? "What can you furnish in the way of entertainment?" asked Hoadley.

"Me?" the flashy young man replied. "I could steer the village miser into a poker game and, believe me, bo, I can make a deck of cards lay down and roil over. What's the idea? What d'ye mean? I got to split with you?" When he declared for the Cooks and Bakers, his choice went down without argument.

Afterwards, we would line up our charges again and distribute qualification cards for sample shoes and hats and clothing. Sizes were limited, and we hadn't anticipated nature's infinite variety in modeling the human form. We made an axiom at the start: the more peculiar the shape, the more particular the owner. "For the lova Mike, mister, I can't wear that coat. Makes me look as if I'd broke me breast bone," said one young man. Another said, "You got to melt me to get me into this." Everybody worked with patience and a desire to be fair, but we had to make both ends meet and as the hours flew by we may have hurried a little.

It was during these sessions that a rotund and good-natured officer gave us a stirring example and prophesied his own future. "You're in luck. That's a wonderful fit," you'd hear him say to a man with a 32-chest lost in a 36-blouse. "You're a perfect 36. Might have been cut for you."

The man gathered a fistful of the excess cloth, stretching it towards the officer, and said, "Cut for an elephant."

"The tailor will alter it so it won't look like the same blouse," responded the officer.

"I'm not saying anything about its looks. All I'm saying is maybe it isn't quite big enough for a good-sized elephant," said the man.

The officer's buttons stretched, as he explained, "If you want to get along in the army, young man, you'll do as you're told. I wouldn't mind wearing that blouse myself."

"But," another officer whispered to him, "you're not quite as big as a good-sized elephant."

The good-natured officer grinned and continued to show us how to make the best of the material in hand. "That hat isn't too big for you," he would call out in his cheery voice. "Gives your hair a chance to grow."

So we struggled on through the days and nights until the first quota was classified and at least partially equipped. And out of that quota came, as related above, 535 recruits—not far from half a regiment. "The men we're to live and fight and die with," someone said.

It didn't turn out quite like that. We didn't foresee the wholesale transfers, the all-night conferences during which officers and noncoms tried to do the fair thing without destroying their organizations. Those dark days of transfers fall more reasonably in another chapter. For at that moment, we were a trifle hypnotized by our growth and our power. We looked along the lines, guessing at the good and the bad—like all regiments, we had both.

The faces we saw were pretty white and the physical frames not, as a rule, powerful. We were a part of the Metropolitan Division and most of our men came from the crowded places of New York. Out of city dwellings, offices, subways, and sweatshops they poured into the windswept reaches of Upton. They knew none of the tricks a boy picks up in the countryside that fits him, after a fashion, for such fighting as we were destined for in the Vosges, on the Vesle and Aisne, in the Argonne, and on the Meuse.

"Will soldiers grow from such material?" visitors asked. From the start officers and men knew the answer as affirmative. Day by day, beneath the bland autumn sun, faces bronzed, chests seemed to expand, and shoulders broadened from the tonic of physical labor. It wasn't all drill. The miracles continued, but there weren't enough civilian workmen available to construct the city, to clear vast spaces for drilling, and to arrange artillery and small-arms ranges. So orders came for the draft men to pitch in and help. Thus commenced the cheerful game of stump pulling. Of our original quota, there are very few that couldn't qualify as expert destroyers of wildernesses. The famous skinned diamond exists as a monument to our skill—the target range is a document written in the passionate sweat of our brows.

During this education, the first effects of discipline were apparent. Faces might darken with rage or whiten from weariness, but in the presence of a superior work they went on without too much painful comment. Occasionally, by chance, you might hear burning opinions of army life in general and stump snatching in particular. At school we had been taught that the average man's vocabulary is scarcely more than five hundred words. The understatement is obvious. Any soldier of the 305th who couldn't apply as many adjectives as that to the common noun "stump" was frowned upon as mentally deficient or as one affecting an ultra-religious pose.

Such tasks were, in a sense, a digging of a pitfall for one's own feet. As the skinned diamond expanded, our drills waxed proportionately ambitious. But the entire process was performing another miracle. Where formerly slovenly ranks had slouched, now appeared straight lines of soldierly figures, heads up and shoulders squared, exuding a joy in things military. "What's all this guff about West Point?" you would hear. "Watch my outfit drill any day."

The veterans of a week or so exposed a most amusing tolerance for newer recruits. The difference between a uniform and civilian clothing created an extensive gulf, but in just a few days it would be bridged. The awkward squad of the day before would face the awkward squad of today with expressions of veteran contempt. The recruits poured in during October: on the 1st, we received 113; on the 9th, 183; on the 10th, 254; and on the 12th, 218. So that by the end of that month we had 41 officers assigned, 18 attached, and 1,313 enlisted men. The 305th was a regiment. All we needed were horses and guns to realize that we were, indeed, artillery designed to throw projectiles at the Huns.

To give variety to our stump-pulling sport, Colonel Doyle called our attention to certain long, low, and harmless-appearing buildings across Fifth Avenue. Still living in the J section, remote from these constructions, the men hadn't wondered about them, but now, it seemed, they were to be our stables. The civilian workmen's responsibility had ceased after they had put up sides and roofs; the rest we must do. We had many fences to build around them, more land to clear for riding rings and paddocks. We were encouraged in our tasks when the government presented us with eight mules on October 18. They were led to the most comfortable stable and treated as honored guests. Quite fittingly, our first veterinarian, Lieutenant North, arrived soon after.

The problem of our missing field officer was solved on October 14, when Major Thomas J. Johnson reported. At that time, no one had an opportunity to know him very well or to judge him competently. It wasn't until we had reached France that we were to realize our good fortune. For on October 26, he was detailed to the School of Fire at Fort Sill. Major Wanvig left for the same destination on November 7.

Without formal battalion commanders, the work of our regiment continued. In view of the lack of equipment, it continued amazingly well. Reserve officers of only a few months training displayed exceptional qualities of leadership. New soldiers wanted to learn. An artilleryman must be able to do more than use a sight, work a breach, or pull the lanyard. The chances of the draft had given the 305th a number of highly-educated specialists for the more complicated work of conduct of fire, and the delicate details of scouting and communication. To that important extent the regiment was already better off than some of the older organizations. By day, the officers instructed and drilled the men; by night, the officers went to school themselves with Colonel Doyle and Lieutenant Colonel Stimson, who did the best they could with the slight material at hand to keep us abreast of artillery developments in the war zone. When we finally got to France, we were overwhelmed to realize all we had to learn.

Now, when we glanced from the slope of Division Hill at the bleak landscape that only a few weeks before had aroused our skepticism, we saw barracks, quarters, and department buildings rising from the ashes of the forest. During October, the regiment moved piecemeal from the J section to its own area. The change was complete on October 24. As we had policed J and its vicinity, so we made our surroundings in M neat and military. Officers and men experienced the fortunate impression of permanence. As long as we remained in Upton, we would have our own home. We felt—perhaps in our own ignorance—that things were going well. Even a band had been collected and could play one or two pieces in public with comparative safety.

During the latter part of October and the first part of November, the officers were brought a little closer to their mission. They were conducted by twos and threes to Sandy Hook to watch the practical working of projectiles and fuses. Forty attended a six-day artillery course in New Haven under the experienced instruction of Captain Dupont, of the French Army, and Captains Bland and Massey, of the Canadian Artillery. On these trips most of the officers saw for the first time the soixante-quinze, the famous 75mm cannon. They admired it as a piece of artillery perfection without being able to guess that it would be their companion for many months, a thing nearly as animate as the men who served it.

What we actually received at Camp Upton at this time was a

single battery of venerable three-inch guns, relics of the 51st Field Artillery Brigade, New England National Guard. Lieutenant Colonel Stimson snared this for us, together with much other useful equipment that aroused the envy of less fortunate organizations that didn't have a former Secretary of War. Certainly, one battery among six was better than none.

When the guns arrived on November 10, the regiment gathered around them, patted them fondly, examined their mechanism, peered down their throats. Pride leaped. "God help Jerry when we show him these!" we exclaimed.

But Jerry never saw them. Perhaps one day, in the dust of some ordnance museum, they may be observed by all the world—precious relics of the extended battle of the 305th at Camp Upton.

Chapter Three

AND BECOMES ACQUAINTED WITH PAPER WORK

Paper work had now become our perpetual companion. Neither by night nor by day did he leave us lonely: he strutted at mess, he paraded across the drill ground, he sat by one's cot through the troubled watches of the night. It became, therefore, necessary to study the creature's habits.

Let us take a fanciful case that everyone can understand, since even in those early days what we called "Corn Willy" was omnipresent. Let us suppose that a mess officer desires some information about this old friend. His impulse might be to dash off a note like this: "To Captain . . . Dear Sir: Having heard that you've made a life study of the subject, it's occurred to me that you might tell me how it is possible to make Corn Willy palatable."

But, if one didn't care to bother the colonel about details of paperwork, Captain Gammell was always glad to put one right. "Not at all, my dear young mess officer," he would say. "Not at all. You must send it through channels."

"I don't think his office is far away. I might just run up and see him," says the mess officer.

"What nonsense, my poor ignorant young mess officer!" replies

Captain Gammell. "In that case what record would exist of this matter?"

So, now you must picture the mess officer in question studying in "Army Paperwork" all about going through channels. As a result, he might turn out something like the following:

Camp Upton, N. Y.

October . . . , 1917

From: 2nd Lieutenant . . .

To: C. O. Dep't of Household Enemies

Subject: Corn Willy

Information is desired as to any known method of making Corn Willy palatable.

JOHN BLANK, 2nd Lieutenant, 305th Field Artillery

That note would occupy some two inches on a sheet of foolscap. A few months later Lieutenant . . . , probably in charge of stables now, might receive a breathless messenger, bearing a huge envelope with his original sheet of foolscap pinned to reams of endorsements. These would run something like this:

1st Ind., From: C. O., Bat'ry . . . , To: C. O., 305th F. A.

Forwarded.

Approved.

2nd Ind., From: C. O., 305th F. A., To: Com. Gen. 152nd F. A. Brigade, with, perhaps, a paragraph or two.

3rd Ind., From: Com. Gen. 152nd F. A. Brigade, To: Com. Gen. 77th Division, with, perhaps, several paragraphs, scarcely ever more than a word in length.

4th Ind., From: Com. Gen., 77th Div., to Adjutant General of the Army.

1. For investigation of record of Private C. Willy.

5th Ind., From: Adjutant General of the Army, To: Com. Gen., 77th Div.

Received.

Contents noted.

No record.

Should be forwarded to Quartermaster General of the Army.

6th Ind., From: Com. Gen. 77th Div., To: Quartermaster General of the Army.

7th Ind., From: Quartermaster General of the Army, To: C. O. Subsistence Division.

8th Ind., From: C. O. Subsistence Division, To: Chief Q. M., Dep't of East.

9th Ind., From: Chief Q. M., Dep't of East, To: C. O. Eastern Subdivision Department of Household Enemies.

10th Ind., From: C. O. Eastern Subdivision of Household Enemies, To: Lieutenant . . . (Through Channels)

1. Received.

2. Contents noted.

3. No method Known.

"What shall I do with it now that I've got it?" asks Lieutenant [. . .]

"What would you suppose?" is the tolerant answer of the expert. "It has become a matter of official record. Consequently, it must be preserved forever, or nearly so. File it away."

"There isn't much room left in our barracks," says the Lieutenant, hopelessly.

But the expert, you may be sure, doesn't let him brood over that very long: "Your morning report was in a shocking state today."

"But I sat up all night, making out individual horse records," says the Lieutenant.

"No excuse. How many horses HAVE you got, anyway?" the expert inquires.

The Lieutenant gulps, saying, "In the stables, or on paper?" And then he retreats, with visions of facing charges.

That matter of preparing charges, by the way, sprinkled with gray the temples of organization commanders, and the scanning of charge sheets made many an enlisted man imagine that his last hour had arrived. Every "whereas" and "in that he did" must be in its proper place; and, no matter how accurately the sheet might set forth the vivid language usually employed by the accused, unless "or words to that effect" capped the quotation the whole business was sent back to the drawer with caustic comment.

In those days, men learned to be expert witnesses and officers became judge advocates, counsels for the defense, and judges with supreme power. But most of the cases brought before the regimental courts-martial were not vicious ones. There really were surprisingly few of any sort. It was inevitable we should have one type of case, for home was very near Camp Upton and passes were not plentiful. A handful of men, when they did get home, found it strangely simple to miss the proper trains back. When they missed too many, battery punishment wouldn't cover the crime and they had to stand trial. Tuesday was the worst day; as a result, such little dramas as the following were not infrequent.

Scene: The orderly room. Battery Commander (B. C.) at his desk, outwardly tyrannous and uncompromising; at heart, fighting a very human sympathy. (Some battery commanders have been known to wish that they, too, might have stayed an extra twenty-four hours with their families). Opposite him the "Culprit" (referred to below as "Doe" by the B. C.) stands, shamefaced, pulling at his hat.

B. C.: "Stop pulling at your hat. Stand at attention." (Culprit snaps his heels together).

B. C.: "Now, Doe, what possible excuse have you for overstaying your pass twenty-four hours? The time was written down. The other men got back. You know what it means, Doe, to be A. W. O. L." (That sequence of four letters suggests the sound of blank walls and firing squads.)

Culprit: (With head drooping, voice thin and tremulous,) "Well, sir, you see me mother-in-law was down already with the rheumatiz. She was that bad that . . ."

B. C.: (Impatiently,) "Go on. Go on."

Culprit: (Less confident,) "And me wife was took Sunday night with the same terrible disease. I was just leaving for the train, too, and I couldn't get a doctor, and . . ."

B. C.: (In an arctic voice,) "That's enough, Doe. Those excuses were old when Noah overstayed his leave from the ark."

Culprit: (With a gleam of tears of disappointment in his eyes,) "I told 'em I wouldn't get away with it, but, hones' to Gawd, Captain, they was the best lies we could think of, and me mother-in-law said the last thing: 'Stick to it, Tim, no matter what your cocky officer says.'"

In an army plentifully sprinkled with men of German or Austrian descent, it was, of course, necessary to be cautious. "When is an enemy alien not an enemy alien?" became for a time the pet riddle of the paperworkers. From month to month, the successful answer appeared to alter and yet, except from the point of view of paperwork, it troubled us little.

There were, however, conscientious objectors—not many, just enough to irritate soldiers who couldn't express their displeasure in a natural, pugilistic fashion without infringing the law. Were most of these creatures nervous or sincere, men asked? Their days and nights in barracks, I imagine, weren't to be coveted. For a conscientious objector, whose sincerity you couldn't question, one might conceive a certain admiration, but with such a war facing a nation the burden of proof, unfortunately, rests on the objector.

Worst of all, conscientious objectors complicated paperwork. From many sources came orders and suggestions as to their treatment. When they flagrantly refused to obey commands, as they had a nasty habit of doing, they had to be court-martialed, and they usually picked the most inconvenient times for their performances, arguing, perhaps, that salvation lay there. We desired to see the last of them. But how? Providence reaches its ends in devious ways.

This is really not straying to another topic. Just then, one of our castles tumbled. We weren't going to live, fight, and die together as we had started at Camp Upton. Specific orders began to arrive that demanded large numbers of men. Up to the end of November, we had lost by transfer two assigned officers, two attached officers, and 346 men. We got in return First Lieutenant Frederick H. Brophy, of the Dental Corps, on October 16; Second Lieutenants George H.

Hodenpyl, Karl R. McNair, and William A. Walsh, on November 12; Second Lieutenant H. Stanley Wanzer, on November 22; and some straggling enlisted replacements.

It is impossible to say where our friends of a few weeks went. They left, more often than not, as casuals bound for some remote division in the South or West. We didn't see them again. This abrupt snapping of barrack ties painted for us more colorfully the serious nature of our new profession. With a sober comprehension, we watched the small bands of casuals, bent beneath blue barrack bags, go lurching down Fourth Avenue to the station—away from Camp Upton, away from us who had more often than not learned to like them, away from the land of passes home.

The philosophy of the average soldier is direct and competent. After one such exodus, a soldier explained to his companions during mess: "What's the use of grouching? That's what war is . . . saying goodbye. Just saying goodbye, fellows. Might as well get used to it now."

These partings, nevertheless, weren't all sentimental. Let us value them at one-third regret and two-thirds paperwork. The orders demanding them frequently slipped into the regimental area during the quiet hours before the dawn. Anything that awakened you was known as a "Trick Order." These Trick Orders seldom came singly: for several nights running they would glide in, lights would gleam from orderly rooms until shamed by the sun, and all those concerned would display at reveille acute symptoms of insomnia. There was no evasion when Trick Orders rustled through the camp. If a battery commander prepared a list against unexpected transfers, Paper Work merely sneered, thinking of the devices he had up his sleeve. At 3:00 a.m., it might be, a red-eyed battery clerk would appear at a captain's cot.

"Sorry to disturb you, sir," said the battery clerk quietly.

Responding with a groan, the captain asked, "Barracks on fire?"

"No, sir. An order's just come to transfer five men," the battery clerk explained.

The captain cries out, sitting up, exclaiming: "Don't you know this is the first sleep I've had for three nights? Didn't I give you a list just so I could get some sleep?"

"Yes, sir," replies the battery clerk gently, "but this is a very tricky Trick Order. The men are to be reported at the station fully equipped, at 5:30 this morning. They'll be equipped, even though the supply sergeant does lead a hunted life for a while. Meantime, I've brought the service records for the captain to initial."

The battery commander surrenders, convinced that no matter how artfully you may dodge it, Paper Work will always tag you around the corner.

I would like to interject here that the preparation of these lists for transfers was a delicate matter. (That's why the subject wasn't changed when we slipped away from conscientious objectors a moment ago.) Clearly, some soldiers could be better spared than others. In fact, there were a few that officers and men desired enormously to get rid of. However, we couldn't picture running along at all without the greater part. It had been impressed upon us that by "men" was meant men of the first quality. At conferences on the subject there developed a touching and sublime faith in human nature, an out-and-out belief that in the very worst of artillerymen resides a mine of extraordinary virtue only requiring the delving of the receiving officer. And, one might add, even in the very most conscientious of conscientious objectors.

Now, to return to our conversation about the Trick Orders, the battery commander glances up from his roster. "Could we," he asks, "spare this man Richard Roe?"

"It would be like amputating a limb," a lieutenant answers, "but it might be managed."

The battery commander grunts, and replies, "Didn't realize he was as bad as that. What the deuce is the matter with him? Isn't he strong and handy? Doesn't he look like a soldier?"

"If you look hard the other way," says the lieutenant.

The first sergeant says in a small voice: "He's a conscientious objector, sir."

"Goodness gracious! I'd quite forgot that," exclaims the battery commander.

"Do we sit in judgment on a man's religion?" someone asks gruffly.

"My dear boy! It isn't a religion at all," says the commander. "It's a state of nerves."

"I don't care what anyone says," the lieutenant puts in, "he's got the makings of a good soldier—if properly handled."

The battery commander, whose indignation is arresting, then says: "Who's been mishandling him here? It's clear someone has, and I'll look into that later. I'm bound every man shall get a fair show. It's clear that Doe isn't getting his here. No matter whose fault now. I'm going to give him his show—send him where he'll get all that's coming to him. Put his name on the list for transfer."

One day, the trickiest of Trick Orders came down. No more conscientious objectors would be transferred. Instead, an attempt would be made to get the objector to return and then a prompt trial would be held for the offending officer. The 305th read the order complacently, glancing down the sturdy brown lines. At that point it held no significance for us. Had we ever had a conscientious objector? No one seemed able to recall. At any rate there was none then, and there was none when we sailed for France.

Now and then the Trick Orders contained troublesome particulars. Perhaps an organization would be called on to furnish a man who was equipped to become a battery or company clerk. Then the committee on transfers would really get down to work, for good battery clerks were as rare as good first sergeants. We would see the members anxiously scanning the well-worn qualification cards again and again. This was accompanied by the shaking of heads and helpless frowns. Then, perhaps, one might speak up victoriously. "Here's the very bird," he would say.

"Read the chief particulars of his qualification card," the chairman demands.

The other holds the card to the light, declaiming in a sing-song voice: "'Ivan Stroffowski. Born in Russia. Occupation pushcart peddler. Education none. Neither reads nor writes English.'"

The members of the committee would glance at each other. Then a tentative whisper filters through the room. But it isn't one whisper, it is a sibilant chorus. "Ivan! Thou art the man!" they shout.

Aside from Trick Orders and routine paperwork, there were family allotments, insurance allotments, and liberty loan allotments. And it mustn't be forgotten here that up to October 28, the regiment had gone into its pocket and subscribed $70,300 to the 2nd Liberty Loan. All of these records figured on the payrolls, providing Paper Work with some of his cheeriest moments. Payrolls, too, gave the men rather more than their share of paperwork. Everybody recalls that spirited lyric, set to the tune of "John Brown's Body:"

All we do is sign the payroll.
    All we do is sign the payroll.
    All we do is sign the payroll.
    And we never get a *blank, blank* cent.

Like much poetry, this was a trifle exaggerated, for on payday (when the long lines formed) there was always some real money on the orderly-room table. Nevertheless, on payday, night groups could be heard intoning another lyric of the war:

The U. S. pays us thirty per,
    Or so the papers say;
    But if you get a dollar ten,
    It's a helluva big pay day.

Yet consider the soldier who gets nothing—the replacement, perhaps, who was entrusted with his own service record and lost it. "Do you know what this means?" the captain snaps at him. "Do you realize your entire record is gone—punishments and rewards, clothing, allotments, everything? Do you understand that without your service record you can't be paid?"

The replacement glances at the paperwork suspended from the walls, littering the table, overflowing to the floor. His lip trembles, as he says, "I don't know much, sir, since I got inta the army." Then, as the captain glances at the paperwork too, the following flashes through his mind: "How much this man and I have in common!"

Chapter Four

ON THE RANGE

We learned things in spite of that curse of military efficiency: simulation. For instance, cold weather found us well along in the standing-gun drill. One battery would get the pieces one hour, another the next, and so on. Caissons simulated pieces and limbers simulated caissons. But we got the mechanics of laying, loading, and firing, and the specialists learned enough to make panoramic sketches of the dreary Upton landscape and to lay telephone lines in suicidal fashion. Stirring in every mind, however, was the desire to hear the crack of a rifle and the rush of its projectile.

That wouldn't be long now, for the target range was progressing. Large signs at neighboring crossroads warned the countryside of danger. "When," we asked, "were we going to justify such violent displays?"

The target range, for the first time, outlined our mission. It is one thing to call out at drill a range of five thousand, it is quite another to walk from a projected gun position to a target five thousand yards away.

The range impressed us as enormous: you could walk across it for hours without ever reaching its boundaries. The broad stretches of woodland and brush appeared scarcely scarred by the months of our labor. But it hadn't been intended, for that matter, that they should be. The plan had been to have the range masquerade as an actual battle terrain. Around the target areas, however, many acres had been cleared and not without a few thrills.

The conflagration happened on a sharp, cold December afternoon. Considering the labor, the hunger, the investigations that accompanied and followed it, it would be an affectation of conservatism to speak of the thing as a mere fire. It began when brushwood mysteriously caught fire at the far end, then sprang to the woods, and developed into a spreading column of flames and smoke that swept towards our mushroom city. Practically the entire regiment worked the latter part of the afternoon and half the night getting the flames under control, and some toiled for an indefinite period trying to fix the blame. That was never done, but for months a shadow hung over suspected spots and Colonel Doyle's lips were often severe. Yet the accident wasn't without benefit. Those who surveyed the charred areas pronounced the range about cleared for action.

However, we still lacked horses. The range was some miles from camp and to get our guns there (with a shadow of dignity or comfort) we would need horses. Because we lacked such vital transport, we could not look upon ourselves as a real Field Artillery regiment.

As always, there had been rumors, but early in December, an amazingly real order came to send details to the remount depot. On December 10, some eighty-seven horses were brought up and quartered in our stables. They didn't appeal to us as at all and were not what we would have chosen for our own stalls. They fell into two classes: cavalry and artillery—that is, individual mounts and draft animals. They were shaggy and unkempt. Some seemed overburdened by the cares of life, while others endeavored to express through vivacious gestures a desire to get at the greener officers and men who hitherto had been mounted only in the tables of organization. Often, while struggling with the curious replacements we received in France, we looked back wistfully to these our first and best animals. From this moment, Paper Work clutched at Department B officers and stable sergeants.

The horses arrived just in time for our first target practice, which was scheduled for December 12. Each of the four batteries had the pleasure of harnessing (with make-shift harnesses) a team of the new, untried animals to a piece and drawing it from the park to the range. Everyone recalls that that day was the first bitter one of an uncommonly severe winter, which instilled in those horses a vaunting ambition. As a consequence, one carriage was nearly upset and the rest were delayed because of cold hands and stiff equipment.

Cannoneers and spare drivers stood in line along Fourth Avenue, between 15th and 17th Streets. The scarlet battery guidons (our flags) fluttered before a frozen wind. Yet, as the first carriage appeared at the top of the grade, there was a satisfied warmth in all our hearts, because at least a share of all the trappings was ours. We could grin and shout *"finis"* to that inefficient monster: simulation.

The carriages rumbled down the slope, swaying from side to side. The drivers didn't look happy. More often than not the near horses were out of hand: some animals pulled, while others ambled, enjoying the prospect. But the carriages did advance. It was city-bred men—having been abruptly informed they were artillery drivers—who controlled untrained stock to that extent.

They got past the difficult turn onto 17th Street. Later, they swung with more confidence into the Middle Island Road. They drew the guns into position and trotted off with the limbers in really dashing fashion. The dismounted men marched out. They were stationed near the guns so that they might see exactly what happens when the lanyard of a three-inch rifle is pulled with a shell in the breach.

Later on we fired as many as eight thousand rounds in an evening, but on this day we expended only nineteen. Many soldiers heard, for the first time, the sharp crack of the piece and the swishing, rocket-like flight of the projectile. They were able to watch a shrapnel burst—a pleasing white ball of smoke, appearing like a pretty cloud without warning.

In order that all this might be appreciated, the target was in clear view from the vicinity of the guns, although indirect laying was to be used.

News of the event had spread: officers of the 304th and 306th came to admire and several officers of marines walked up the road. Where they had come from no one knew, and there was too much else on hand to bother about finding out.

Lieutenant Colonel Stimson was to fire the first problem. He walked with Colonel Doyle and a knot of officers to the observatory a few hundred yards ahead. Everything seemed to be ready. Lieutenant Hoyt, in charge of the range party, was said to be down near the target. As soon as he announced that the range was clear, crews from each battery were to fire in turn. The battery commander and his executive ran here and there, giving final words of advice to the gun squads, examining the sights, inspecting the bores one last time. The telephone officers and their details struggled with the primitive system of communication. There were no switchboards; all lines were party lines. There would have been no wire if a friend of Lieutenant Colonel Stimson's hadn't presented the regiment with sixty miles or so of heavy, twisted pair. Yet we were quite proud of that net. We had done our best according to the sacred precepts of Volume III [of the Army manuals on artillery gunnery]. One shudders trying to conceive what [our amateur efforts] would have done to us at the front.

Anyway, the system worked most of that afternoon in the winter peace of Long Island. We had limited faith, however. On nearby crests, lonely figures etched against a sullen sky the broad strokes of the semaphore code. We had even erected two wireless stations using Lieutenant Church's homemade set. They didn't work particularly well, but they looked as if they did. (It took an expert to know one way or the other.)

The men, standing waist-deep in the underbrush, shaking from the cold and probably a little from the excitement too, craned their necks in the direction of the target, two thousand yards away. The white flag on the hill continued to flutter, advertising that there was no firing and that the range was safe. We knew that until the white flag was replaced by a red flag nothing of interest would happen to us. The gray afternoon waned. Cannoneers blew impatiently on their hands, while the ranks in the underbrush stamped their feet and waved their arms, setting up a crackling like the advance of a vast army. Little groups ran up and down the road to keep warm. Whispers lost their stealth, became audible, burst into an impatient chorus: "Why don't they shoot and let us go home?"

Then, through the mysterious army channels of rumor drifted down a fact: someone had been seen on the range nearly an hour ago but now they weren't able to find him. "Where's Hoyt?" an officer inquired. "Why doesn't Hoyt get him off?" In response, a soldier explained, "Hoyt's on the range, seeking the cheerful villain."

The executive strode to a man stretched beneath a shelter tent with the receiver of a service buzzer at his ear. "Get me the observatory," he demanded. (Or, didn't we call it "B. C. station" in those ignorant days?) After a time, the operator passed him the receiver and transmitter, and told him: "It isn't altogether clear, sir."

"Observatory?" said the executive. A pause, then he repeated, "Hello! Hello! Hello! Observatory?" With a low voice, he said again, "Hello! Hello! Hello!" With a high voice, he said "My God! Hello! . . . Hello! . . . Hello!"

The telephone officer stood by watching and made a gesture of disgust. "Don't say 'hello!'" he offered. "It's meaningless. It only wastes time. It never gets you anywhere." If a telephone officer has ever talked to you like that, when you held a dead instrument and big things were afoot, you need precisely no analysis of the executive's emotions.

The executive sprang up, casting the offending telephone parts from him. He glanced dangerously at the telephone officer and he, as they say, collected himself. "I've said 'hello!' all my life," he muttered, "and I'll admit it's never got me less than it has this afternoon."

"Oh, don't get sore," the telephone officer said breezily.

The executive confided, quite in private, "If you do as well as this at the front, the Huns will court-martial the first man that hurts you."

"Buzz it," the telephone officer said indifferently.

The executive, who chained his wrath, replied, "I'd rather give it to you straight. Want any more?"

"No, no," said the telephone officer pityingly. "I mean your message. The buzzer often goes through when the voice won't."

"Oh!" the executive exclaimed, turning to the operator. "Tell them it's nearly dark, and ask what the deuce is the delay." A whining buzz came from the shelter tent—it lacked conviction. "Your man up there got the Saint Vitus dance?" the executive wanted to know. On the nearest crest one of the lone figures was now etching, with eager and excited strokes, a response. "Says," the private in observation read off, "Colonel . . . Doyle . . . wants . . . to . . . know . . . why . . . wire . . . communication . . . has . . . ceased . . . func . . . tion . . . ing."

"Test your instrument," the telephone officer called to the man in the tent, "and you," he ordered another, "get out on the line." A stooped figure threaded the underbrush, letting the wire run through his fingers. In a few minutes he was back, saluting. "Line was cut, sir, not fifty yards out there," he reported.

"Probably one of your cannoneers," the telephone officer complained to the executive. "They must learn that wires are sacred. Court-martial offense—carelessness with wires."

"Speaking of court-martial," the executive whispered, "remember that the Hun that hurts you will be tried by some bigger Hun."

"Got you the first time," the telephone officer grinned.

Behold! The white flag fluttering down! The red flag streaming up!

"Lieutenant Hoyt is back," came over the wire. "The range is clear."

The chief actors became rigid and expectant. "Cannoneers posts!" someone shouted. The men sprang to the pieces like a football formation jumping into play. From the shelter tent the operator commenced shouting out the firing data that drifted over the repaired wire from the observatory: "Aiming point that bare pine tree five mils to the left of the left hand edge of target. Deflection, six-three hundred and fifty. On second piece open ten. Site three hundred. Korrector thirty. Battery right. Two thousand."

The sights and the tubes responded to the febrile motions of our amateurs. The executive repeated the commands one by one. You fancied that through the taut atmosphere came their echoes from the far target. A captain ran along the line, verifying the laying. There was no longer any stirring in the underbrush, nor any movement on the road. A branch snapped, as Lieutenant Norman Thirkield, the recording officer, balanced in a tree and precariously raised his glasses. The brown cloth of the shelter tent bulged and the voice of the operator ran with awed vibrations across the tight silence. "Fire when ready!" he cried.

The executive raised his hand and brought it down with a sharp motion, bawling out: "Fire!" The section chief of the first piece repeated the gesture and the command. The silence was destroyed. It seemed to fall away before the snapping concussion of the discharge and the departure—invisible but fairly sensed—of the projectile.

The operator cried, "On the way!"

The first shot fired by the 305th sailed majestically over Long Island. In succession, the other pieces followed and far off, in the general direction of the target, one by one there appeared (after several seconds) the white smoke balls. The stirrings in the brush-wood recommenced and a great sigh went up—it resembled an exclamation of childish wonder. A feeling of relaxation followed, for it was as if with that first shot we had altered from an inert, incoherent thing into a body abounding with an ordered and flexible purpose. We sensed it as we swung back through the sharp, early dusk. The rumbling of the carriages behind us expressed it. And ahead the lights of camp twinkled at us with a new appreciation. We had made a crossing.

That night we said goodbye to Lieutenant Colonel Stimson. He had been ordered to report to the port of embarkation at Hoboken for transportation to France. That firing of the first problem was his last duty with the 305th until he rejoined us nearly six months later, when we were training in the south of France. After that, until a few days before we sailed, Colonel Doyle was the only field officer with the regiment.

On [December] 13, we took our materiel to the range again and fired ten rounds at the same target, with Captain Gammell conducting.

Glancing back from our veteran viewpoint, it may require a difficult focus to see those pitifully few rounds in their just perspective. Each one might have been a priceless jewel released by some patriotic collector. It took the better part of two afternoons to sprinkle their contents on the target, or near it. They were responsible for hours of discussion in preparation, and evenings of the same. Every burst became the subject of orations. Each was recorded on special forms, and the War Department in general and the ordnance people in particular were told all about it. Temporarily, one of the shell cases was mislaid. The dark eye of suspicion rested on possible souvenir hunters. Those in any way responsible were frowned upon as unique criminals, because that was in the days before the Regular Army got over its ritual attitude towards ammunition. Only when the case had been found did the atmosphere clear.

We had trained for more than three months before firing the precious twenty-nine, and we were to wait more than three months before firing another. But—for one must focus—they taught us what a rifle would do if rationally treated. Each gun squad had had a chance. Incredulity, as to sights and scales and instruments of precision, had been demolished by the men's own labor. To that measure they had already become artillerymen.

It was of even greater advantage that those nineteen rounds had let us measure the results of our training. We could judge ourselves and each other—we could see that, on the whole, we were good. The various details had had practical experience. Operators had actually transmitted—over lines laid by their own hands—words of the highest importance. Twenty-nine rounds at $25 a round! They did more to make our regiment find itself than millions of dollars spent in other ways.

Chapter Five

HOLIDAYS AND RUMORS

During these thrilling days, the powers of administration had not by any means neglected us. They ordered twenty-five officers from the Second Officers' Training Camps to descend upon the regiment, on December 15. The proportion of first lieutenants made at the Second Camps was greater than at the First Camp. A number of our young second lieutenants had been recommended for promotion sometime before, but when their commissions finally came through they were dated later than all the commissions given at the Second Camps. In other words, the men who had set their hands first to the tasks, had struggled with raw beginnings, had molded regiments, were outranked by these youngsters fresh from three months at school. This amazing fact is mentioned in passing because it created a situation a trifle delicate and not without humor.

It is simple to say: here are captains and first lieutenants, give them the authority and responsibility that goes with their rank. It is quite another to project instantaneously into their brains the necessary practical experience our officers, junior to them, had acquired during four hard months. The problem was solved by detailing, temporarily, these superiors as assistants to their veteran juniors.

"Please do this and that, captain," a second lieutenant would have to say.

Or at retreat—to which the new ones, thirsty for things military, always turned out—a second lieutenant would make his assignments, wondering what was wrong with the world. "Please take the first platoon, captain," the second lieutenant would say. And he would distribute the other platoons in a dreamy way among the group of first lieutenants. "At ease!" or "Attention!" the shavetail would roar, and the silver-decorated shoulders would droop or straighten obediently, but in their eyes would appear inevitably a light of something out of the way.

These were excess officers, so a rearrangement of quarters was necessary. No longer could everyone have that little rough sanctuary so essential to concentrated study. The juniors were doubled up to give the new superiors each a room to himself. The majority of these officers were merely attached and remained with the regiment, receiving valuable experience only until its departure for France.

During this period, however, we received a number of officers who did become a part of the organization. Among them were: Second Lieutenant Ellsworth O. Strong, who came to us on December 10; First Lieutenants Wilfred K. Dodworth and Paul G. Pennoyer, who reported on the 17th; Second Lieutenant Edward F. Graham was assigned on the 20th; and First Lieutenant Albert R. Gurney on the 27th. Just before the Christmas holidays, Captains Anderson Dana and Alvin Untermeyer were also attached to the regiment. They had trained with the Second Battery at the First Plattsburg Camp and had been held as instructors for the Second Camp.

Except for a brief period, Captain Dana remained with the regiment during the remainder of its history. He came, of course, as an old friend, since he had known and trained with most of the officers during their novitiate. A few days after his arrival the powers transferred him to the 306th Field Artillery, but when Captain Devereux was promoted and transferred to the 304th, Captain Dana came back, definitely assigned to the command of Battery A. (That change was made officially on February 4.)

Captain Untermeyer, on the other hand, remained attached to our regiment as adjutant and acting commander of the First Battalion, until just before we sailed for France.

After target practice, our minds turned to the holidays. They were heralded by a series of lectures by British officers who had survived some of the bitterest fighting of the war. We heard at first hand of tanks, and machine guns, and gas, and discipline. We gathered from these few intimate talks more knowledge than a library of books and months of reading could have given us. They reminded us of what lay just ahead. They told us of the nasty effects of phosgene and mustard gas, with which we were to have too close an acquaintance later on. From Colonel Appen's stirring talk on discipline, we carried away an unbendable belief that in discipline resided a defense almost as powerful as ordnance. We resolved to equip ourselves with that weapon.

In spite of such grim reflections (or was it because of them?), the holiday spirit captured us excessively. There was a strengthened pleasure, a trifle pathetic, in the holly wreaths and mistletoe and tinseled evergreens of home. That classic tinkle "For Christmas Comes But Once A Year" was in our minds. What changes, we wondered, would come to pass before another year should bring its unique feast. It was, roughly speaking, twelve months later that the regiment held its first memorial service in a sodden meadow of the Haut-Marne.

According to the Paper Work, every officer and man (except victims of discipline) could have either at Christmas or New Year the period between Saturday morning and Tuesday night at home. Some fortunate ones got both holidays. The crazy specials pulled out of the terminal with eager youths overflowing to the platforms; and always fresh columns marched up, were inspected, and passed through the gates. At the Pennsylvania Station, a civilian was a somber piece of driftwood in a restless, muddy sea. We gave all New York a brown tinge that Christmas. In clubs, hotels, on the streets, and in nearly everybody's home khaki was a perpetual reminder of war and of approaching departures. When we returned to camp, we found that the few left behind had not gone cheerless. There had been turkey and mince pies, and the mess halls were still green and red from brave and abundant decorations.

The return from New York on New Year's night we put down (without dissent) as "Horrors of War No. 2." They had had us out at fire drill Saturday morning and a few frozen ears and fingers had warned us that the frost king was after new honors. The journey up, through a lazy snowstorm, had been suffered patiently because of its warm destination. But the mercury continued its ambitious ways, and it was always colder at night than by day. Towards midnight of New Year's, to anyone standing on the platform at Jamaica, it was obvious that records had been broken.

When the train finally came along, we crowded eagerly to get in. Strong soldiers shrank from the open door and hoarse voices called on regions of perpetual warmth. But the strongest and the hoarsest had no antidote for steel coaches that were fresh from the yards and unheated and unlighted—save for a single candle in each, burning high, suggestive of a votive light in an Esquimau's tomb.

Compared with the atmosphere in these coaches the outside air was warm. We had to remain where we were, crouched on seats or in the aisle with our feet on suitcases or on each other, while the train crawled, while we counted the minutes, while the air froze tighter. Gems of advice slipped from one to another. "Don't go to sleep, Edward. They says they never wake up," said a soldier. "Better try it. Be a dashed sight warmer where you'd go, Benny," said another. And, "Move your legs, boy. Keep 'em moving," instructed yet another. "If you freezed in that position they couldn't get you out of the car till the spring thaws." The last piece of advice offered was: "I heard that if you thought anything hard enough it would be so. I'm going to think I'm warm." The response to this: "Tell that to the Baptists, George. I'm a Shaker."

And that night because of these things the railroad, too, suffered a little. In some cars the metal floor was discovered to be an excellent bed for a fire and the wicker seats passable as fuel. The combination resulted in discussion between headquarters and the railroad barons.

From that moment thoughts of home receded. The bitter weather lasted, and there was a famine of coal in the land. These facts added, probably, to our improvised heating arrangements, caused special trains practically to become extinct, and passes nearly so.

The first warm weather brought a new complication. At best it had taken delicate handling to get an automobile, without prematurely aging it, in or out of Camp Upton. Spring altered rock-like dirt roads into unnavigable morasses. For a time, the railroad was our only practical means of communication with the outside world. Fortunately, the coal situation had improved by then, and our erratic fires had been forgiven. Special trains ran again, and the days of generous passes were revived.

Although the cold weather had cut into our drills, there had been plenty to keep us busy. More horses had arrived, along with another veterinarian, First Lieutenant John J. Essex, assigned on January 14. Grooming occupied a lot of time, and the care of harnesses and carriages a lot more. The liaison schools worked so hard with theory and practice during the cold days that a Regular Army inspector was lost in admiration to the point of saying: "Regular Army, National Army, or National Guard, I've never inspected details as well instructed as these." No matter how cold it was, unless snow or fog made the visibility bad, Colonel Doyle took the officers and portions of the details to the hill above the infantry practice trenches, where he instructed them in the Fort Riley method of conduct of fire. We fired problem after problem from imaginary guns, while Lieutenant Hoyt, at the targets a mile or more away with erratic smoke bombs, made us feel how bad we were.

In February, Dame Rumor stole from her winter quarters. One day we were going to France on a moment's notice; the next, we would be lucky if we ever got there. The third rumor: our boat was in the harbor and we'd have to hustle to get off.

Some of the saner-minded weighed the matter. We couldn't fight the Huns with our one battery, our few horses, our insufficient harnesses, our incomplete instrument equipment. Moreover, a number of our battery commanders were at Fort Sill for instruction, and others were scheduled to go. If proposed departures should be cancelled, and the absent captains recalled, we would begin to put our affairs in order, for it was clear we couldn't go on marking time perpetually at Camp Upton.

In some measure, Washington's Birthday cleared the air. It fell on a Friday. We began to speculate our future when we were informed that on the holiday there would be a parade and a monster Division ball that night in the armory of the Seventh Regiment. And, as many of us as possible would be given passes between Thursday evening and the following Monday's reveille. "Looks like a farewell show and a last chance for a good visit home," was the common interpretation of the events. This view was strengthened when we struggled to town Thursday night and word passed through the train that the absent battery commanders had been recalled.

The parade was solemn but it had an exotic touch. American soldiers had never looked quite like that before: the men wore their new winter caps instead of the familiar campaign hats, and a blanket of snow that fell became a part of the uniform. The spectators gazed with a sort of wonder at city youths—broadened and ruddy-faced and clear-eyed—in a setting that placed them all at once, as it were, in a different world.

It was almost entirely an infantry affair. In spite of the highly technical nature of our branch, our lack of equipment even at this late date barred most of the artillery brigade from the column. Among the entire three regiments, there were still only our four venerable rifles. The honor of parading these fell to Battery A, in command of Captain Dana. He was the first officer of the brigade to have a chance at entraining and detraining a battery. It spoiled his holiday, but it was good experience.

The crowd cheered that single battery as it crunched through the snow past the reviewing stand. Little Wing, the Chinaman, on one of the lead horses, illustrated with unconscious pride the democratic, the universal power of our army.

At the Division ball that night—somber with brown figures and gay with the evening best of mothers, wives, sisters, and sweethearts —stalked an oppressive succession of hazards. What did it all mean to these cheerful brown figures and these smiling women who danced the night away, together?

Two days later, in the Cohan and Harris theater, Lieutenants Sage and Roesch staged a monster benefit for the regiment. Our own talent was supplemented by a glittering array of Broadway stars. The show made enough money to pay off the debts owed by the regiment to members who had gone into their own pockets to buy what the powers had failed to provide.

On our return to camp we waited for the words of verification. It came on Tuesday morning. The acting division commander, an infantry brigadier, desired the presence of every officer that could possibly be spared from duty in the Y. M. C. A. hall on Upton Boulevard. The noncommissioned officers ruled the regiment during that pregnant hour. A huge theatrical success wouldn't have filled the hall more uncomfortably: infantry, artillery, machine gunners, medicos, the trains, they were all there.

This was not like previous gatherings for advice or reproof. Suspicious individuals stood at each entrance, scanning the arriving officers. Certainly we were going to hear secrets. The usual laughter, gossip, and calls to distant friends were replaced by a dreary and unnatural silence. It was as if we had aged unexpectedly. More than the customary smoke curled towards the rafters. The brigadier entered and faced us with countenance and attitude sterner than the ordinary. "Are there any enlisted men present?" he inquired. Verily we were to hear secrets!

After we had heard everything, we questioned if the enlisted men didn't know nearly or quite as much, and we wondered why they shouldn't. As the discourse developed, we discovered that while we were sailing soon no definite date had been set. All we could do was to equip and train the new men who would be arriving. In order that the enlisted men might be kept in ignorance of these things, we were to tell them carefully there were rumors we might leave.

The officers filed out of the gathering and wandered back to the regimental area chatting together softly. In those first hours, it seemed inevitable we should go almost at once.

When organization commanders faced their men, they gathered that the men knew where they had gone and why. A recital of the rumors seemed superfluous, for in the faces of the men, too, there was a solemn sense of imminence.

Chapter Six

THE AGES OF GETTING READY

We failed to sail within a fortnight, or within several fortnights. Perhaps it was just as well that we lacked transportation, for there was much more preparation needed than we had suspected. Lieutenant Walters left us and Lieutenant McKenna came into his own. That is, McKenna was assigned to the command of the Supply Company and before long his promotion to the rank of captain arrived. From constitutionally reluctant quartermasters he tore supplies with the same cheerful energy he had displayed in the days of recruit fitting. Yet the more we got the more we appeared to need—the lack of artillery harness was not the only hurdle between us and the docks.

While McKenna hustled we entered two new phases: one might be labeled "The Age of Gas" and the other "The Age of Equipment Checking." In my memory, the second looms larger. "A complete check of personal property will be made before retreat," was the order that faced us day after day on the threshold of the afternoon. It meant we had to lay out on bunks all of the issued equipment, according to an intricate pattern. And it meant a review of every piece, checked against an official list of equipment C.

Someday, a Regular Army quartermaster may divulge to us the

structural secrets of those lists. For our part, we never quite understood the logic of reversing, sometimes mutilating, the descriptions of familiar and intimate articles of clothing. "Bags, barrack," it began. Why, in the name of abused commas, wouldn't "Barrack bags" have done as well? Then there was: "Breeches, O. D.," "Socks, winter," "Gloves, riding," "Poles, tent," "Razors, safety," and "Tags, indentification." It ran on something like that, and we followed, if reservedly. We only revolted at: "Shirts, under," "Drawers, under." Perhaps an obsessed clerk, typing the copies, was responsible for that.

This is how one spent one's time in the age of equipment checking: in the somnolent barracks you arranged your equipment according to the intricate pattern. Everybody had a different idea as to some of the more esoteric details of the pattern, and you compared notes until you didn't know whether you would be passed, arrested for distortion, or praised for acute originality. Then you endeavored to keep awake. If you were an officer, you took your lists, tried to get the cunning pattern through your head, and wished to heaven you could smoke on the job.

A noncommissioned officer slams into the sleepy room, singing out: "Attention!" The officer then walks in. It probably isn't severity that gives his face that peculiar expression, you decide. It's more likely a stifled yawn. "Rest!" he croons. "All except this first man." He checks the articles on the cot, and demands, "Where, is your fifth pair of socks?"

The warrior blushes, and says, "On me pusson, sir."

The officer reflects on this. This time his frown isn't wholly concealed, for the orders were absolute, everything must be seen before being checked. The soldier, meanwhile, stoops obediently, removing his legging and probably murmuring in his mind: "I'm not trying to put anything over on you, and I'd wear them in the army whether it was my habit or not, because your issue shoes aren't exactly plush."

"Where's your other O. D. shirt?" the officer asks. But he catches himself and explains, "I mean, Shirt, O. D."

Again the soldier displays emotion and responds, "In the laundry, sir."

Once more the officer reflects. It seems expensive, unjustifiable, and meat in the mouth of Paper Work to issue this man (and all the other cleanly men) masses of equipment to be turned back on the arrival of their laundry. On that point there should be something definite, so he seeks the captain for a ruling. The responsibility is great, so the captain seeks the battalion adjutant. Then the battalion adjutant seeks the regimental adjutant, who in turn seeks the colonel. Beyond that the chain is vague, but in a few days a ruling comes down that for the present equipment in the laundry may be considered as present and accounted for.

The checking officer, meantime, makes out a painstaking little list for each soldier. "Private so and so has in laundry . . . " The list is long, and those who hear it decide that "so and so" is effete. The conversation in the room has developed from tentative whispers following the officer's "Rest!" into comments, exclamations, and arguments, centering around the flow of well-known raconteurs. The checking officer hears all this, grows a trifle absent-minded, and then has to make alterations to his neat lists.

"I pasted him in the jaw, honest to Gawd I did, and he didn't have no come back," brags a soldier. "You saw the bout, Jim. If I hadn't caught my shoulder in the ropes, he'd never have knocked me out. Ain't it the truth, Jim?"

"If you ask me," Jim replies evenly, "I think you had horseshoes hung all over you to last as long as you did."

And from a group of three serious-faced young men, two of whom have just returned from the third R. O. T. C.: "Germany's financial structure is as restless and insecure as a house built on sand." To which one of the three responds, "That's logic, but logic and the truth are often bad friends."

"Oh, Lord," groans the checking officer inwardly, making another mistake with his lists. And, to cap the climax and spoil an entire sheet, he is interrupted by the following exchange.

"Billy told me about it," the recruit began. "If the Y. M. C. A. could have seen him then! Nellie had him up to tea Sunday. Least he thought he was drinking tea. Looked like it. You know a martini and tea are the same color. They put cocktails in his cup instead of tea, and he smacked his lips and drank four cups, and all the time the poor simp thought he was drinking tea."

A deep voice cuts the air, snorting and booming and shouting: "The hell he did!"

The sergeant tries not to grin, while the checking officer says, passionately, "Attention! Sergeant, if another man speaks put his name down, and I'll take care of him later. At Ease!" He turns back to his checking, aware that what he had really wanted to say was: "Men! This job has got to be done. It hurts me more than it does you."

Sometimes we checked and were checked at night, too. Whose fault was it, this ceaseless repetition that carried us each time only a trifling distance forward? It must be admitted that in some measure the blame was our own. There were a number of men whom you could check at two o'clock and find, with the exception of allowable deficiencies, up to the mark. At three you might check them again and learn they had lost within the hour such prominent objects as tent poles and shelter halves. One little bandsman was suspected of an appetite for tent pins, as his disappeared so rapidly and regularly.

But we weren't to blame for the futile effort of checking and rechecking, for a complete check could only be made with every soldier in his place and each piece of equipment in view. At one time the stable sergeants and the grooming and feeding details would be at the stables. Check or no check, the horses had to be cared for. At another time the cooks were scattered on various duties. Naturally the men couldn't be checked at the price of starvation. And every day at headquarters and in the orderly rooms soldiers of clerical ability bent before the sacred shrine of Paper Work and couldn't be torn away. So the "Age of Checking" was prolonged through March and April, and even up to the day we sailed.

The "Age of Gas," while less irksome at that time, was rather more unpleasant. Lieutenant Mitchell had taken a course from a Scottish noncommissioned officer—so he was looked upon as an expert now and we were content to pin our faith to him. But one night we were summoned to hear Mitchell lecture. He sprinkled bright little stories among statistics, depressing and, we fancied, a trifle exaggerated for our good. We drank in extended figures of casualties caused through carelessness or ignorance; of casualties, on the other hand, scarcely to have been avoided. He had the house at his feet. In a fashion, he beat the English lecturers at their own game. He'd found out about some new gases that shriveled you up all at once or got you with a delayed and terrible kick long after exposure. And instead of a cheerful Christmas time just ahead, there was actually—gas.

He asked us to listen to him again the next night, and when we obeyed we found a table piled with masks. He showed us how to put them on and take them off. We gasped in the strange, uncomfortable, stinking contrivances. We laughed—not uproariously, you understand—at our own appearance, abruptly converted into something monstrous.

Noncommissioned gas officers were appointed. The men spent a definite period at gas drill each day, holding competitions, running courses. They looked like types of a new race, born of some dreadful catastrophe.

We were introduced to the gas house—a wooden shack near the machine-gun range—where the Scottish sergeant was heard to say: "We got to ha' a wee bit o' luck this afternoon. We carried out thu-ree corpses this marnin', and they only allow me fower for a full day." This prompted Lieutenant Mitchell to say, in a stage whisper, "Laugh, or you'll hurt his feelings." So we laughed "ha, ha, ha" at his joke, but it was more like a cry for help.

The procedure for the gas drill was explained by a captain of the Medical Corps, for that was before the powers gave gas to the Engineers. "I'm going to let loose a killing mixture of chlorine," he ended, "so it would be as well to inspect masks carefully." We hoped he was trying to impress us, but the ranks, one noticed, took a long time over the inspection of face pieces and canisters.

We were ready finally. The medico then put on his own mask, entered the shack, and sealed it. Through the single window we saw him turn the escape valve of a cylinder tank. He opened the door, stepped out, and removed his mask.

"Come close," he said, "so you can smell the stuff. Then you'll know I'm not putting anything over on you." When we obeyed him, our lungs refused to breathe the sickly air. We donned our masks and filed in and the door clanged shut behind us. We were imprisoned for ten minutes, half expectant of catastrophes. Through our goggles the air had a bluish appearance, but in our lungs it was pure.

We escaped at last, relieved to be able to breathe naturally again and to know that the masks were really good. Afterwards we were treated to a lachrymatory mixture that hurt our eyes. After that we were permitted to march away, cracking gruesome jokes for the benefit of those whose ordeal still awaited.

We took gas in the stride of our work of preparation for war, which continued with slow sureness. Day after day Captain McKenna opened the regimental storehouse on newly collected treasures, and each organization sent details to bring home its share. Then followed hours of fitting and issuing and checking again, until we realized that the regiment was nearly equipped.

Each officer and man was given twenty-four hours at home to attend to his personal affairs. (This brought the war so much nearer.) On March 18, a review and a dance of the brigade was held in the 69th Regiment Armory, which meant we had from Saturday until Tuesday morning at home. "And this time it's surely 'so long, Mary,'" one heard going up on the train. There was, indeed, an atmosphere of climax about the whole affair. For March, the weather was warm and Lexington Avenue and the side streets, as we came up, were nearly blocked by restless spectators. They lacked the air of a crowd at a parade—their brief cheers touched formality, they were restrained, they vibrated with a quality a little choked. Suddenly, we realized that the men and women seemingly unrecognizable in the night were those that loved us.

Automatically, we recalled stories of the departures of regiments from New York for the Civil War. Always such pictures were set in sunshine with a ring of quaint costumes and a brave show of flags and music. We had looked forward to something of the sort. There was music, all the more brassily insolent because its source was unseen, and, lost in the shadows, we knew our flags shook in the tepid air. The rest was wholly contrast. The columns, swinging up through the dark, pushed back the restless shapes. The door of the armory opened and the shapes slipped through. They had to traverse a broad band of light and, as we looked, I think it came to all of us quite abruptly that it was simpler to be of the offering than among those who tended the altar.

On our return to Camp Upton we entered the "Age of Packing"— a most complicated and laborious epoch. Every day, and until far into the night, the mess halls resounded with a new activity. Battery carpenters hammered on packing cases and painting details striped them with maroon and white, the division colors. Packing details filled them with instruments, and ordnance, quartermaster, signal, and engineer property—and paperwork. From duplicate lists clerks checked everything in.

Typewriters clattered on the tables. In one corner two men bent over typewriters tap-tap-tapping numbers and names on identification disks like a new race of Nibelungs. In another corner an exchange had been established and brisk bargaining over odd sizes of equipment imposed on the general pandemonium a shrill note of wheedling or invective. Any harness we had was draped from uprights and hanging from the ceiling beams were rows of blue-barrack bags, still wet and splashed with the white and red of division markings.

Black days of unpacking and repacking, to meet some new Trick Order, followed and the checks continued. One Saturday a check of the harness disclosed the fact that two sets were missing from the regiment. That time the men were more fortunate—the columns of pass holders could march down Fourth Avenue as usual. But later an edict came from the colonel that no officer, whatever his remoteness from harness, should leave Camp Upton until the missing sets, or a reasonable explanation, had been found.

By night the amateur detectives—and everyone had joined the

quest—saw their last theories crumble. Every inch of the area, they swore, had been searched. No one had escaped a bitter third degree. The harness, to all appearances, had dissolved. We were released, but the shadow of the mystery long hung over us; and through the shadow, after a time, gossip stole. You may accept it or reject it, but it might be well to picture a couple of officers and a few men gathered in an orderly room. On studying their faces, you might decide they gaze with horror on the result of some red and impulsive work their hands have just accomplished. Or you may think that the souvenir of some murderous indiscretion has unexpectedly risen from the past to challenge their content. For their faces are not without horror—a helpless, desperate horror, and one does gasp: "Great Caesar's ghost!" But there's really no ghost, or any crimson relic—nothing exceptional at all in the plain little room except one perfectly good set of artillery harness.

In the little room, an officer flings his hands above his head in a gesture of despair. "Surveyed! Finished with! Bunches of paperwork on its grave! Where in the name of kind heaven did you find it?"

"In the stables, sir," one man responds, "covered up by accident in a manger . . ."

The desperate hands go higher. They now also express supplication. "It can't be found! My God! It can't be found!" cried the officer.

"You're right," one agrees, "because according to Army Regulations it has ceased to exist. To try to bring it to life again might take years of investigations, valuations, boards, I guess it would stop the war."

"Probably," says another, "it would put G. P., meaning general prisoner, on the backs of most of us."

"Drather find nitroglycerin," a man comments.

A murmur crystalizes the thoughts of all: "If it were done away with quietly, dispassionately, without cruelty?"

You can't depend on this idle gossip, for the set was never heard of, at least publicly. One of the conspirators was seen in friendly conversation with an officer of the Supply Company. Perhaps a stratagem was found. Maybe there's something in the story after all.

Days of doubt descended upon us. For some time, each weekend at home had been treasured as our last but we didn't move. "An order has come from General Pershing," McKenna informed us, "that no artillery units are to sail without their full equipment of harness." But a word might alter that. If we could go without guns or caissons or horses—for gradually it had become clear our animals would be left behind—why all this fuss about harness?

And the division was moving. Headquarters stole out of camp one early April night. Not long after, we were awakened by the shouts of many men and the wanton splintering of barrack window glass. The sky reflected many bonfires. The next morning the area of one of the infantry regiments was empty. Machine-gun battalions followed. Then another infantry regiment. Each day we expected our orders. During this period of suspense several changes occurred. A special order from the War Department arrived giving Captain Untermeyer an extended leave of absence. In his place arrived Captain Henry Reed who had received his commission at the First Niagara Training Camp, had instructed at the Second Camp, and during the winter and early spring had been just across the hill instructing at the Third Camp. He was assigned to the regiment as adjutant of the First Battalion. Major Wanvig returned from Fort Sill. Lieutenant John W. Schelpert of the Dental Corps came to us on March 24, and remained with the regiment until August 19, when he was transferred to the Ammunition Train.

Then the blow fell. A very high officer was heard to say with a laugh at the Officer's House: "The artillery? They won't get to France before apples are ripe." And on top of that came the order that seemed to confirm him. An infantry regiment that was moving at once was short of men. The artillery brigade would fill it up.

By that time we had developed that organization spirit that is just as essential as it is delicate to breed. To take fifty or sixty men from each battery seemed a destruction of the greater part of all that we had worked to achieve. As a rule, men who had trained during seven months in the ways of artillery resented being transplanted all at once into a branch of the service to which they were strangers. Nor did their officers care to see them go.

"Good men! Good men!" was the cry.

By that time we believed there weren't many that didn't fall into that class. But somehow the lists were made up, the victims equipped, and the dazed exiles marched away to a new formula, to strange companions. It happened once more just before the last infantry regiment departed. As the result of those two orders, within a few days of our sailing for France as a combat regiment, 698 men had been torn from us. The Headquarters Company lost 50; the Supply Company, 27; Battery A, 93; Battery B, 119; Battery C, 113; Battery D, 95; Battery E, 116; Battery F, 82; the Medical Detachment, 2; and the Veterinary unit, 1.

At Camp Upton the artillery alone remained, and we stared (as if threading the mazes of an unpleasant dream) at half-filled mess halls and skeleton ranks. Then troops began to pour in from the south. Upton, we heard, was to become an embarkation camp. Our area, however, would remain sacred to us.

The vast German offensive of the spring of 1918 was dangerously under way. We could understand a pressing need for infantry and yet, we argued, infantry in such a war isn't very valuable without supporting artillery. How could Europe furnish enough of that? "We won't move before July," was the general cry, and on studying our shattered regiment that was easy to believe.

The changes—the incredible changes of army life!

Coming back from town on the night of April 14, it was surmised that October was the most likely date of our departure. As it turned out, that was to be our last Sunday home before sailing. On Monday morning the October guess continued good: a new smoke-bomb range had been designed and miles of wire laid; we were instructed to unpack a great deal of equipment; and elaborate schools were planned for the warm, favorable weather.

On Tuesday whispers slipped apparently from nothing. On Wednesday the Supply Company awakened to new activity. From it escaped the significant news that we would get harness at once.

Nothing more was to be unpacked and all that we had taken out was to be put back again in the cases. "But," we objected, trying to stick to logic, "they wouldn't have stripped us this way. We can't go without men, and you can't take green men and train them on an ocean voyage."

Can't you, though? We were to find out about that, for on Thursday the officer in charge of arriving casuals conferred with us. From him we learned that trainloads of men from the West had been gathered at Camp Devens and would come to us at once. We grasped at every comfort—if these replacements were from the West, they'd probably know something about horses.

Selected officers and noncommissioned officers were awakened at two o'clock on the morning of the 19th. The trains were about to arrive. There was a chill in the air and a mist—pearl-colored about the lamps—veiled the dreary similarity of the barracks. The trains crawled in with a stealth harmonious with the secrecy of all these movements. The throbbing of the locomotives was discreet, as if the mist sought to muffle it.

Out of the cars they poured—sleepy-eyed, struggling ineptly with barrack bags, not at all voluble as soldiers in groups usually are. Our seasoned men lined them up with a gentleness designed to destroy their attitude of strangers, bashful and apprehensive. We counted them again and again to be sure we were getting all we were entitled to. We marched them off in groups through the fog. The fog seemed friendly to them, for at that time they were without personality to us —just so many things, counted and recounted, to fill the ranks of a regiment about to go to war.

Taking up the march again, after a rest during which two groups had got a trifle mixed, an officer counted his objects and found one missing. He and his noncommissioned aides ran up and down through the mist.

"I'm shy a man. Have you got an extra man? Count up," said the aide.

"What's he look like? Know his name?" an officer asked in response.

"How the deuce could I? Doesn't make any difference. All I want's a man. Anything'll do," said the aide.

After many counts he was supplied, and the nameless things, taking up their barrack bags, stumbled on through the mist.

It was four o'clock when we reached the area, but lights burned in the mess halls, and mess sergeants and battery clerks were about their tasks. The odor of coffee was prophetic.

Each barrack swallowed its quota. The seasoned men, who neglected the sleepy, half-frightened expressions of the recruits, stared at the amazing variety of hat cords. Only on a very few hats did the red of the artillery show, on the rest were the colors of the infantry, the Signal Corps, even the Medical Corps. With sinking hearts, we remembered how our artillerymen had gone to fill the ranks of the infantry. By what curious chances during those days did a man find himself here or there? By what devious contrivances was such a circle drawn?

Despite so many men in the mess halls they were curiously silent. The drone of voices reading service records or questioning increased an atmosphere of somnolence. There was the familiar variety of names and accents and countenances. Most of these men were, in fact, from the West and many of them had had experience with horses. That would help.

The mess sergeant placed steaming cans of coffee and tins of corn bread on the counter. His voice sang out cheerily, "Come and get it!" The inert and drowsy groups in the hall aroused themselves. A rough line was formed and passed stolidly by, each man taking his share without words. As they munched, the men stared at the bare walls and the pine tables and the windows beyond—where through the mist an indifferent dawn illuminated the unfamiliar waists of Upton.

A sergeant cried with rough good humor, "We're not going to bite you. What's the matter? Talk up! Haven't you got a song?" On some of the sleepy, grimy faces a grin struggled. There was no song, but sporadic conversations sprang up here and there and then died away. One man's head rested on his arms, which were stretched across a table; his snore disturbed the silence. Other snores followed with varying effect. There was a laugh or two. In a corner a little fellow, bronzed from the western sun, sat before his untasted bread and coffee. He didn't laugh with the others—his expression altered, his mouth twitched uncontrollably. For a moment we thought he was going to laugh, too, but instead he silently and with difficulty began to cry.

Chapter Seven

GOODBYES AND THE SUBMARINE ZONE

In retrospect those who got home may wonder at the quiet force of the regret that crowded those farewell hours. As that philosopher of ours had said, "War is saying goodbye." And goodbyes are seldom easy. Since most of the regiment couldn't go to town, families came down; and wives, mothers, and sweethearts don't speed their nearest on to battle with dry eyes. These final farewells were given, as far as practicable, a just proportion of the last rushed days. From morning to night, the hostess houses were filled with soberly clothed women who knitted and, for the most part, sat silently, glancing up each time a brown-clad figure hurried in.

Towards the end they learned the way to the barracks, where they sat in noisy, cluttered mess halls. At each opportunity their men would sit with them. One marveled at the lack of words—there seemed nothing left to say except goodbye. At night, in the dusk of the station, this unnatural repression would be momentarily destroyed; shattered, as it were, by an unavoidable release of emotion too long subdued. The long trains filled slowly, for the passengers, as a rule, waited until the last minute, huddled in the pen-like enclosure beyond which soldiers might not pass. From it arose a perpetual monotone, like a wind in heavy pines—the last effort at repression, the farewells of those who only dared whisper.

Guards and railroad officials urged on the unwilling civilians with: "See here, you've only got a minute! Want to miss the train?" Then, almost always as the dark mass began to move, fighting back upon itself, the monotone would rise as the wind in pine trees rises. And like a knife in the heart of the whispering stillness would flash a cry: "My boy, my boy! Oh, my boy!" These last goodbyes weren't said until a few hours before our departure.

On April 22, Lieutenant Arthur A. Robinson was assigned to the regiment from the Depot Brigade. He had been with us for a few days in December, coming down from the Second Plattsburg Officers' Training Camp. The powers had taken him away almost at once, but there had lingered an impression of an exceptionally pleasant and efficient personality. When the regiment found itself a second lieutenant short at the very last moment, it got Robinson and gave him for the time being to the Headquarters Company. Lieutenant Robinson's career was unique in a number of ways. He was, as you shall see, the only officer in the brigade to be awarded the Distinguished Service Cross. He served with more organizations of the regiment than any other officer. As soon as we got to France he went from the Headquarters Company to Battery E. After a few weeks Battery B got him. In the Lorraine sector Battery C was short and had to have a competent officer, so Robinson was shifted and fought through the war as executive. McKenna got him for the Supply Company in the celebratory piping days after the Armistice. Everybody wanted Robinson; when he left us so tragically on the journey to the embarkation center, he left a gap that couldn't possibly be filled.

Colonel Doyle and Captains Dana and Starbuck went to Hoboken on Tuesday, April 23. Major Johnson returned from Fort Sill that day and, in the colonel's absence, took command of the regiment. Under his friendly and easy guidance, the task of getting away seemed simple.

We were to leave at two o'clock on the morning of April 25. On Wednesday morning, immediately after reveille, the straw from the bed sacks was dumped in huge piles and burned. The flames rose high above the buildings. Men, waving their empty white bed sacks, danced around the fires. The picture had a ceremonial air and before long only ashes remained.

We policed barracks and quarters, which were as empty as when we had first invaded them and ready for the next recruits to come. We wandered about bare places that appeared all at once unfamiliar to us. We were homeless. We had only to count the minutes while we reviewed details. At midnight we had a supper of sandwiches, cakes, and coffee.

Paper Work alone enjoyed himself, altering not at all his ways. In Regimental Headquarters the clerks still toiled.

The organizations were formed on the parade ground, and each man placed his pack in his place, so that when the command to "fall in" for departure should come we could be off in a minute. The imaginative among us busied themselves with the manufacture of placards that they nailed to the barrack doors. They said, for example: "This house to rent. Owners spending the warm season in France." Or "Good-by, Upton! Hello, Berlin!" And, "Wipe your feet. We're off to kiss the Kaiser, and can't do it for you."

Out on the parade ground Pullen's bugle blared. The lights in Regimental Headquarters expired. Paper Work went to sleep for the night. We listened for the orders—"Fall in! Hustle it up there! Squads right! March!"—and moved off through the darkness, turning to the left on Fourth Avenue. It was beyond belief, for we were walking away from Camp Upton. Feet shuffled as if trying to dissipate a dream, but it was real. We were actually marching and our destination was the front.

There was a precision about that movement that augured well. We found our trains waiting at the railroad station. The column was divided and the proper number of men placed in each car without delay or confusion. We had scarcely finished being packed in when the trains started. Through the dawn we approached Long Island City. As the first green flashed by from trees and bushes, we wondered what the spring would be like in France. We were under strict orders not to open windows, not to call to people on the roads or at the stations, not to sing. Early morning passengers at the stations watched with a dumb curiosity these trainloads of soldiers silently gliding by.

At Long Island City we crowded our way onto ferry boats that took us around The Battery to Hoboken. The city was scarcely awake; only here and there did a man wave his hand carelessly from a park or a wharf. There was nothing glorious about it. We were only interested in what boat we would get, and wallowing up the North River we saw that a number of big ones were in the harbor. We nosed towards Hoboken, where the *Northern Pacific* and the *Von Steuben* (the former *Kronprinz Wilhelm*) lay. The First Battalion was destined for the one, and the second for the other.

We poured off and formed in the odorous dusk of the pier. The place was crowded with a feverish activity reminiscent of a factory—a huge factory, greedy for material, which it belched forth, ready for the front. Red Cross men and women trundled little carts along the lines, offering us hot coffee, buns, and cigarettes. We ate greedily but we couldn't smoke, because it was forbidden in "the factory."

While we munched, Paper Work awakened. But we had him well in hand—our passenger lists were right, as were our accommodation lists, our service records, and our inoculation cards. We were permitted to embark and went up the gangplank in single file. We were counted off, assigned to space, and then they stopped bothering us for a while. When examining our temporary home, our hearts sank a bit. The bunks were in three tiers crowded close together. There was an odor of disinfectants, of departed meals. The top bunks seemed safer on the whole.

But we were fortunate, for the *Northern Pacific* and *Von Steuben* were better than a good many other transports. And they were fast. Anyway, there wasn't much grumbling, because whatever came was a part of the game. Yet that day and the next were hard—more difficult than storms at sea or the conscious dodging of submarines. For during that period, we lay docked at the pier and watched the ferryboats go by, answering the fluttering handkerchiefs or the few cheers and, all this time, forbidden to step from the transport, we watched the smoke curling above our homes.

We took refuge in our only antidote: we wrote letters and signed safe arrival cards. These bore on the back the printed legend, which we were ordered not to alter: "I have arrived safely in Europe." Yet when those cards came through to be censored there were few that didn't carry something else—about love. It didn't do any harm. Probably the final censors thought so.

Naval officers seemed to have lost their voices. We had no idea when we would cast off. And there was a strain about this waiting, chained within sight of home. At five o'clock on the afternoon of April 26 the strain broke. The fuel barges moved away, men hauled in the gangplanks, and they began to cast off the moorings. The boat slipped into the river with only a discreet blowing of its whistle.

Everyone was ordered below decks. No uniform showed outside except the blue of the navigators on the bridge and the brown of the officer of the day dashing importantly here and there. And the world outside seemed oddly indifferent. We crowded to portholes and windows, hungry for a last glimpse. At dusk the companionways were opened, and we climbed to the decks. We were through the narrows and ahead lay the gray, empty sea. Behind us, far in the distance, resembling details of a mirage, the towers of New York penetrated the haze, then were lost.

The following seven days shared a drab, uncomfortable similarity. Aside from a half hour's sketchy physical exercise and abandon ship drills, there was no effort towards concerted work. The limitations of shipboard life decreed that.

Abandon ship drill, our most serious occupation, began on Saturday. Everybody had a blue life jacket—we grew so accustomed to those life jackets that they seemed a part of the uniform. They were light and not uncomfortable, which was fortunate, for after the first four days, when we reached the danger zone, we wore them at all times. We were no longer, in fact, permitted to remove our clothing at night. We slept in boots and breeches and blouses, with the blue life jacket over all.

At first the drills fell at anticipated hours, and we would get our belts and be ready when the bugle blustered. We received at once assignments to boats and rafts. There weren't very many boats, but there were a lot of rafts, so that the great majority of us examined the floats and the open lathe work between, and speculated on methods of launching, wishing we had been lucky enough to get boats. The rafts would simply be flung overboard, and we would go down rope ladders and get on them as best we could. It looked hazardous, but we believed it could be done if we had a system. So we developed one and tried to account for everything.

We resented the advice of a fortunate individual assigned to a boat—and it wasn't merely a boat, it was the captain's gig. "It's well enough for you to talk," we said, "you're in a boat. You're lucky." Our hearts were full of envy.

"I thought I was at first," he admitted, "but I'll swop with any of you. Somebody's reminded me of a thing I'd forgotten, and I'm trying to duck that boat."

"What is it?" we asked. "You're crazy."

"Oh, no. Not at all," he replied. "You see the captain's the last man to leave the ship."

No matter where you were, when the bugle cried for abandon ship drill you had to rush to your bunk and wait there in the dusky and close hold of the ship until the gong sent the long lines worming at double time up the companionways and to the deck. It was a good deal to ask a man to leave the air and the sun, in an emergency, and to fight his way through narrow, insufficient passages to the stifling hold, but we could see it was the most efficient way.

As the days passed the drills became more ambitious. They came at unexpected moments—often in the middle of the night. In the turmoil of getting ready, you would hear the men shout: "Shake it up there! Get to your place! Don't block that passage! Hay, Brown, where did you get the molasses on your shoes?" We were never quite sure whether it was a drill or a dangerous actuality.

It was forbidden to talk at abandon ship drill. That was difficult, as sometimes it was nearly an hour before the recall blew. So men talked, and when they did strange punishments were invented. You might see a forlorn individual standing in ranks with a placard hung about his neck, informing all the world: "I talked at Abandon Ship Drill." Or another, at the head of the companionway, singing out "I got to learn to hush up when it's orders" over and over again, like a man reciting some frantic litany.

The necessity of such precautions, and this severity, were clear to the dullest among us. Because of their speed, the *Northern Pacific* and *Von Steuben* had no convoy. They crossed side by side—two little specks in an endless waste of water. But there were places in that waste where it was necessary for us to go, where submarines lurked. We would eventually be picked up by destroyers, but only a day or so out of Brest, France.

Sometimes the boats were so close together that with glasses we could recognize friends part of the other battalion. One was tempted to shout across. And through this narrow lane one night, with the whole sea to accommodate him, a tramp steamship blundered. There was something of the miraculous about that escape. We conducted abandon ship drill more earnestly after that.

The crossing wasn't all abandon ship drill; the weather occupied us quite a bit, too. After the first two days the sea rose and the boats showed us how they could roll. Familiar faces disappeared. By Tuesday there was a really high sea running and preparation for morning inspection of quarters became an ordeal. Instructions were to get every man on deck unless he was literally too ill to be moved.

"What's the matter with this man?" an officer asks the first sergeant, peering into a clearly occupied bunk.

"Says he isn't seasick," the sergeant answers with a cruel sneer.

"Not seasick . . .?" the officer interrogates.

A voice very weak but firm arises from the bunk: "No, sir, not a bit."

"Then what's the matter with you?" inquires the officer.

"I think I got the . . . the . . . the grippe," says the suffering soldier.

"Up then, and get where the air is fresh," demands the officer. "It's what we're prescribing this morning for grippe." Thus caught the invalid does get out but not without leaving awful souvenirs of his prevarication.

There were some, heaven knows, that didn't lie. When an officer asks why another man was still in bed, the first sergeant says, "We can't move him, sir."

"Feel better if he'd get up," advises the officer. "Now what's the matter with you . . .?"

"Oh! Oh! Oh! Oh!" replies the man miserably.

The officer then asks, impatiently, "Answer up. What's the matter with you?"

"Oh, my God! Oh, my God!" cries the man.

"That's nonsense," the officer tells him. "Do you good to get on deck. Seasickness is all imagination." The officer looks around him quickly. His own words fail to comfort him, as a lurch of the ship throws him against the bunk of pain. If he doesn't come up for air pretty soon himself his end is clear. "All imagination," he insists weakly. "Get out of here."

With the aid of the first sergeant, the officer gets the man out. Swaying and clutching at the air, the man moans, "If I ever live to get to France, I'm going to stay there and become a frog."

"Excuse me, sergeant," says the officer vaguely, "be right back. I've got to report . . ." He then staggers up the ladder.

"All imagination, did I understand the lieutenant to say?" grins the sergeant.

Then the complaining voice of the invalid: "Honest, sergeant, they wouldn't treat a dog so."

"What you kicking about? Didn't you see the officer had all he could stand?" the sergeant explains.

The invalid's voice, suddenly strengthened, replies: "You ain't foolin'? Honest, sergeant? Ha, ha, ha! Damfi don't feel better."

Tuesday was our day of greatest casualties and wicked was the wit of the survivors. If the quarters were bad, the mess hall made them seem very pleasant by comparison. As long as they could manage it, men went there because—except for a few crackers and such things at the commissary—it was the only place on the boat where they could get anything to eat. And somebody had started the abominable lie that eating is the best cure for seasickness. The food was good, too. Let that be put down.

The mess hall was formerly the first class dining saloon. It was so far down that with any sea running at all no portholes could be opened. Here and there traces of its former luxurious decorations survived, but in place of mahogany one gazed on deal mess tables crowding each other. An ancient square piano was lashed to the end wall. By the main entrance were the tubs and cans of the cleaning detail. It is no wonder that the grease of one meal couldn't be cleaned from the mess kits for the next. Meals nearly overlapped each other, as organizations had to be fed in turn. In the corridors were processions of men wondering if they could last until they got in, or if they could manage to get through if they did. And the odorous ghosts of many vanished meals illustrated the transient nature of the one in progress.

For one on the edge, the atmosphere inside was nearly unbreathable. The floor was awash with greasy, coffee-colored water. Kitchen police in those days should have gotten citations. On the wildest night the old piano broke its lashings and went drunkenly fraternizing with the tables. It lost a leg and then permitted itself to be led back and tied up again. It furnished a humorous interlude that helped some men, who asked that the piano be allowed to perform thusly every night.

The guard and the soldier lookouts had to do their jobs, seasick or not. The captain of the ship had offered a prize to any soldier who spotted a periscope. It kept the lookouts wide awake and it didn't do any harm to the flotsam and jetsam that were reported as periscopes. There were rumors every night that submarines had seen us.

On Thursday evening, when we knew we were well within the danger zone, the bugle called us to abandon ship drill. There was an element of strain present. The naval officers had looked glum all day. It was whispered that submarines had been reported near us, that we weren't far from the French coast, that our escort of torpedo boats ought to have picked us up that afternoon, and that the skipper was crowding the air with demands to know where they were. And so a feeling grew that this wasn't a drill at all. We all came tumbling down to the close hold, which was only lit by an occasional blue globe, and stood attentively at our bunks. When the gong rang, we jumped up the stairs with no more than the prescribed hurry. As the last light faded over the water, we waited patiently for whatever might follow.

We could see that both boats were taking a zigzag course, strengthening our belief that there were submarines about. As the minutes slipped by, the recall didn't come and the presence of submarines was accepted. We strained our ears for an explosion. From the bridges of the two ships signals flashed out. After a long time, when it was quite dark, the recall blew. The men gathered about the decks in whispering groups. No one regretted the experience. It had shown that the crowded boats were at the pitch to behave just so if the real thing should happen.

That night (or early the next morning) a story went on the lips of the most conservative that we had, towards midnight, actually run into a submarine nest and two torpedoes had been fired at the *Northern Pacific*, and one at the *Von Steuben*. Judging from the letters written home, it was accepted generally as a fact.

We knew we should be in by Saturday and everyone was glad. It was growing irksome to sleep with one's clothes on, to carry the blue life jacket everywhere, to stumble about at night in the insufficient green light, unable to read or play cards.

When we went up on deck Friday morning, we saw five destroyers low in the water, their sterns piled with depth bombs, their hulls and superstructures curiously camouflaged. They chased about us as if in pursuit of each other, tearing along our sides, doubling about and dashing perilously beneath our bows or stern. The sight cheered everyone. The sun was unclouded, the sea had gone down, and we began to pack.

Early the next morning thick fog shrouded us. We were summoned to an abandon ship drill—and when we glanced at our compasses we saw that the boat had turned around and we were headed west. Was it a flight? We were not released from the stuffy hold until nearly noon, when the white pall of fog had thinned and we had gotten back on our course.

Because of this delay, we didn't pick up land until after luncheon. There was no dramatic abruptness about our first glimpse of France. In the beginning there was just a shadow on the sea far in the southeast, and then little by little it deepened and lifted itself above the water. Nearly without words, we crowded the rails and watched the thing grow. Out of the somber, low clouds protruded details: a suggestion of green wavered then spread along the water. It ceased to be nebulous, defining itself for us as the bold headland of Finistère.

France, we thought, where war's happened for four years and flames now, waiting for us! That was the reason for the nearly motionless silence along the decks, for the eyes fixed on each detail that seemed a little sacred. The outlines of trees and houses traced themselves before us. We had left America just struggling from the sober cloak of winter. Now spring had done all it would for France—the coast appeared abnormally green and gay.

Airplanes whirred overhead and a dirigible, catching the sun like a placid planet, came to meet us, swung about, and escorted us in. The white and brown cliffs closed around us, like a welcoming embrace from the land. We felt ourselves drawn to a smiling serenity, a drowsy and remote content. Yet all the time we knew it was nature's masquerade and it changed nothing for us. We were in France, which for nearly four years had submitted to the scarlet and voluble shock of a perpetual disaster.

Chapter Eight

SOUGE AND FIRST CASUALTIES

The Chinese, we soon realized, would be an irritation, for we wouldn't be allowed to loaf here. At once we were prepared for the fulfilling of our destiny. First of all, we were equipped like the artillery regiment we were. Six batteries of *soixante-quinzes* were delivered to us in the spacious gun park. Sleek and lithe with an iron grace, they stuck their noses from their painted shields. They looked terribly competent, a little snobbish, too. They seemed to remind us that they weren't three-inch guns, and that we had a lot to learn before we would really be fit to handle them.

Limbers and caissons were of an unfamiliar pattern. We gathered about the gray *fourgons*—a cross between a gypsy van and a prairie schooner. They looked sturdy and faithful, and they turned out to be so. Telephones, switchboards, wire, wireless sets, goniometers, scissors—they all came streaming in. Except for horses, we were fully equipped within the first few days. (When the horses began to arrive, they would breed dissension almost at once.)

We didn't have much time to admire all this, as we were put to work to learn something about it before we tried it on the *bosche*. The course was announced as eight-weeks long. After the first day we glanced at each other hopelessly and thought: What had they done with us at Camp Upton for seven months? How could we absorb all this strange, fascinating, and fundamental knowledge in a few days?

At first officers and men went to standing gun drill. The officers followed with terrain-board work and specialist instruction, and spent the rest of the day at general lectures on conduct of fire, orientation, communication, materiel. We were given elaborate range tables and taught about stripping ranges and transport of fire, and about "D zero" and "K zero." Our heads buzzed from all the technical artillery jargon. "If I have to figure all these things before I shoot at the *bosche*," someone said, "the war will be over before I get my first shell off."

The sun grew hot and the sand more clingy, reminding us we were in the south, as we trudged to classes or walked many kilometers with plane tables and instruments for orientation exercises.

During this period of education, the regiment more or less ran itself. Officers and men went to different classes; the hours didn't coincide. Often for drill there would be no officer present. Yet the work didn't slump and discipline maintained its old standard.

We were the first National Army division in France, so our instructors had been drawn from the few Regular Army and National Guard divisions that had preceded us. They had a little experience in what might be called "parlor trench" fighting. We grasped at it, and tried to emulate their easy command of the finer points of French artillery specialization. It was invaluable training for us.

Frequently we got to Bordeaux for a weekend of relaxation. The neighboring villages of Martignas-sur-Jalle and Saint-Jean-d'Illac offered us smiling hospitality. For less adventurous spirits, there was a collection of booths just outside the gate, where one could sample French cookery and wines. Then, during the second week measles appeared and for a time all passes were stopped.

As part of our training, we had solved the mechanical puzzles of the *soixante-quinze* and something of the mysteries of orientation and modern conduct of fire. On May 27, we went to the range to shoot. There were just enough horses at that time to draw one battery out— the Second Battalion got them for E Battery, which had won the gun drill competition, and had been selected to fire the first shots on the range with C Battery.

C Battery tried trucks: they got the pieces and caissons as far as the macadamized road went, but there remained, perhaps, a mile and a half of sand. The trucks wouldn't touch it. The cannoneers looked at the deep ruts and the heavy pieces and said to each other, "We have been honored with this first job to fire." They put their shoulders to the wheels. They kept talking about that honor and wondered why they had ever gone into the artillery to be so appreciatively singled out. Although a little limp themselves, they managed to get the carriages to the position in front of Observatory 3, where others had dug emplacements and sunk trail logs. The details located the guns, got the aiming sticks up, and ran wires to the observatory and into the range telephone system.

Captain Roger D. Swaim, of the New England National Guard, was the First Battalion's firing instructor, and Captain Kelly, of the same organization, was the Second Battalion's. They met us at the observatories at 7:30 Monday morning, and we started.

We had so much ammunition that we forgot to gaze at each shell as if it were a precious pearl being cast before swine. The projectiles went away in quick salvos, and after the first few we knew that while we weren't perfect we could bracket a target and get real effect on it. Then the instructors criticized us, the colonel did the same, and the majors usually had their say. Those who hadn't fired looked at the man conducting smugly. Yet always sooner or later they got, as one phrases it, theirs. This was the beginning of endless hours in the observatories. We averaged four hours firing and eight hundred rounds a day.

One recalls that our first day on the range saw the opening of the great German offensive across the Chemin des Dames, through Château-Thierry, and nearly to the gates of Paris. After the thrust at Amiens and around Ypres, the *bosche* had lain quiescent, and his startling initial successes carried a vivid shock to us in the midst of our schooling. We guessed our plans would be altered, for more artillery was needed—and a cry went up for every available man. Yet when the change came it was no less of a shock than the great battle. The schedule was published at the end of the week. We would start on the range at seven o'clock and get back in time for a hurried bite of luncheon. From then until five o'clock, we would have terrain board and specialist instruction, and gas would have to be worked in between five o'clock and supper. From supper time until nine o'clock, we would listen to lectures on ammunition, fuses, and other various subjects. Then, if we liked, we could attend to our routine organization work and study. Then, if there was any time, we could go to sleep.

Indeed, the emergency was grave. We even heard rumors that the government had moved from Paris to Bordeaux a second time, and we went into town that weekend apprehensive of too many figures in frock coats and silk hats. After a few days, the news was better but it didn't affect our schedule. During the afternoon classes, after nights of insufficient rest and mornings of intricate calculations and eye strain on the range, we struggled against sleep.

During this phase, Lieutenant Colonel Stimson returned to the regiment, after having visited several fronts and taking the course at the staff college at Langres. Lieutenant Mitchell, in spite of his experience, was not named regimental gas officer. That position went to Lieutenant Gilbert Thirkield. Our gas drill consisted of exercises in speed and walks or runs while wearing the masks. We tried to accustom ourselves to goggles that always clouded, to mouthpieces that left us a trifle choked, to headbands that exerted a painful and increasing pressure.

Into the midst of this earnest endeavor the horses came, and time had to be found to take care of them and to wrangle over them. They weren't very good horses, but they served to arouse that passionate gypsy instinct that informs all lovers of animals. There was sharp trading and devious scheming to get the best of each lot.

A new batch would arrive from the remount depot. It couldn't be assigned to one organization without giving the others a fair chance for its best. An order would come around that organization commanders might exchange the choice of their individual mounts for anything that caught their fancy in the new lot. The horse fair would then begin. It was usually held in the deep sand by the stables. Officers and men would form a ring around a row of shaggy beasts held by self-conscious orderlies. Critical eyes would run down the line, taking in the choices: the badly used thoroughbred, a thing of possibilities; the narrow-chested overbreed; the useful animal of poor but honest ancestry; and the pitiful crocks. Arguments would spring up as to the virtues of some particular beast. You invariably weighed the reverse of an expressed opinion. Faces would grow red and voices grew hoarse from reiterated convictions.

"I'll swop for this one," a captain says.

"All right," from the officer in charge of the fair. "Bring on your best mount."

The captain strides away. After a time, the circle parts to admit him and his prize—a spring-kneed, mangy cob from the hospital. It takes two orderlies to support it. "Whoa!" cries the captain, and pats him gently as if to persuade him not to cut up. He points to the new horse he has chosen and instructs his orderly: "Lead that fellow out. I think I'm getting stung, but I agreed to swop, and I will." The orderly leads the invalid, glancing back as if to make sure he hasn't toppled over.

Raising his voice, the officer then says, "Like the deuce you'll swop. What did you bring that hat-rack here for?"

The captain's expression is of innocent surprise. "To trade with you as the order directed," he says.

Sneering, the other says, "Thought you'd made a mistake and believed I was running a soap factory, or maybe you want to borrow a detail to dig his grave."

"Very funny! Very funny!" the captain responds. "That's one of the best horses in the regiment."

The orderly puts in gravely, "It's a real hardship to see him go, sir. He's just a little sick."

"My interpretation of the order," the objector says, "is that you can trade your best individual mount. If that's it, your battery will walk."

Gesturing, the captain says: "Orders are orders. You've got to trade."

A very superior officer intervenes. "Gentlemen!" he cries ". . . Or maybe I ought to say gypsies . . . We can't do business this way. We'll get an interpretation that will give everyone a square deal. Meantime, put the horses up."

And the red disappears from the faces of the wranglers, and they walk away arm in arm, good friends until the next fair day.

Sharp trading was necessary. Not only were many of the horses bad, but they died in large numbers and replacements weren't simple to get. Major Johnson was largely instrumental in holding casualties down and in conditioning the survivors. He was also a bulwark between us and the gypsy desires of other organizations, as the horse-trade fever swept the entire brigade. "I thought they might court-martial me today," he would say, after an hour or two at the stables or Brigade Headquarters with higher ranking officers than himself, "but I've held them off our horses."

The remount men watched the bargaining and smiled. They had their own axe to grind, and they liked to see a favorite animal well placed. They were capable of diplomacy when officers of higher rank than the one chosen threatened to interfere. "Sure, a beautiful horse, sir," the remount man might say to the very high ranking officer. "Few better in looks have come out of the depot. You might go farther and fare worse." He winks at the junior officer for whom that horse is destined.

The senior officer glances up and asks, "What do you mean? What's the matter with him?"

"Matter! Who said anything was the matter?" says the remount man. "Of course, sir, all horses have their little foibles."

"I thought so. Talk up. What's the matter with this one?" the senior officer demands.

The remount man gazes at him admiringly. "No fooling you, sir!" he tells him. "But I don't go back on what I said. A beautiful animal, and he might give you good service if you took chances and had a little luck. I go on the principle that no horse is hopeless, but this one is a genuine bad actor."

Exit high ranking officer.

We had practical uses for our horses now, as some of the gun positions and observatories were five kilometers or so from our quarters. It often took hard riding to snare a bite of luncheon before the first of the afternoon classes.

Lieutenants Hoyt, Montague, Gurney, and Church, who had been delayed in America to bring over casuals, joined us early in June. Shortly afterwards Lieutenants Hoyt and Norman Thirkield were sent to balloon school and Lieutenants Jones, Montague, and Gurney to aviation instruction. Soon thereafter Lieutenant Hoyt was ordered by G. H. Q. to the liaison service and the regiment said goodbye to him regretfully.

We had gotten into lateral and bilateral observation by this time. Often the guns were several miles from the officer conducting fire, but communication was always open, and the result of these exercises plainly told us that we were nearly ready for the Hun. Before this war it would have been considered an absurdity to try to train an artilleryman (even in the old fashioned methods) during so brief a period. But here we were—good. The regiment felt it and, a little later, the Hun felt it, too.

Our first casualties came to us on the range at Souge, on June 20, when we were registering for an intensive barrage that would mark the close of the course. The two battalions had established command posts at some distance from each other. Each had put in elaborate schemes of communication, practically independent of the range system. Major Johnson had received permission to locate the pieces of the First Battalion according to the technique of actual warfare. He got camouflage nets that, with the natural cover, hid the positions so successfully that an airplane photograph taken for our instruction was innocent of warlike indications.

The first platoon of Battery B was scarcely more than fifty meters from Major Johnson's command post, Observatory 1. The pieces were echeloned, each under its own camouflage net. The registration progressed, as registrations do, to a precise and dreary measure. Without warning and with no unusual noise, Battery B's No. 2 piece was shattered by a premature burst. For a moment a cloud of smoke obscured it. As it drifted away we saw that the camouflage net had disappeared, that the caisson was blackened and smoldering, that the breech of the piece had gone. The crew went from being an ordered group to a thing scattered and incomplete—men stumbled oddly as they ran out of the cloud and some seemed to be missing.

"Cease firing!" Major Johnson ordered. "Where's the surgeon?"

The operator passed the word over the telephone. Flames sprang from the smoldering caisson; shells were evidently bursting there. Major Johnson ordered everyone from the observatory and, followed by his adjutant, Captain Reed, and Captain Ravenel, walked forward and threw sand at the caisson. Unasked, volunteers sprang from the ranks into the danger zone. In a few minutes the fire was extinguished.

Those on the outskirts questioned: "How much damage? Anybody hurt?" And from the group around the smashed piece came back the quiet answer: "The gunner and No. 1 killed." Everyone had guessed that would be so—sitting on either side of the breech, the men never had much of a chance.

Following the event, the director of the school came, a board was appointed, and the evidence taken. We had learned to fear long fuses, but the damage had been done by a white fused shell, and No. 2 had looked through the bore, so that the blanket verdict of faulty ammunition went down.

On that day, an ambulance dashed up and backed towards the group. Two covered forms were lifted into it, then it clanged a swift away towards camp. "Brace up!" an officer called with kind brutality. "You'll see plenty of other men killed before you get through with this war. Get on the job now. Firing will be resumed." The men responded, shaking themselves rather as dogs do after an unexpected immersion. By that afternoon there was a new piece firing from the destroyed gun's platform. The gunner and No. 1 did not flinch. The day's work went on with a noisy rapidity. "Yet," as someone wisely remarked later, "it can't be like seeing men killed in battle."

Privates Jeremiah S. Lynch and Harry J. Posner were buried the next day. Chaplain Sheridan conducted the services, and Mrs. Gariessen, of the Y. M. C. A., who had a short time before lost her own son in action, tried as best she could to take the place of the mothers. Lynch and Posner received full military honors. Men from every organization attended the funeral and saw more distinctly in the bland southern sunlight the vicious and amazing shadow that is war.

Chapter Nine

HUSTLED TO THE FRONT

The regiment went about its business with its former eagerness. We were told that our first rolling barrage was worthy of veteran troops. It certainly made enough noise and black smoke. The second, with the guns of the two light regiments in a long row, was as good. We admired the dust clouds half obscuring the quickly sliding tubes and the changing black curtain drawn across the range. "No one," we told ourselves, "could get through that." Our instructors admitted that there didn't seem to be any holes.

Such perfection wasn't reached without delays and adventures. The weather had grown steadily warmer and there had been scarcely any rain, so consequently the range was abnormally dry. When the 306th got its 155mm howitzers and opened fire with practice shells, these factors produced worse conflagrations than we had had at Camp Upton. They stopped our work and instead sent us to warm and uncongenial labor. Towards the climax of a delicate adjustment it was distracting to hear someone say to the instructor: "Isn't that smoke over there sir? I think it's a fire on the range." The instructor always looked through the binoculars and nearly always, in a tone of helpless disgust, called to the operator: "Cease firing! Fire on the range." Then the battle roar would die before a threatening silence.

We never learned. We always hoped until the last minute that the flames would burn themselves out. But always the small smoke ball with its red center would grow and spring into a black fan with a flame fringe, sweeping before the wind that always blew in that place.

Then the colonel, or the brigade commander if he was there, would call for trucks and men until the greater part of the brigade and the ammunition train was on the range, starting counter fires or using picks and shovels to clear the ground before the flames. It usually meant an afternoon's hot work at the expense of specialist instruction.

Our instruction had about run its course anyway. The days had slipped into weeks and towards the end of June we knew we were as nearly ready as Souge would make us. Our departure waited only on transportation. We speculated as to where we would go—our infantry had trained with the British in Flanders, so for a long time we thought we would fight there.

Tours wanted to know which regiment would volunteer to hold itself ready to move at a moment's notice. The 305th offered itself. We entered a new age of packing. We had more equipment but we also had more experience, and we got ready with little of the neurasthenic hurry of Upton. Our carefully studied organization, however, was to be shattered. Other artillery brigades were coming to France and they would have to be instructed. Under orders from the chief of Artillery, the Souge instructors were asked to choose a certain number of officers from the brigade who they considered had shown aptitude. They would either remain behind now, or be called on later to teach artillery.

We felt our regiment had been unduly complimented. Captains Reed, Delanoy, and Ravenel were to leave us at once. Lieutenants Camp, Church, and Fenn might be called from their organizations at any time. Lieutenant Colonel Stimson went to G. H. Q. to accomplish the release of the three captains. Lieutenant Camp was made acting adjutant of the First Battalion and Lieutenant Fenn acting adjutant of the Second. Lieutenant Montgomery took command of Battery B. Sometime before, Captain Fox had been made personnel adjutant, so

Lieutenant Kane was the commanding officer of Battery C. With these radical changes made, we were ready to go into action.

From day to day we waited for word from Tours that our transportation was ready. The Fourth of July was near. The general commanding the base section wished the brigade, if it had not moved by the holiday, to take part in a monster parade in Bordeaux. That ceremony kept us anxious for a number of days: in the morning the parade was a certainty; after luncheon there wasn't a chance that we would make it; and the next morning there was no question we would make it. It wasn't until July 3 that we knew, and then we were told that we would leave, mounted, immediately after luncheon, camp at a race course outside Bordeaux, march in the next morning, participate in the parade, and come all the way back before night. On July 5, the regiment would start entraining.

It looked like a difficult program to fulfill. Our drivers had had very little road experience and the regiment had never before been mounted as a whole. We were afraid of our horses. Could they do it? Was it wise to make them do it, when they would have to stand immediately afterwards in box cars for three days?

Just before we left, Major Johnson's promotion to the rank of Lieutenant Colonel came through. It cast another shadow, because we knew the powers wouldn't let us have two lieutenant colonels.

After luncheon the regiment gathered in the gun park. The teams were brought from the stables, protesting at the unusual exercise. The drivers reproved them with harsh voices. A fog of dust arose and settled over the place. Through it, you caught glimpses of prancing horses, struggling men, yellow harness; out of it, came a chorus of commands, entreaties, threats. Guidons flashed red like a gleam of sunlight through the rolling mist. The sunlight grew, the mists rolled away. Wheels, swings, and leads stood in their places and behind them the yellow and black carriages rested expectantly.

"Prepare to Mount! Mount!" someone shouted. Drivers sprang to their saddles, the leading battery moved out, and the others followed. As they were leaving camp, the column twisted a little and wheels slipped into the sand on either side of the avenue, but the column kept growing until it stretched into a string incredibly long and business-like and military.

Road discipline came to us, as it were, instinctively. There were no stragglers. Drivers mounted and dismounted precisely at every halt. We took a narrow country road and on a curving hill—as difficult a place as you could choose—we met a supply train coming up. We got our carriages into the ditches, wormed by, and nothing was upset. On the jammed roads at the front, we found nothing much more puzzling. Right then and there we began to take pride in the regiment mounted. Self-satisfied, we listened to the heavy rumbling of the carriages. From every turn we glanced back at the struggling horses, the sleek pieces, the caissons low and awkward. The whole had an appearance of grotesque beauty.

The Stade Bordelaix was green and trimmed like a huge formal garden. We camped by the steeplechase course. After parking and pitching tents, we, for the first time, faced the problem of watering on the march. We found the familiar lack of facilities, the accustomed waste of time in going long distances with a few horses. But it was experience that we needed, and we saw that it was a good thing we had come.

A few fortunate ones got passes for Bordeaux. The rest, after mess, lay about in fresh-cut hay and tried to realize it was their last experience in the S. O. S.—the Services of Supply.

The next morning our apprehension vanished. The First Battalion took one road to town the Second Battalion took another. "We'll rendezvous all right," the commanders said confidently. And we did. In spite of the apparent confusion in the city, every element fell into its own place in the column and the parade started.

Bordeaux gave us a gracious welcome—masses of citizens threw flowers and confetti from bunting-embellished buildings. They liked the looks of the American Artillery equipped with their own *soixante-quinzes*. They were glad to see the Americans. Turning onto the Place de la Comedie the band blared out the "Sambre et Meuse." The closely packed mass of French people burst into cheers, flung hats into the air, and madly waved banners.

A tribune had been erected in the Grande Place, where local celebrities and French and American generals stood. Opposite was a line of veterans, some with missing limbs, who held flags decorated with the names of breathless battles. When they dipped these flags, our bright new colors bobbed back. By painting our work for the first time with sentiment, the parade did us good. It was our first touch of the spectacular side of things military, a thrill that war lacks.

We paid a small price: only one piece was put out by unmanageable horses; only one man on that piece was hurt; and only one man was thrown from his horse—Dr. Parramore, who was tearing back to attend the victim of the accident. The crowd was interested.

Immediately after the parade, Regimental Headquarters and organization commanders hurried by automobile back to Souge to prepare for the movement to the front. The regiment, in command of Lieutenant Colonel Johnson, returned to the Stade Bordelaix, where they were watered, fed, and messed, and afterwards made the long march back to camp. On the way we had one lesson that impressed on us the necessity of close liaison even in the smallest column. At a crossroads, another regiment cut our line of march and the Second Battalion followed in its wake. There was a good deal of time and energy lost in finding the three batteries, turning them around, and getting them back in line. We pulled into Souge at dusk—tired, dirty, and with a lot of grooming and rubbing before us—but on the whole triumphant.

The next day the movement commenced. The Headquarters Company left the railhead at Bonneau, where less than two months before we had detrained, uninstructed and unequipped. Nearly everyone, it might be said, thought that we would be billeted behind the lines for several weeks of the road work we so much needed. That took a little of the seriousness from the journey.

Regimental Headquarters and the Supply Company left the afternoon of the sixth, and First Battalion Headquarters and Battery A that same evening. During the next three days the other batteries pulled out, while the 304th and 306th waited their turn. We said goodbye to Captains Reed, Ravenel, and Delanoy without knowing when we would see them again.

Entraining a battery mounted was a new experience for all our captains except Dana. The entire regiment had arrived in one train. Now each organization had a train to itself, and was forced to crowd a little to get everything on. These artillery trains were all of a pattern: there were the Hommes and Chevaux for animals and men, a combination first and second class coach for officers, and a string of flats for the carriages.

At Bonneau there was a loading platform. In some places we found none, and used instead clumsy moveable ramps. Yet methods varied little. With practice we achieved some of the skill of circus men. The different tasks proceeded simultaneously; an incoherence seemed to prevail. Then all at once the groups would scatter, and you would see that the job had been done, that the train was either loaded or unloaded.

None of our organizations needed the three hours allowed them for entraining at Bonneau. The carriages were little trouble—squads ran them from the platform to the flat cars across heavy planks, fitted them into the constricted space allotted, and lashed them there with cleats. The drivers, however, struggled with the horses, for the latter never got to like the Hommes and Chevaux. They rose on their haunches, at times crying out their disapproval. The men tugged at their halters and persuaded them from the rear. A horse already in the narrow, shadowy car would look out and shake his head. It was often quite difficult to combat such friendly advice.

The stallions in particular were a problem. If you put them together, they gossiped about old scandals and ended by fighting jealously. If you placed them with lesser beasts, they expressed their contempt with tooth and hoof. "Get 'em in so tight they can't fight," crystallized the advice of most of the men, and it worked fairly well. Eventually, we got to know which horses liked to travel together and that simplified matters.

From the moment the train was loaded until it was unloaded, one lived in a racket like the beating of countless bass drums. Noiselessness in a horse was a symptom of extreme illness. Sick horses were, in fact, a problem. Unless an animal was practically in rigor mortis we took him along. Sometimes one died in route then we had to telegraph ahead and make arrangements to evacuate him. Sometimes the sick survived the journey and died on the picket line afterwards, infrequently they got well. It was the best we could do with animals as scarce as they were.

When a battery had finished loading it looked a good deal like a circus train. The heads of horses appeared through the open doors of some boxcars. Men sat dangling their legs in others. The *fourgon* always appeared gigantic on its flat and behind it stretched the sleek, inquisitive noses of the pieces and the stubby bulk of caissons and limbers. Usually the watercart and the rolling kitchen were on a flat next to the men's cars. Brown figures were busy about the kitchen and a promising smoke belched from its chimney.

It was on that first journey that we learned to know and love the clumsy, sooty rolling kitchen. On the road it was incredibly noisy, and it had a habit of shedding its parts, yet it stood frequently between us and hunger and cold. It was our best friend against evil weather and too much physical labor. On these train journeys it gave us hot food, and it made us independent of the very unsatisfactory coffee stops.

There were certain stations that were announced to us as "coffee stops." The train paused at them, usually at inconvenient hours, long enough for the men to line up with mess cups that were filled with a black liquid from unappetizing pails. They were supposed to be a convenience, but they seemed to possess also a routine element. An interpreter would rush up to the officers' car sometime before reaching one of these places and announce: "Coffee stop in an hour. You will want coffee there." It was not a question, it was a command. The train commander would shake his head at this, pointing to the black cloud rising from the rolling kitchen. He could grin at the surprise and disapproval of the interpreter. Corn Willy, too, it ought not to be forgotten, loses much of its agony when warmed and disguised with some less dreadful substance such as canned tomatoes or stewed carrots.

Eating from the rolling kitchen introduced a sporting element into our travels. The mess sergeant gambled on having his meal ready for a suitable stop. The train commander hazarded leaving many men behind when he ordered them to descend from their cars and form a line by the kitchen, for you couldn't tell much about the length of halts anywhere except at coffee or watering stops.

The train would pull up, let us say, at noon. The mess sergeant would announce himself ready. The train commander would confer with the *chef de gare*. Sometimes the train commander would know French, more often he wouldn't. "Ici!" he would say. "Combien de temps?"

The *chef de gare* would look at him, puzzled. Then a gleam of pleased intelligence would light up his face. "Oui. Fait beaux temps— tres sec," he would say.

The train commander would look at him doubtfully. Did that mean much or little? The word *sec* had a brief sound, but one had to make sure. He would point, therefore, to the train and then with his hand indicate motion. He would display his wristwatch and wheedle: "Ici! Beaucoup or petit?"

The *chef de gare* would smile in a friendly fashion, and reply, "Oui, Mon Capitaine. Beaucoup des Americans. Les boches seront malade."

The captain's face would usually express an emotion bordering on tears—an eloquent emotion, which usually interpreted everything for the official. His face would brighten. He would look at his own watch and, realizing the futility of further words, he would carefully indicate two points on the dial. "Quarante . . . minutes," the captain would say. "Get them out with mess kits," he would call to his aides.

Tumbling from their cars the men would come and form a feverish line. Details would carry pails of food forward to the drivers. The captain would watch with a smile. "You know I'm picking up a lot of this lingo," he would boast contentedly. Then the locomotive whistle would blow, and he would exclaim: "That can't be for us!"

But the *chef de gare* would think otherwise. He would come

running, waving his arms and shouting, "En voiture! Vite! Le train partira." That's always easy enough to understand.

"Quarante minutes. Vous—dit," says our captain, protesting.

The *chef de gare* would be through with argument. The engine driver, never having wasted words on the subject, would simply start the train and out of the kindness of his soul would hold the pace down at first. The men would tumble back into the cars with their half-finished dinners. The details would come scurrying back with their pails. From all directions soldiers who had gone in search of water would tear back, their clusters of canteens tinkling pleasantly. Usually everybody got aboard. Word would come back to the captain that the men had been checked. Then everyone would comment pleasantly on the customs of the country. As a rule, we got fed and it was good, very, very good.

Sometimes the long halts for meals allowed us to water the horses. But the schedule for a troop train is not a constant thing, and these halts often came at bad times. They were not troublesome affairs as a rule, for beside our siding were usually a number of taps, so that the job seldom occupied much time. Sometimes we could wheedle hay from the American officials, and sometimes we couldn't. Yet on the whole those summer changes of stations were not unpleasant or too troublesome. The weather was fine, the men were not crowded, they sang—that's the best indication you can have that things are going well.

Up through Bordeaux, Périgueux, Limoges, Châteauroux, and Auxerre we journeyed towards the front. We expected our definite orders at Is-sur-Tille, but at noon on the 8th, when we paused at Nuits-sous-Ravières, we received a telegram that changed our route and promised us orders at Chaumont. We got them in the evening there: we would detrain the next morning at Baccarat. It rained that night, so it was in depressing and gray weather that most of the regiment reached its destination. Exactly as the entraining of one battery is much the same as another, just so the arrival of each organization at Baccarat differed only in the hour.

Escaping from sleep, we glanced from the cars at a strange France.

The change was due to more than the dull sky, the drifting rain, and the deserted appearance of the little station. Opposite stretched a row of depressing stone barracks, oddly scarred as if they had been neglected for a long time. Nearby a group of gaunt walls suggested a devastating fire. A large sign hung from the front of the station: "Shelter for forty men," it stated. There existed about that place an air of stealth and imminence. One felt the proximity of the *bosche*— any man set down there unexpectedly would have taken one look and known he was in the war zone.

We asked the officer in charge of the yard if we could have breakfast before unloading. He looked at us as if he suspected our sanity. He glanced about with nervous eyes, and said in a low tone, "Get this battery out of here as quickly as you can. *Bosche* planes come over all the time. You don't want to get caught, do you, with your whole outfit in this yard?"

We went to work without argument. It seldom took a battery, under those circumstances, more than an hour to desert its train. The horses were hustled down the runways, the carriages were lowered along ready planks, and the teams were harnessed— the battery stood ready for the road.

We glanced often at the dull sky, with our ears alert for the whir of airplane engines or the crash of bombs. The air remained free of menace, but the sense of imminence persisted, and we were glad when a French guide appeared and told Colonel Johnson he was to conduct us to our bivouac. The column started. Colonel Johnson paused to confer with the colonel commanding the French Artillery brigade, which our brigade was to relieve. The colonel said a *coup de main* was planned for three days later. Colonel Johnson was then determined to win permission for some of our artillery to take part in the preliminary bombardment, so he dashed ahead to Neufmaisons where Infantry Brigade Headquarters had been established.

The column, meantime, left Baccarat. The order was for a fifty-meter interval between carriages, so if *bosche* bombing planes appeared they would do a minimum of damage. There were a number of ruined buildings along the road, souvenirs of bombardments and bombing attacks. We turned onto a forest road that breasted a hill and rested at the top behind a heavy screen of evergreens. The first sounds of actual warfare reached us there. We all felt that we seemed too near the front for road training. The men fell silent; faces were serious.

A good deal of that firing was undoubtedly from infantry grenade and small arms ranges, but we didn't know that then. Our minds absorbed the bark of cannon and the hateful stutter of machine guns as special menaces for us. We visualized ourselves as just behind the frontline.

We finally reached a thick forest on the slope of a hill, where scattered among the trees were Adrian barracks and huts constructed of small logs and trees, of a pattern we had all seen in pictures of fighting in the Vosges. This was the Bois de Grammont on the main road from Bertrichamps to Neufmaisons. We learned that the Headquarters and Supply Companies were located in the woods by Bertrichamps, and the Second Battalion would bivouac near them. Both these woods were too peaceful for war time. In their shelter even the firing we had heard earlier fell away.

But, as Colonel Johnson told us, it was "a bad place for gas."

"We're as close as that?" someone asked him, thinking it was rather near for a bivouac.

Colonel Johnson smiled and whispered: "Not a bivouac. It's our echelon."

Such peaceful woods didn't meet with one's preconceived notion of an echelon—a station of extra carriages, animals, men, and supplies just behind the lines. How could such a place be so peaceful and attractive even on a gray day.

"The first platoon of Battery A," the colonel said to Captain Dana, "will go into position tomorrow night."

It brought the war very close. But those who got that first word also had the impression that the movement would be a temporary one, and that the battery would come out again after the *coup de main*, and that we would somehow get some road work. The colonel, however, shook his head at this. The batteries would go into position as soon as possible after their arrival. The French would remain for a while to show us the ropes, but the task of supporting our infantry was now to be our own. How would the men accept such news in its naked unexpectedness?

The National Army was a good deal of an experiment: it contained every type, race, and temperament. Had its brief training fused these uncongenial elements into a serviceable whole? Each battery commander asked himself this, when he made his abrupt announcement immediately after his arrival, before his men had had an opportunity to forget the fatigue of their three days' journey. One such scene answers for the whole.

The day was about done. In the chilly shadow of the woods the battery stood in line. The men, with shelter halves draped over their arms, waited for the order to take interval and pitch tents. Except for a pleasant rustling of wind in the tree tops, the forest was silent.

The captain faced his command and called, "At ease!" There seemed to be something unfamiliar in his tone. The dead leaves of the forest carpet rustled with the restless movement of many feet. And, serious, expectant eyes answered the battery commander's stern regard.

"Men," he began, "I have an announcement to make. I know you have looked forward to a period of road training before going into action. My announcement is that you won't have it. You're going into the line. The first platoon of this battery will go in tomorrow night. The second platoon will follow the night after. That's all. Battery attention! Count off!"

Heels clicked together and then the numbers "one, two, three, four . . . one, two, three, four" ran crisply down the line. You may have heard a number of organizations count off, but it's doubtful you've ever heard anything like that outside of the National Army in France. The serious expressions didn't alter particularly, but the heads snapped around with a rare precision. The voices were big and hoarse with a sort of helpless effort. It was as if this odd assortment of men were all trying to tell their captain the same thing and, because they wanted to tell him so badly, couldn't quite get it out.

Chapter Ten

MAKING THE HUN DANCE

That same evening the expected blow fell upon us—rather sooner than anyone had anticipated. Major General Duncan, commanding the 77th Division, sent for Colonel Johnson and took him away from the regiment, assigning him to G. I. at Division Headquarters. That loss is hard to estimate—the regiment missed his understanding and the inspiration of his ambition. He never lost his interest in the 305th, but his influence came from afar, for he was no longer a part of us. For the difficult moment, Captain Dana became acting battalion commander. Early on the morning of the 10th he took his acting adjutant and his battery reconnaissance officer and set out to reconnoiter the position Battery A would take up.

There are all sorts of reconnaissance, and we experienced most of them between Lorraine and the Meuse. Some are pleasant and not particularly hazardous; some are dangerous in the extreme; and some are not fit to write about because of their labor, their anxieties, and their lack of result. This was one of the first sort, for it was always more or less pleasant relieving the French. Both battalion commanders can tell you the same story of the Frenchmen's kindness, helpfulness, and hospitality utterly at variance with one's notions of life at the front. We never ceased to marvel at the easy and efficient control the French had of their work. Things that seemed most dreadfully complicated and difficult to us at first, they took with a smile and a careless gesture. They impressed us as having assumed a habit of war that obliterated all the past, that assumed until the end of the world a continuation of disagreeable and morbid events that must be made the best of.

During the reconnaissance, we trotted towards them through a succession of bivouacs of troops who were either resting or waiting to go up. At Lorraine we came upon our first "shell screens"—rows of dead cedar branches, or dirty sacking, stretched between poles. At frequent intervals were lines hanging overhead from which branches were suspended. These shielded the road from aerial observation.

Regimental Headquarters had been established in Neufmaisons, a village of perhaps a hundred houses nestled in a fold of the hills. The French for the present were standing by and rather teaching the child to walk. They gave us our destination: the group headquarters in Pexonne, a mile and a half nearer the enemy. We discovered that the road beyond Neufmaisons was more carefully screened. Ahead at last lay a village that even from a distance had the appearance of something dead and corrupt. There wasn't a house that hadn't suffered from shell fire: many were heaps of rubble, or the facade would be gone. You could see into the intimacies of some houses— clothes hung against a wall, a row of bottles in an open cupboard, a tumbled bed. In the choir of the church yawned a hole large enough to take a column of squads.

There were doughboys in the streets keeping close to the walls with furtive movements, as if they expected someone to catch them at an indiscretion. Engineers indicated the presence of nearby dumps. One of the guards posted in the village stopped us near the church. He seemed to think we had lost our way and wouldn't let us pass until he had learned our mission and had scanned our identity books. Just beyond we found the French group headquarters in a large dwelling that was reinforced with splinter screens constructed of logs and sand bags and was comparatively unhurt.

We had been told to ask for Captain Nicoll, the acting group commander. It must have been after seven o'clock by that time. We knew the captain had been warned the night before of our coming. Our minds were full of ourselves and the serious nature of our errand. We felt as if the war might depend on what we were doing that morning. War for us was a matter of perpetual wakefulness, of extended hurry and effort, whether useful or not.

There was no sign of anyone stirring in the headquarters. We knocked, we pulled at a broken bell handle, and we glanced at each other in amazement. "Is it possible," we asked with the innocence of amateurs, "that they are still in bed?" Apparently, it was possible. After an interval a shuffling step from within became audible, the door opened, and a sleepy, half-dressed soldier appeared. At first he gazed at us as though examining a collection of unexpected specimens, and then he overcame his astonishment, and led us into a dining room that was tastefully paneled in dark wood. From there we heard reluctant stirrings upstairs and before long three lieutenants appeared. Their astonishment (or perhaps disapproval) was smothered behind greetings and an undreamed of hospitality.

The captain, they explained, had been occupied until very late the night before but our affair was quite simple. One of the lieutenants produced a cobwebbed bottle from a cupboard in the dark paneling. "It is forty years old," he said, pouring a white liquid into glasses. Coffee soon appeared. These officers were in no hurry to discuss our affair. We experienced a sense of guilt while we waited for them to come to business. Our restlessness grew, for we wanted to be doing something.

At first that was the attitude of the average American soldier towards his job. Experience taught him eventually to take the day's work a trifle more sanely. But on the whole he was in a hurry, and in quiet sectors he was up and at work earlier than the French. He took about one-fifth as much time for meals as they did and went to bed a good deal later (but seemed seldom to have had enough sleep). Until he learned something of the tricks of war, he was always surprised at the end of a day to find that the French, while apparently loafing, had accomplished a good deal more than he had done.

When the coffee was finished our Frenchmen were inclined to smoke and chat. Since we were in their hands we could only hint at our anxiety. They pointed out the paneling of the room, and one said, "The house belongs to a rich man. Your soldiers call him the Count of Pexonne." Another picked up the dusty bottle, and explained, "He had a taste for such things. You haven't seen his cellar. You know in French a cellar is a cave, and a cave has come to mean a shelter from bombardment. When we saw the Count's cave we decided never had war led us to such a shelter, and we didn't care how long the *bosches* kept us there, for it was filled with such bottles as these. They're about gone now, for the town is to be abandoned, and since there is very little transportation for the civilians, the count has sold his treasures to the French and Americans for almost nothing."

We told them we were astonished to learn the town was to be abandoned. "Yes, as you can see," responded a Frenchman, "it is under constant shell fire, but the principal thing is the gas. They can fill it full of gas in a moment. You will notice that all the civilians carry gas masks, for the gas comes in frequently. In a few days the village will be deserted."

We moved at last. We descended first to the famous cave—a damp, vaulted cellar—that was the heart of the group's system of communication. A telephone operator crouched before three four-direction switchboards against the front wall. A number of wires came through an opening and meshed like an untidy spider's web across the ceiling. "You can communicate with the whole army system from here," one of the lieutenants explained. "That will make a little difficulty for you at the start, because since the village is to be abandoned, you will have a new command post. You will have to arrange a new telephone central there."

One of the other officers got his horse, and we mounted and rode from the village at last. We hadn't expected to be able to continue our reconnaissance mounted, but most of the road, our guide explained, was defiladed and on such a dull day the *bosches* weren't likely to be troublesome.

We left the dying village by a country road and after a few hundred meters we reached the first of the battery positions. The pieces were placed in casemates constructed in the high bank of the road. The whole was extremely well camouflaged and impressed us as a perfect position. The road did away with the danger of fresh tracks and simplified the bringing up of ammunition. Then we noticed many shell holes on both sides of the position and close to the guns. "Yes," our guide explained, "the *bosches* have located this position. It would be well for you to leave this camouflage up and locate your guns somewhere else." So we casually examined a number of possible positions, but that morning we were chiefly concerned with the location of Battery A's guns, which were to fire in the proposed *coup de main*. The French had decided on their approximate position near one of the French batteries in the thick woods of La Haie Labarre.

As we climbed a hill, the sun appeared from behind the clouds and we were captivated by the beauty and apparent peace of this rolling, wooded country of the Vosges foothills. Between groves of birch and hemlock, the fields were yellow with ripe wheat—and flashing out from the yellow, like elaborately set jewels, were turquoise-blue corn flowers and poppies of vivid scarlet. What firing there was that morning was far off and troubled us not at all. Except for our mission, there was really nothing to remind us we were at the front and well within range where we could be opened on at any moment.

We rode down a slope along a narrow path that was nearly obliterated by overhanging branches. Here and there among the trees appeared French artillerymen, one of whom took our horses. The forest was full of a quiet but intense activity: some figures lifted stones and great blocks of cement with difficulty; others moved among the trees, bearing iron beams and logs, heavy and unwieldy; and many stooped and rose rhythmically, as they dug spades into the earth, making a crunching sound and then a thud each time the dirt landed on dead leaves. Our guide took all this in with a sweeping gesture and explained, "We have already got the new battalion command post well started here. You have only to install yourselves and complete it as you go along."

Nearby we found the battery under the tutelage of which our Battery A would be placed until the final relief. Captain DesVignes, the officer commanding, took us over the position. We marveled at the neat and efficient arrangement of the positions and the ammunition dumps. We had never imagined such trail logs as the French had here. The captain showed us the temporary position suggested for Battery A, not quite four hundred meters to the right. There was plenty of natural cover—just to the rear sloped a steep wooded hillside, perfect for the construction of dugouts, and at the edge of the forest was a rough road that men and carriages could track safely. Captain Dana was satisfied and returned to the echelon to arrange for getting the first platoon up that night.

It was understood that morning that the French group would remain with us for a week or more. On their departure we would leave the temporary positions for the ones they occupied now. All that was altered the next day and, except for the first platoon of Battery A, the guns of the regiment went directly to the French emplacements.

It was noon and the French habit of a long midday break was imposed upon us. Why, the captain wanted to know, shouldn't we lunch? The one officer of Captain DesVignes appeared, Lieutenant Riveau—the executive, reconnaissance officer, telephone officer, "department B" man, and *popote* (as the French call their mess officer). A table, situated in front of a round white tent and beneath the pine trees, had been laid with cloth napkins and china. It wasn't war . . . it was a picnic. A copy of the *Mercure de France* lay nearby. We didn't talk of war—the only reminder was the distant mutter of guns.

The Americans tried to wheedle the chatter back to the things that obsessed them. "Do you French always run an orienting line?" one of us asked the lieutenant.

"Always," Riveau answered languidly, "in theory, never in practice." He steered the conversation quickly away from war talk. "I have been reading some of your American books," he commented.

The captain, meanwhile, sipped his *pinard* (the French issue wine) as thoughtfully as if it had been a rare vintage. At the end of the meal, with a ceremonial air, he produced a nearly priceless bottle of liqueur from the tent. But the minds of the Americans were on orienting lines and gun positions. Lieutenant Riveau surrendered to our interests at last and accompanied us to a jog in the woods of La Haie Labarre. We had a plane table with us, which Riveau set up. We removed our helmets so as not to disturb the needle, while Riveau oriented his board with a declinator compass. We shot a line across the map from our location through the registration point and drove a stake on the continuation of that line in the wheat field. We drove another stake beneath the plane table.

A rocket went up, but we scarcely noticed. Then it suddenly came to us that we were locating the first piece of the National Army at the front. Lieutenant Riveau, of the French Artillery, had his hand in that, along with Lieutenant Camp, acting adjutant, and Lieutenant Brassell, Battery A's reconnaissance officer. That was the climax of the afternoon; everything was ready for the guns.

We returned to the echelon and were met with the news of a change (necessitated by Colonel Johnson's departure): Lieutenant Colonel Stimson had been given command of the First Battalion. He brought with him from Regimental Headquarters his old Camp Upton adjutant, Lieutenant Klots.

Battery B had arrived during the morning reconnaissance and Battery C came in that afternoon. The movement commenced that night according to schedule. It was not a relief. That started the next night after it had been announced that the French would depart, leaving us to work out our own salvation.

During the afternoon, Captain Dana had sent a detail of men to La Haie Labarre to prepare the emplacements. At eleven o'clock the horses were harnessed to the carriages, the drivers mounted, and the platoon moved out of the black woods and down the road. There was no nervousness—these men went about the job with the efficiency of veterans. It would have been impossible to suspect that they faced the enemy for the first time. There was only one thing particularly noticeable: everyone was unnaturally quiet, as if the Hun might hear. The rumbling of wheels on the hard road surface was disquieting, until you stopped to compute how far away the enemy actually was.

It was a dull night. Except for some firing on the left and an occasional star shell [used to illuminate the battlefield], there was nothing to startle us. Neufmaisons had gone to bed. From the country road, the star shells were plainer but the woods were peaceful—and black. We were to learn to use such darkness as cats do, but that night was the regiment's first experience.

Anyone that flashes a light at a battery position is either a spy or a fool. (The discipline is nearly the same in either case.) Delicate tasks must be performed by the sense of touch, by a special instinct that an artilleryman has to develop. The pieces must be accurately placed, the trail must be nicely fitted into the trail log, and you have to pile ammunition according to the law. Your camouflage must be perfectly arranged so that the first gleam of daylight will find everything covered. The only lights that are ever allowed at a battery position are the shrouded bulbs at the sights, the tiny slits of the aiming stick lamps, and the hidden gleam of a candle in a dugout where the battery commander or the executive figures new targets. All of these, if properly arranged, give away nothing.

The green men of the 305th accomplished their tasks in the brief time they had. No. 1 piece was set directly over the stake the reconnaissance party had driven that afternoon. No. 2 piece was twenty paces to its left. The platoon was ready to fire before daylight.

With the departure of the French announced, a more extended reconnaissance was made on the morning of the 11th. Lieutenant Colonel Stimson went ahead to Pexonne in the sidecar, while the commanding officers of Batteries B and C had their first touch of the front that day. Our little party was welcomed—as we rode into Pexonne, eight shells fell in the town, followed by a noisy and thick barrage from anti-aircraft guns. We glanced overhead and saw among the white bursts directly over the ruins eight Hun planes, flashing white in the sun. We dismounted hurriedly at the command post. Our guide of the day before came running from behind the splinter screen by the door. "Get in here quick!" he warned. The officers responded, while the orderlies trotted the horses off to a comparatively safe stone stable.

We waited inside while the anti-aircraft barrage drove the planes higher and higher and finally back to their base. Then we settled down to the business of arranging the relief. It was complicated . . . and it required a delightful luncheon, moistened with some survivals of Count Pexonne's cellar. This irritated the Americans who felt they were wasting time. However, there was far more to be learned from that luncheon than appeared on the surface. In spite of our impatience, we absorbed sector gossip that would scarcely have come to us from a study of plans of employment or the terrain itself.

Our infantry, we gathered, was having greater losses than we had expected from the normal activity of that portion of the front. One battalion had been caught during a relief and had experienced many casualties. A few nights before, the Huns had placed a box barrage around a platoon, had come in with gas and a new type of grenade, and had practically wiped out the command. An officer from our infantry Battalion Headquarters dropped in for coffee and told us a story of the affair.

"[The officer] who was in command of the platoon, you know, got hurt . . . lost his foot, in fact. That's tough luck . . . in a way. Looking at it in another way he'll go home and maybe be decorated.

"By the way, he had a little Italian in his outfit. I remember the fellow well. Utterly worthless, that's what we all thought. Couldn't speak English. Rotten soldier. On kitchen police most of the time. [The officer in command] had tried to transfer him, but nobody would stand for it. So the Italian was in the trenches with the platoon when the show started. The barrage Jerry treated 'em to plastered the whole works. Then he threw in gas. Shriveled some of 'em up. Then he came himself with these new-fangled grenades and mopped things up. [The officer in command], as I say, was hurt. He lay on the floor of the trench, and a Jerry officer and two or three Jerries were around him, going through his pockets. He heard something and glanced up. There at the turning of the trench stood the Italian who couldn't speak English, who was just about perpetual kitchen police, and who [the officer] had tried all along to shake. His gas mask was off, his face looked different—it expressed a decided disapproval of the whole proceeding. The little fellow's lips set. His rifle, bayonet fixed, rose slowly to the charge. He leaned forward. The [wounded officer] saw and called to him: 'Get back, you idiot! For God's sake, get back!' But the Italian, single-handed, ran to the rescue of his officer. He charged the lot of them."

The narrator paused, as if he was all finished.

"Well?" someone asked.

"Oh! What do you suppose? One of the Jerries tossed a hand grenade and blew the little Italian to pieces."

The story elucidated something for us.

At that luncheon, too, we heard of the various barrages we were supposed to fire under a variety of conditions, and why some positions in the sector were better for the work than others. Captain Nicoll, it developed, had an exceptionally complete dossier that contained: plans of the telephone, wireless, and optical liaisons; careful maps showing the barrages and the Operational Camouflage Patterns; an extended plan of employment; and infinite orientation data. It made us rather dizzy. It seemed incredible that any human mind could digest the voluminous contents of that folder.

We examined the positions recommended by the French. Battery A would move into the French emplacements occupied at present by Captain DesVignes's battery. Battery B would go to a fresh position in a wheat field a kilometer and a half to the southwest of Pexonne. Battery C would take over an old French position on the edge of Ker Arvor woods. Its platoons would be separated by a hundred meters. To balance this inconvenience there was an elaborate system of dugouts and a quarry offering dead space close to the back wall. Lieutenant Kane at first established himself here, but the menace from gas was great, so he moved to a dugout on the hill. Lieutenant Montgomery chose a tumbledown farmhouse near his guns for his command post.

We learned that day that we would not have perfect observation.

The battalion observatory, in a fringe of birch and hemlock between two fields of standing wheat, offered a good view of the left of the sector but nothing of the right. It was called Nénette and the command post went by the name of Rintintin. It was our first introduction to this interesting pair.

During our stay in Lorraine we were always reconnoitering for a more satisfactory observatory. We became convinced that it didn't exist. Most of our barrages would have to be fired blind. Rockets from the right of the reference point—the ruined church tower in Badonviller—would have to be relayed, always a dangerous and uncertain expedient.

Battery C had an eventual barrage in front of the left of another army. There was an observatory at a place called Pierre-Percée from which Lieutenant Kane could register his guns for this mission. The dossier recorded a forward observatory. When it was examined, it was found to be well in front of the normal position of our front-line platoons—that is, in No Man's Land. The French advised against making use of it, for it is a serious thing to place artillerymen in danger of capture needlessly. They know too much.

The situation in the Baccarat sector was unusual. The front was so thinly held that one was always apprehensive of a surprise attack. There was a line of resistance and everything forward of that was provisional. Patrols moved cautiously through a maze of abandoned trenches. At night "Cossack posts" crouched in shell holes or at trench corners. Often Americans glided inside the Hun outposts. The reverse, of course, was inevitable and there were desperate, little combats in the dark. It was troublesome to get the wounded back. Such conditions created too anxiously anticipatory an attitude.

In case of an attack in force, these outposts were to fall back on the line of resistance where the real stand would be made. That necessitated an extreme care in the system of rocket calls for barrages. (You will see later how it worked out.) It made us all the more dissatisfied with our observatories. Yet we established only one new one that was in no way superior to Nénette: we built a platform in the tops of several birch trees on the edge of a wood, which gave us something to fall back on in case we were shelled out of Nénette.

About three o'clock that afternoon, on July 11, Captain Dana, Lieutenant Brassell, and Lieutenant Camp were at Nénette locating points in the sector from the battle map. Lieutenant Colonel Stimson appeared. Captain Dana wanted to register, but Stimson was anxious to avoid stirring the enemy up. But the platoon was in, the guns were ready, and the effect on the men of a few rounds was worth considering. So Stimson consented and Captain Dana telephoned the data down to the battery. The registration point was a corner of a Hun trench at a range of 5,500 meters.

We heard the cry, "Fire when ready!" The crack of the gun reached us. We heard the projectile rushing over our heads towards Germany. The first shot of the National Army artillery was on its way. The shell was normal charge, high explosive. Considering the range and the nature of the terrain, it was quite reasonable that it should not be observed. The captain called for high-burst shrapnel and not long after we heard the projectile's swishing flight, we saw a pretty white ball of smoke appear near the corner of the trench. There was an error of only three mils in deflection and less than a hundred meters in range.

For the shot that put the National Army artillery in the war, Corporal Andrew Ancelowitz laid the piece, Sergeant Fred Wallace gave the command to fire, and Private George Elsnick pulled the lanyard. "Guess," said someone drily, "they heard that shot in Berlin." Certainly, it was the first note of the music to which the Hun danced back to the Rhine and defeat.

Chapter Eleven

CONSOLIDATING IN LORRAINE

The Second Battalion followed close on the heels of the first: Major Wanvig and his staff arrived in Baccarat with Battery D at midnight on July 10; Battery E came in on the morning of July 11; and Battery F that afternoon. Major Wanvig established his echelon near the Supply and Headquarters Companies in the woods above Bertrichamps. The major—with Lieutenant Fenn, his acting adjutant, Lieutenant Church, acting telephone officer, and Captains Starbuck, Storer, and Mitchell, commanding the three batteries—made his reconnaissance on July 12.

The reconnaissance efforts for the relief of the French, as has been said, all shared the same surprises and the same hospitality. However, the conditions the Second Battalion found differed in some ways from those met by the First. To begin with, the French group had only two batteries in position. It was decided to place Batteries D and E in their emplacements. A new position was chosen for Battery F, which was to the right of the Neufmaisons to Vacqueville road.

The group command post was in Vacqueville, a pleasant little village that shell fire had spared. Major Wanvig moved into the Frenchmen's quarters and offices. Scotland was the inherited name of the command post and Godfrin of the battalion observatory. Here, too, the question of observation offered no perfect answer, for Godfrin was not better than Nénette, nor did it have as good natural cover. It was an overgrown hole in the ground—covered with a sheet of elephant iron—in front of the woods. Because of its vulnerability, it was used only for observation of the sector. For conduct of fire, each battery had an observatory of its own, but none of them approached perfection.

At the start an unexpected task faced the Second Battalion. There was a battery in their portion of the front of two ninety-millimeter and two ninety-five millimeter howitzers (sector property). Lieutenant Pike of Battery D was given these guns with nine men from each battery of the regiment and told to find out how they worked, to register them, and to fire them on demand. He and his makeshift crew solved the mechanical and theoretical mysteries of the strange guns and fired with the rest of the regiment.

The relief of the French, meantime, was well under way. The second platoon of Battery A and the first platoons of Batteries B and C went in on the night of July 11–12. The remainder of B and C followed the next night. Two guns each from D, E, and F moved up on the night of July 13–14. The rest of the Second Battalion completed the relief on the night of 14–15. We escaped without a single casualty. Either the Huns hadn't got wind of the change or else they had guessed the wrong roads.

It is, nevertheless, always a nervous business going into position over main highways that you know the enemy has registered, and when you are well aware that his intelligence department is performing miracles to learn the exact hour of your relief. All you can do is to leave wide distances between your carriages, but often the roads are too crowded for that. The whir of every airplane is a warning to take cover and, of course, you can't leave the road.

The chief danger lurks at the position itself. The pieces to be relieved must remain in their emplacements ready to fire on call until the relieving guns are at hand. Consequently, the guns are jammed in a small space and many men and horses are crowded in and about the pits, working in the dark. It is at such a moment that a shell gets the maximum confusion and the greatest number of casualties. In the Baccarat sector, the Huns shelled and dropped bombs at the wrong moments—something we could laugh at, for we were in position and fairly well protected, and we were ready to back up our own infantry.

Now we faced for the first time the problem of organizing a position. That is an irritating and endless process for a green outfit. During the three weeks we spent in Lorraine, we learned more than months of school could have taught us.

The Second Battalion, with the plant it had inherited from the French, settled itself with less trouble than the First. Lieutenant Colonel Stimson moved at once from Pexonne to the new command post in Haie Labarre woods and, with details drawn from the batteries, hurried the work on the dugouts that the French had started. Until some of these dugouts were completed, the headquarters was quite unprotected. And that was only one of the tasks—a new system of communication was still needed and both battalions had to organize their observatories and arrange their liaisons with the infantry.

We had realized all along we were short of officers, but we had felt we were plentifully supplied with men. This abrupt concentration of work, even in a quiet sector, taught us that the artillery tables of organization made in America had not foreseen all the demands of this type of warfare.

At the front, the Headquarters Company could no longer be treated as a unit. Regimental Headquarters, the two Battalion Headquarters, and the echelons were separated from each other by several kilometers. At the start, then, the three details were divorced for tactical and administrative purposes. That raised new problems of subsistence and transportation. Each detail, moreover, was subdivided. Men had to be at the echelon to care for the animals and to draw and transport rations. After the specialists had gotten to the command posts, we found it necessary to supplement them by drafts from the batteries. The batteries, then, complained that that left them shorthanded. The telephone details were woefully small, so we had to shift scout and instrument men into communication. We tortured the dignified tables of organization until they were unrecognizable, but the result was something that could wage war.

At the start, let us review what we did with communication, for that was the first problem we had to solve. A regiment at the front without practical means of communication might just as well be in America. It is out of action. Telephone officers and men, therefore, must lay and maintain wires, no matter how heavy the shelling. They must keep every portion of the organization in touch with the others and keep the whole in talking radius with neighboring units.

In that sector, the 305th had about a hundred kilometers of wire to lay or maintain. We took over many lines from the French, but a good deal of their wire might have been a souvenir of the First Battle of the Marne. For no apparent reason beyond senility it would go dead, and that type of trouble is difficult and hungry of time to locate. A great deal of the new wire issued us had insulation that cracked easily, and because of its color and texture, the wire shielded its faults jealously. As a consequence, we had to lay it with extreme care. The weather helped. It rained very little—so with the heavy twisted pair given the regiment in America, we could supplement our poorer stuff and keep communication always going.

The cellar in the Second Battalion command post at Vacqueville made an ideal central switchboard, and the few new lines necessary for the command were quickly run. The First Battalion completed a small dugout the first two days in, and set up its switchboards there. It made use of what French lines ran to Captain DesVignes's old position, but for the most part it had to run new ones to its various units.

Two men were on duty in the centrals for shifts of twenty-four hours: one man sat at the switchboard while the other could sleep or read or write letters. They could change about as they pleased. But it wasn't as simple as it seems, for at times those boards were busier than the busiest central switchboard in the stock exchange. Often there was more necessity for speed on the front than in the commercial world and high ranking officers as a rule are less patient than the tired business man.

Then, there were the unforeseen complications. For instance, we all knew that code names were used at the front. That was natural. It was impossible to shout names and organizations over wires when the enemy was almost certainly listening in. But we hadn't suspected how quickly such customs of secrecy cast a net of fascination for even mature men. In Lorraine, nearly every officer devised a code name for himself, and until higher authority interfered, guarded it jealously. This produced a clenching of hands and a tearing of hair among earnest operators. Here is a sampling of the code names: Colonel Doyle was "Hub" and his adjutant went by the tinkling sound of "Mess Kit;" Lieutenant Colonel Stimson was "Night Gun" and his adjutant, with perfectly straight hair, was "Pompadour;" and Major Wanvig was "West" and his adjutant was "Kansas," which at least suggested an origin.

The operators took it seriously enough—they had to. Their mispronunciations were due to phonetic idiosyncrasies rather than any humorous intention. Rintintin, for example, became a staccato "Ra-ta-tin" and Nénette often became "Nanny-et." So one might hear: "Pump A Door's busy Mess Skit." Or, "I can't get Night Gown."

Such stealth had its more critical side for the telephone men. The infantry had those who listened in, who spent their days and nights trying to catch operators talking "in the clear"—that is, using the numbers of organizations or the names of places or well-known individuals.

One day a terrifying document reached the regiment. One of our operators had been heard using the names of places. The infantry brigade commander, we were informed, was extremely angry about it. There must be no more talking in the clear. Word then went around that the operators were to put no one through unless they asked for organizations and persons by their code names. That same evening the irritated general wished to speak to one of the command posts. His adjutant got the switchboard and said, "Any officer will do." The youthful operator, faithful to his job, not being able to guess that the infantry didn't know the local trick names of the artillery, replied: "Can't put you through unless you ask for the officer you want to talk to by his code name."

Drama!

Persistent diplomacy alone spared a breach between the two branches of the service. But the operators couldn't get it straight. "If you talk in the clear," they said, "you get the deuce, and if you refuse to talk in the clear you get the devil." But generals, as well as the men, learned from practical experience with such inevitable inconsistencies. And Division Headquarters stepped in: it published a list of those officers who ranked code names. No others would be authorized or tolerated. But some habits aren't broken easily, and often over the wire sighed the eccentric nicknames of the lowly.

The operators did a good job and, even in that sector, a hard one. Lines could go out due to shell fire, weather conditions, traffic, and bad wire. The panels were tested every hour. The operator would plug in, and if he got a response from the other end, he simply said: "O. K." This meant he was testing and was satisfied. If he got nothing in response, or ground noises, he reported to his telephone officer that the line was out and two men were sent to find the break and repair it. They went in pairs, so that if a man should be hit in a lonely place help could reach him.

The hauls were long in Lorraine, so you had to carry a telephone for testing. You would go along for a few hundred meters, scrape the insulation from the wire, hook your telephone in, and call central. When you failed to get a response, you knew the break lay between you and your last testing point, and you examined that section of wire until you had spotted the trouble.

There were alternative talking routes to all stations. When the operator found a line was dead, he got the other end through a different line and warned the operator there to send men out. The other fellow didn't always do it, and one pair of men might have to walk five or six miles to find the trouble—this happened a number of times—in the other fellow's switchboard. That didn't make for the best of feeling among the details, but such irritations were temporary.

Then, there were always curious things happening to the lines. For instance, we had a grounded circuit to Pierre-Percée, where there was a French central switchboard The fact that the line had a ground return indicated that it was not used much. It was, in fact, only important in an emergency. Still, in view of that emergency, it had to be kept working and it was perpetually going out. One day Corporal Caen and an operator went through the lonely, wooded country that separated the two central switchboards About half way they came upon a party of French telephone men who were stringing a wire that looked suspiciously like a remnant of our Pierre-Percée line. (A gap nearly a kilometer long existed in that.) Corporal Caen spoke French and he could gesture, too, like a Frenchman, and he knew some of the most powerful French phrases. But when questioned, the party shrugged his shoulders.

"You could be shot for this," stated Corporal Caen bluntly.

"Ah, oui," said the Frenchmen indifferently. But finally he consented to explain. "Our officer told us to run a line to the infantry, *coute que coute*. We didn't have enough wire. It's only cost a kilometer or two of yours. What are you scolding about? Don't we, like you, have to obey orders?"

The corporal didn't crave international complications, so he trudged back, got more wire, and bridged the gap. But there was a curse on that line. Another day he found a party of Americans from a neighboring unit playing the same salvage trick, and those fellows he had on their knees, begging him not to court-martial them or have them shot at dawn. Tampering with a line in the field is likely to do incalculable damage and thus is a very serious offense.

There was one line that some of the men thought was bewitched. It played its tricks on a very rainy night, with Coheleach on the switchboard. When he ran a test about ten o'clock, instead of calling his customary "O. K." he looked puzzled and said something rapidly in French. "There are frogs on this line," he announced.

"Impossible, because that line runs to Battery B," someone replied.

"Sure, and I can hear the B operator talking across the frogs," said Coheleach.

It looked like there was a cross in the line. The French line had probably been blown from its supports and had fallen over ours. The wet weather and faulty insulation would account for the rest. Only one man left from one end. In half an hour, a small voice came over the wire, reporting. Through his uncertain words we could hear French flowing. The conversation had an astral quality. We could not interrupt it. The groping demands of our man somewhere on that line in the wet, dark night had failed to dam it.

"The line," we distinguished above the queer conversation, "has been tied into close to the road." This seemed impossible. We asked the startled linesman if he had traced the wire. "Where does it go?" we inquired.

"That's just it, sir," said the linesman. "It isn't natural. It goes to a dark dugout."

"Maybe Huns with a listening-in set," someone conjectured. But even the puzzled linesman didn't believe that, for over the wire came a weaving of French phrases that meant it was a bitter night for those who fought, a bad night to die.

Our man wasn't afraid of Huns with a listening-in set. That meant a fair fight, but he didn't like that dark dugout with such a conversation slipping from it over a wire. He hadn't followed the wire in and disapproved of attempting it. A direct command was necessary. But he was so long reporting after that that we became uneasy. Perhaps there had been something he couldn't control—such as too many Huns talking French.

The B drop fell at last and he was on the wire. His voice was conversational again—rather more agreeable than usual. "Spooks? Quit your kiddin,'" he chided. "Who said anything about spooks. Frogs. Line looked as if it was tied in, but it wasn't. A cross. One of these frogs was *couchayed*. Other got lonely and was chinnin' with some central. He had beaucoup *mangay* and after we chowed he came out with me and helped fix the line. O. K. now. Goodnight."

We always had that fear of Germans tapping our lines. There were spies about, the conditions favored them. In Lorraine, most of the inhabitants speak German and there are many German names. The mingling of Americans and French helped the Hun spies—for a

Hun in American uniform among the French or one in French uniform among the Americans was likely to go unquestioned.

A line to one of our advanced positions was interfered with several times. Switchboard operators were called by men whose voices they failed to recognize. These men asked carefully formed questions designed to draw military information. An investigation disclosed that there were tiny breaks in the insulation, such as a listening-in set might make. We placed patrols on that line and, one day, close to the infantry, they caught a fellow fumbling with the wire. He couldn't give a clear account of himself, so he was turned over to a higher authority. What became of him we never heard, but that form of annoyance on that line ceased.

We were particularly anxious about our wires to the infantry. In order that the artillery may properly support the infantry, it must know the doughboys' needs, where his frontline is, where his advanced patrols are. In Lorraine, we kept telephonic communication open fairly easily. In other sectors it was, as you will see, a more difficult problem.

But you must have something besides that. Artillery officers must live with the infantry commanders, explaining the possibilities and limitations of artillery fire, acting as go-betweens, as it were. Regimental Headquarters kept personally in touch with Infantry Brigade Headquarters. An officer was usually sent to infantry Regimental Headquarters. Always a lieutenant went from each artillery battalion to the infantry battalion commander in the frontline.

Lieutenant Edward F. Graham went from the First Battalion to the infantry and Lieutenant Karl McNair from the Second. Each took with him half a dozen enlisted men to act as runners and forward agents. The first day down some of these men were taken on a tour of inspection forward of Battalion Headquarters. In the smashed village of Fenneviller, they were caught by *bosche* harassing fire. They dropped into a ditch by the side of the road, but they saw a medical captain and a doughboy seriously injured and another doughboy killed. They found the ditch comfortable until the *bosche* had had enough.

When they reported at headquarters that afternoon, with a message from the infantry, their attitude was prophetic. They had flung off the shadow of the disaster they had witnessed. They were elated because they had received their baptism by fire. Little Michel of the First Battalion came up grinning, calling to his friends: "Say, boys. This chicken's been under fire. Gee! It was great." Indeed, a spirit of frivolity colored the triumph of the little party. A soldier removed his tin hat, pointed to a deep dent in the side, and said, "Pretty close one that!" This was met by a snort of disgust from one of his companions, who said, "Saw him myself. He took an axe to it."

Chapter Twelve

BARRAGES AND RAIDS

In Lorraine, however, liaison with the infantry was never the bugbear we had feared. One had to be diplomatic. The gravest danger lay in a slip there.

We had, as a matter of fact, forward guns nearer the Hun than infantry Battalion Headquarters. We were ordered to place these as soon as we were in position. They were called pirate guns and their code name was, appropriately, "the goat." Their mission was to deliver harassing fire, to snipe at fleeting targets, and to safeguard the battery positions from sound and flash ranging by making it necessary to fire only barrages from them. In other words, the pirate gun went into action with its eyes open. The Hun could spot them by sound or flash ranging. And the Hun did, so those guns were always shelled more or less.

Battery A sent in the first pirate gun for the First Battalion under command of Lieutenant Ellsworth Strong. The emplacement was an excellent one in the cellar of a ruined house in Fenneviller, which was heavily casemated. To guard against emergencies, it was necessary to keep the limber and teams at hand in a stone stable. The Second Battalion had pushed its gun forward to a French emplacement in a piece of woods. Lieutenant Watson Washburn took it up. We wanted to keep an officer with each of these pieces, but we had too few. Instead, it was necessary to put them in charge of noncommissioned officers. But this was a good thing, for the results increased the confidence of the officers in their enlisted assistants.

Both of the aforementioned positions were shelled. Fenneviller got it nearly every day. It was the custom when the music started to take the men off to a flank and keep them there until the concert was over. Later, the Second Battalion put out a piece from Battery F.

Another phase of organization concerned the observatories. To be serviceable, they had to run according to a perfect system. Conduct of fire was only a short side of their usefulness. Rocket signals from the infantry were relayed through them. Scouts sat at the instruments all day watching for signs of enemy activity and for fleeting targets. Through careful watchfulness, the scouts often located enemy positions and observatories, which indicated (to a certain extent) the enemy's immediate intentions.

There was always an officer at each battalion observatory, and at the regimental one, which was far back behind in Neufmaisons. At each, battle maps were carefully marked and dead space and visibility maps were made, as well as elaborate panoramic sketches. Anything observed on the terrain could be reported by its coordinates.

Our organization was good, but the question of rocket signals disturbed it always. It seemed simple enough in the beginning; heaven knows why it wasn't always. We placed at each observatory a circle on which the limits of our sector were fixed. When a rocket went up, an indicator was turned so that it pointed to the burst. That showed us at once whether or not the rocket was intended for us. The rocket guard was always on duty.

There were very few rocket signals—one for each of the various barrages, one for short firing, another for gas, but among the higher officers there seemed to be a diversity of opinion as to which signal should indicate what. It gave the men on the frontline lots of fun guessing as to which signal to use in an emergency, and the men in the observatories an equal pleasure in gambling on what was wanted when a rocket appeared.

The system was altered frequently and that's where the confusion lay. One morning during a reconnaissance of the frontline, a captain of infantry asked our advice. He ran through a batch of orders and memoranda and then flung up his hands. "If I should need a normal barrage tonight," he said, "I honestly don't know what I ought to send up. Any one of three rockets might be right—or wrong."

Such a situation could not be tolerated, so our officers in liaison with the infantry did what they could, as both branches were equally anxious. There's enough danger in a rocket signal anyway, and that is no reflection on the doughboy. An inexperienced, noncommissioned officer with a small squad in an exposed and lonely place, when he becomes aware of danger or fancies it, wants help in a hurry. In his anxiety, he may send up the first thing that comes to hand or everything he's got. Or, in the dark he may easily mistake a rocket. The artillery must sense such mistakes. When signals are changed too frequently, it requires a clairvoyant.

A new order came down that settled the matter. There would be a rehearsal of the new signals that night. The telephone officers arranged a system of barrage calls by projector with the infantry. While the rehearsal was in progress, that and the telephone were the only means open to the infantry to cry for help. The Hun didn't catch on and attack. The rehearsal proceeded peacefully—it was rather like a pleasantly conservative display of fireworks. The telephone system was given every conceivable test. Runners were sent breathlessly from organization to organization, and to and from the infantry. Bicycle messengers tore along the dark roads. Everything worked. Towards midnight, we talked it all over and went to bed with a sense of security that we had hitherto lacked.

That's the way things go in war. Within an hour we were awakened by our first real emergency. And there was plenty of confusion that the night's display had not accounted for. It began when Lieutenant Colonel Stimson's telephone buzzed. The officer at the First Battalion observatory was on the line. A red rocket, he said, had just gone up from the infantry. He had repeated it to the batteries. A red rocket under the new system called for a barrage on the line of resistance. It was not, therefore, to be fired without confirmation by telephone. Yet those at the observatories were under the impression that they had been told to pass red rockets directly on to the batteries. Our line of resistance was full of men who were happily asleep.

It was one of those times when our switchboard was busier than the one on the stock exchange during a panic. Lieutenant Colonel Stimson got to work. He put in calls for all three batteries at once. He wanted infantry Battalion Headquarters, too, to find out what the emergency was, for certainly there had been a mistake in the rocket. Our officer with the infantry, for his part, wanted Stimson, too. The battery commanders had judgment. They also wanted Stimson, to find out why the red rocket had been relayed to them. Regimental Headquarters wanted the battalion commanders. The regimental observatory was in the same case. It was necessary to report to three outside stations that a barrage was to be fired.

In the meantime, while waiting for the battery commanders to respond, we listened, apprehensive, for the sound of our own guns firing that short and murderous barrage.

As I said, the battery commanders had judgment and they didn't respond to the signal. It appeared that our operators were good enough for the stock exchange on a panic day. They got the calls through and put the battery commanders on Lieutenant Colonel Stimson's wire as they came in, until he was talking to all three at once. That situation was saved. But what the deuce did the infantry mean by firing a red rocket? They wanted something and they wanted it in a hurry. It might mean anything from a small trench raid to the attack in force we always felt was a possibility in that thinly held sector.

A captain back in pleasant Neufmaisons evidently sprang at the worst. We had no time for him. Lieutenant Graham was on the telephone, calling from infantry Battalion Headquarters. They hadn't been able to get anything from their frontline. Infantry brigade took a hand. As far as we could understand, the infantry seemed to want the

"Negre" barrage—a barrage similar to our so-called "normal," but to the right.

The battery commanders had the receivers at their ears. Lieutenant Colonel Stimson called the single word "Negre." The guns spat, and their rapid fire filled the night woods with an evil, staccato, crashing sound. And although it has taken moments in the telling, that response came in an amazingly swift period after the red rocket had awakened us.

Infantry Brigade Headquarters took a hand again. They had had enough of the Negre; they wanted the normal. Two batteries were about through and finished before shifting. Another, a trifle behind, shifted nonchalantly in the midst of its firing.

The Second Battalion was in much the same case. Its central boiled, too. Neufmaisons informed it that the Huns were breaking through the center and cried for the "Chamois" barrage. One battery was ready to respond to the red rocket, but was stopped in time. Another fired the "Grand Bois," and then shifted to the Chamois.

"By gad!" the infantry said afterwards, "it was bully to hear those shells ripping over. Sounded efficient and safe somehow." We smiled in a superior fashion. Why had they sent up their red rocket anyway? No one ever found out. As far as we could learn, it spoiled the evening for everyone except the Germans. They seemed particularly peaceful that night.

While the firing continued, the details were armed with the few rifles available and runner relays were got out. But most of the men agreed with the infantry—it was worth staying awake to hear such a superior noise.

When quiet descended upon the woods, except for some distant firing, a call came through from Battery C for an ambulance and the surgeon. Three men had been struck by shell splinters. That was our only material damage. But the night's work disturbed us, for there was a vagueness about the whole proceeding. It intimated that the infantry was not in that close liaison with us, the type of liaison that we saw as necessary for success. And other sectors would offer nastier problems.

Only one unpleasant incident followed this affair: a charge of short firing against one of our batteries. It was not pressed because of the strain under which our under-officered brigade was working.

In view of the generally peaceful nature of the sector, sleep was surprisingly scarce in Lorraine. We tried to do everything at once. We felt that a multiplicity of endless conferences was necessary. A man needs a clear head, especially when he is new at the game, to figure complicated corrections for modern artillery.

Nor should it be forgotten that Paper Work had taken a new interest in us. We had foolishly imagined he would be left behind when it came to killing Huns. Absurd dream! He stalked into our midst with a new confidence and destroyed friendships, threatened reputations.

The morning report and the sick book were complicated by the fact that each organization had men in two or three places. The firing battery, for instance, was at the position. The drivers and extra cannoneers were at the echelon several miles away. Communication between the two was seldom good. A few men would be at the observatory, at a pirate piece, with the infantry, or on detail at Battalion Headquarters. Yet reports on these men must be consolidated and at Regimental Headquarters at the usual hour.

There were reams of extra paperwork. The war diary became a bogeyman. If, the men asked, they had to have anything of the sort, why not do away with all the other reports. For the war diary brought everything together—positions, men, animals, casualties, rations, forage, ammunition. At the front, where we had less time than we had ever had, repetition haunted us. The information on that little war diary form had to be collected from many sources, and the batteries had to have their figures together by five in the morning, for Battalion Headquarters wanted them by six, and Regimental Headquarters insisted on receiving them by seven. That meant somebody had to sit up nights completing the form, and usually it was the battery commander.

The figures didn't always come through on time. They couldn't understand that in Neufmaisons. One makes no excuse for these delays. Those at the front were engaged in the biggest and most dangerous war in history. It is incredible, perhaps, that they should have been more interested in hurting the Hun and sparing their own men than in compiling innumerable neat figures that scarcely changed from day to day. It took some harsh words from Neuf-maisons to bring them to their senses. Paper Work had to be fed, for Regimental Headquarters had many people whose only duty was to look after the thing. And Brigade was voracious, and Division was unappeasable.

Then there was an observatory report in code to go down at 5:30 a.m.; a munitions report at 6:00; another at 11:00; a third during the afternoon. There were firing reports and supplementary observatory reports, too. In spite of all this, we did manage to annoy the Hun at times, and after a while we got enough of a system to run the thing after a fashion.

Another ideal was shattered in Lorraine. At Souge they had told us that while supplies might be difficult to get there, at the front we would only have to telephone the echelon to have anything we needed brought up the same night. Our instructors had been at the front during a period of stabilized warfare with only a handful of Americans on whom our entire service of supplies had been concentrated. Conditions had altered by the time we got in: there were more Americans and warfare was no longer stabilized. Also, echelons were further back and roads were not as well protected as they had been. Actually, the materiel didn't exist to satisfy everybody. And yet we were absolutely dependent on equipment. We learned, therefore, to be economical, to improvise, to salvage.

Camouflage was one of our chief needs. We got enough flat tops to take care of the batteries, but we needed protection for ammunition dumps, wireless stations, the observatories, and the entrances to command posts and positions. We made a careful study of camouflage in Lorraine, and the experience we had there was invaluable in Champagne and the Argonne.

When we were left in complete possession, we found a number of fresh tracks that had to be covered up. The springs were danger points—water is heavy and men want to carry it by the shortest route. We covered such places with fresh-cut foliage and established penalties that kept us all in the desired ways. For larger work, such as entrances to positions, we used small trees to supplement our insufficient nets. The engineers helped all they could, but they had many organizations to look after. They gave us what materiel they had for our dugouts, which progressed day by day. We needed gas-proof curtains and got them somehow. A sly spirit developed here and there. A man who got much needed materiel usually went around with an expression that connoted: "Ask me no questions, and I'll tell you no lies." And one watched carefully what one had got.

While all this work of organization continued, we paid some attention to our more strictly military affairs. One does not recall the number of supporting barrages we figured for one purpose or another and never fired. It was splendid practice, but the futility of it depressed us. Things didn't always come off as one planned in Lorraine. The show for which Battery A had been rushed into the line never got beyond paper. And that wasn't the only instance.

No one in the regiment is likely to forget Sunday, July 28. We figured a box barrage for a raid that day. We were a good deal concerned about it, for we had been told it would be a daylight raid whose object was the bringing back of prisoners. Captain Barrett of the infantry would be in command of fifty men. It seemed a hazardous undertaking to us. We knew that the most accurate fire would be necessary. That noon we were informed that we would not fire—yet the raid would go on.

Between two and three o'clock we heard machine guns and the popping of grenades. The rest is history. Eighteen men, we were told, came back and just two of them were unhurt. Captain Barrett and the rest were killed or made prisoners. Evidently the secrecy that had eliminated the artillery had failed to mystify the Huns.

Many other raids were projected and died. There was, too, the usual crop of rumors. You would hear after nightfall that the Huns were going to attack before dawn and told to hold yourself in readiness. You sat up all night waiting for the first guns and, as a rule, nothing happened. Sometimes, as you waited sleepless, you almost wished for the real thing. Our officers in liaison with the infantry were lavish with these rumors. It is inevitable that the infantry should get its wind up, and one must take its imaginings as seriously as its facts.

False gas alarms were more annoying than anything else. You can't fool with a gas alarm, for discipline's sake, even though your judgment tells you the presence of gas at a given place at a given time is impossible.

You would hear far in the distance towards the frontline three rifle shots in quick succession. They would be repeated nearby. Steadily they would drift back, exigent, uncompromising, accompanied usually by the jarring screech of gas horns. Weary men would turn over and groan. Our own alarm would belch and you would struggle into your stifling respirator, giving up all idea of sleep until you got the all clear. It may as well be said now that although gas was infinitely plentiful, we weren't so conscientious as far as gas alarms. We had too much work to do and formed the habit of trusting our own noses.

After one of these alarms the whole world would seem to lie awake and ask for trouble. A screech owl would set a dozen alarms going. A runner would tear in from the infantry, gasping in his mask. He'd got a whiff of something on the road, he told us, and the wind was blowing in our direction.

The men at the echelon usually wore their masks in the alert position when they came up. That was proper protocol, and they had to put up with it for only a few hours at a time. These men had become strangers to us. We often envied the more comfortable conditions under which they lived. They appeared at the front only by night when they brought up rations and ammunition. No one coveted that side of their job, for registered roads don't make for contented travel.

The drivers announced their approach by shouts and a cracking of whips. The details rushed to the entrance of the position. The contents of the G. S. carts or the *fourgons* were unloaded and carried away with anxious haste. The drivers would chat with the cannoneers for awhile. Now and then a nearby battery would cut in and the horses would grow restless. Then the drivers would mount and rattle off again to the remote and desirable woods they inhabited. That's the way our rations came to us. Food, too, brought its new problems at the front.

A book might be written in praise of the army cook. His name, as everyone knows, is no stranger to the casualty lists. His devotion to his work was nearly fanatic. Others might falter or straggle by the road, but the cook clung to his rolling kitchen or his field range stove with pathetic devotion. And always, quite naturally I dare say, he craved to build fires. Flame became to him a sort of god and its resultant smoke was incense from an altar. The rest of us couldn't look at it that way, as smoke was as dangerous as the flash of our guns. For the enemy it was a banner advertising our positions.

As long as wood was dry, we could manage to keep the cooks at their devotions, and not without benefit to ourselves. But in damp, chilly weather the wet wood was too much for our experimental smoke screens. It was frequently necessary to scatter and extinguish fires while the cooks stood by with an air of witnessing a sacrilege. Fortunately, it didn't rain much in Lorraine, and we were sufficiently far back to make fires practicable most of the time. We weren't destined to fare so well again until the close of the war.

Nor did we dream we would be left in Lorraine for long. The fighting was taking a new turn, that destined to be its final phase. We had been rushed into the line. So, it developed presently, would we be rushed into the hottest battle of the war at the war's supreme strategic point. As the truth faced us more and more frankly, we reviewed our slight training, our mistakes on this front, and we asked ourselves if we were ready. The powers, meanwhile, were in no mood to consider such things meticulously. We were a regiment and we could shoot and so we were needed.

We had erected our wireless station on the hill above Battalion Headquarters, and from it the communiques slipped down to the command post, unofficially but vividly. Newspapers (a trifle stale) came up at night from the echelon, so after a fashion we kept in touch with the vast workshop of the Western Front. We could see there roughly the modeling of our immediate future.

We read of the Hun's last great offensive on the side of the Château-Thierry salient. We shrank from a repetition of the anxious days at Souge, when Paris had been menaced. That menace seemed to exist again and uglier than ever. But all at once the spirit of the news altered. Foch's brilliant counterattack was under way. And as green American troops had stood with smiling ease and confidence on the defensive against those vicious thrusts of May and July, so now they were tearing forward with the French, laughing and singing as they went, killing Huns, and dying with a courage superb and indifferent.

Château-Thierry came back to France and many smaller towns. The Huns were going out of the salient like water from a pressed bulb. Fère-en-Tardenois, their base of supplies, was threatened and had been entered by American troops. The allies stood in front of Fismes, then were in the city. Except for a few outposts, the enemy was between the Vesle and the Aisne rivers.

Rumors thickened into fact. We were to move almost at once. No one shirked the fact that we would probably be thrust into that vast, sanguinary, and decisive battle. But Battery B offered a complication. On July 15, Corporal Samuel W. Telling was sent to the field hospital in Baccarat and back to us drifted the dreaded word "typhus." The battery would be quarantined and the most involved sanitary precautions would be taken throughout the rest of the regiment. Except for its officers, Battery B was passed through the delousing station and placed in shelter tents in the woods near Baccarat. Yet the battery could not conceivably abandon its share in the missions assigned to the regiment. A detail of cannoneers, drivers, and telephone men were sent from each battery to Lieutenant Montgomery and, in his stride as you might say, he welded them together so that his work suffered no interruption.

At the echelon, however, things didn't go as well. There were very few officers there, and this influx of green drivers added much weight to their already great burden. When the regiment finally pulled out, some property was left, paperwork was involved, the colonel was annoyed, and there was a good deal of harsh language expressed. From a broader point of view, however, the meeting of this emergency by Battery B was an extraordinary accomplishment.

Around this time, the situation was relieved somewhat by the arrival of two officers fresh from Saumur: Second Lieutenant Charles F. Wemcken was assigned to us by order of July 10 and was sent to Battery C; Second Lieutenant Charles F. Perry was assigned by order of July 20 and was sent to Battery B. At the same time, Lieutenant Robinson was shifted from B to C, with whom he fought with pronounced success until the Armistice.

Another form of encouragement came in a telegram for Lieutenant Colonel Stimson. An extraordinary exception had been made in response to his plea. We would soon have Captains Reed, Ravenel, and Delanoy back. On the other hand, we lost First Lieutenant Watson Washburn, who was transferred to a staff job at Army Corps Headquarters, and First Lieutenant Paul Pennoyer, who, while on a temporary mission from Souge, had been given a Army Corps staff job, too.

The Hun had probably gotten hold of some of our rumors. At least he was extremely attentive during our last days in Lorraine. Nearly every night now we got some kind of an alarm from the infantry, and we retaliated by planning many *coups de main*, ordered by Infantry Brigade Headquarters, few of which materialized.

One morning towards the end we were awakened by a heavy bombardment. Shells were bursting close to the First Battalion command post. Either the Hun was registering to transport directly on us or he was after Nénette. Lieutenant Brassell was at Nénette with Corporals Tucker and Goldberg and Private Braun. Lieutenant Brassell telephoned down while we snatched a bite of breakfast and, to all appearances, dismissed our uncertainty. "I think they're bracketing Nénette, sir," he said plainly.

We settled our tin hats on our heads and climbed the hill. The arriving swish of the shells and the noisy bursts were not comfortable. With each burst close at hand, little volcanoes of jet black smoke sprang out of the pretty wheat field. Thirty odd projectiles fell over and short of Nénette and to either side. There the show ended, for Nénette had not been touched. We tried to assure ourselves that all we had got were "overs"—fire intended for an anti-aircraft battery near the Pexonne Road. Yet, Nénette was always an anxious place after that, and we held ourselves ready at the first alarm to shift to the alternative observatory among the birch tops. And, we endeavored again to find other points suitable for observatories near the frontline.

During one reconnaissance, Lieutenant Colonel Stimson came upon an observatory unique in conception and treatment. It is doubtful if the war produced anything of the sort more admirable.

We were on a defiladed road immediately behind the infantry frontline. To the right was a hill, thick with tall pine trees. A fundamental protection of the place was its patent antagonism to terrestrial observation, as you can't observe through pine trunks and heavy foliage. But the French had got around that. They had gone above the foliage but didn't use the common expedient—a platform in the trees —which sooner or later gives itself away. They had instead constructed a huge new tree. They had a tower raised from similar trunks and covered it with the same foliage. You could stand within a few feet of it and remain unsuspicious of its existence. You had to climb many ladders to the observatory at the very top, and once there you had a sense of Peter Pan come true. You swayed in the breeze and you looked almost directly down into the Hun lines.

The infantry was in possession and we went back to Nénette and our poor makeshift post. Then, on July 30, French officers appeared at the command posts and informed us they were going to take over, beginning the next night. These men had just come out of the great battle. We, suspecting an immediate entrance into the battle they had left, listened breathlessly to their talk of unheard of artillery concentrations, of long casualty lists, and of a supreme exhaustion.

"*Formidable!*" was their favorite word. "You've never dreamed of the noise and the effect of their barrages," they said, "*Formidable!*"

One of the French officers glanced about our pleasant woods. He sighed contentedly, and said, "It is tranquil here. A sector for un père de famille." In spite of ourselves there was a little envy in our hearts.

Certainly the bosche guessed something was going on. We had known all along that he had control of the air in Lorraine. His planes were constantly overhead and the bell-like note of the Archies (our name for anti-aircraft gunfire) was with us much of the day and night. There were nearly always white clouds in the air (unrelated to the weather). Still Jerry had not been very aggressive. French planes had been up and given us one or two réglages undisturbed. On the night of July 30, however, the Huns came over in force loaded with bombs. Unquestionably, they fancied the relief was under way that night.

A huge ash can dropped down beside the Battery F position. The force of the detonation knocked Lieutenant Derby down and spattered a dozen men with dirt and twigs. By incomprehensible good fortune, the hot, ugly pieces of metal touched no one.

Another big one landed in the field back of Nénette and sprinkled fragments all around the observatory. That was near enough to the command post to make it advisable to get the men in the dugouts. Then the planes turned and went back to Germany, sprinkling their foul droppings as they went. We escaped, but there were casualties close by in Ker Arvor woods.

The next day our formal orders arrived. Two pieces from each battery, except B (whose position would not be taken over by the French), would be relieved that night. The whole of B and the remaining guns of the other batteries would go out on the night of August 1–2. The first guns out would go to the echelon and wait there until the next night, when they would join the last guns that will proceed without stopping at the echelon to the division regrouping area two marches away. The Huns were evidently satisfied with what they had done a night too soon. The relief was undisturbed.

Battery B again presented a special problem. Since its position was not to be taken over by the French it was necessary that its plant be kept intact until the last moment. Yet it could not delay to the point of losing its place in the column. There were miles of wire to salvage and much equipment to be packed at the very last. Lieutenant Montgomery managed it and pulled out on time.

Lieutenants Camp and Fenn remained behind with the two French groups for twenty-four hours to induct them into the mysteries of the sector. The French weren't exigent, so half a morning served to organize them completely. Again, one was forced to admire the way they achieved the completest results with a minimum of effort.

The first night of the relief Lieutenant Colonel Stimson left the regiment and would never return to it. A telegraphic order had reached him that afternoon, instructing him to report to America for duty with the Field Artillery there. We watched him drive away that night with a sense of grave loss. Afterwards, we heard that he had been made a full colonel and given command of a new regiment in training at home. The Armistice came before that regiment would sail.

Chapter Thirteen

THE FIRES BEYOND CHÂTEAU-THIERRY

Our movement from the Baccarat positions was not as simple as we had expected. For its entire length, the road was perfectly visible to Hun airmen, so it was advisable to march at night. The column was late starting and it crawled, as such columns do, on traffic-laden roads. Our schedule called for a bivouac at Magnières during the daytime on August 2, but it was long after daylight when the regiment arrived, anxiously glancing aloft. By the time horses and men were settled the hour of departure was at hand.

Again the roads were packed and progress was snail-like. It was nearly noon on August 3, before the dusty and tired column entered its regrouping area on the Moselle. We hadn't imagined the movement of a single division could be so complicated and tedious. The march, however, was not without its valuable impressions. For the most part it lay through the district of Lorraine, which had been destroyed by the Germans during their retreat after the Battle of Grand Couronné de Nancy, the eastern phase of the Battle of the Marne. The smashed villages were now sketchily inhabited and the fields were under cultivation again, but around this resurrection still clung an appearance and an odor of death.

Our own area was just beyond high tide of the Huns. For us, after that journey it was impressively undisturbed and peaceful. We felt that our ugly carriages parked in fields along the Moselle were out of place in such a landscape.

Regimental Headquarters and the Second Battalion were at Bainville, and the First Battalion was at Mangonville, two kilometers to the south. The Headquarters and Supply Companies were in and around the charming Château de Leménil-Mitry, three kilometers to the east of Bainville. Significant changes were announced here. Among them was the transfer of General Rees, who had commanded the brigade, to other duties and the appointment of Colonel Manus McClosky, soon to be made a brigadier general, to replace him. For a few days Colonel Doyle, as senior colonel remaining with the brigade, was in command. There was a feeling in the air that the changes wouldn't stop there.

Captain Dana, of course, was again in temporary command of the First Battalion. Captain Reed reported back from Souge on the first day and took over his duties of battalion adjutant, while Lieutenant Klots went back to the Headquarters Company. These two officers set to work with determination, to get the battalion ready for the serious work just ahead. Meanwhile, Captain Mitchell was transferred from Battery F to the Field and Staff as adjutant of the Second Battalion and Lieutenant Derby took command of Battery F.

We had expected two days in this regrouping area, but it stretched into four and no one was sorry for the delay. It was pleasant there and we had a great deal to do. We settled down to straightening out the tangled paperwork situation and completed the divorce of the three details from the Headquarters Company. Men, animals, and equipment were reported to Regimental and Battalion Headquarters and were assigned to organizations for travel and rations. The Battery B men, released at last from quarantine, reported back.

We were ready when the order came to march on the mornings of August 6 and 7. Regimental Headquarters, the remnant of the Headquarters Company, and the Second Battalion proceeded to Charmes on August 6, where they entrained. The First Battalion and the Supply Company entrained at Einvaux on August 7.

This movement was unlike the one from Souge, when a brigade had a week to entrain. Now, from a small section, an entire division was going out practically in a single day. While there were a number of points of departure, the congestion at each was such that a careful schedule had to be made and followed. Therefore, each battery broke park and took the road at a stated moment and arrived at its entraining point at a given time. It fed and watered according to the clock. We passed large parties of our doughboys maneuvering in the fields while they waited their turn at the trains. They interested us and we intrigued them. Their glances followed the long, overladen column from which the sleek snouts of the pieces—that escaped from burdens of forage and equipment—peered at them encouragingly.

The Supply Company was off first. Battery A commenced entraining at two o'clock and was completely loaded at 3:30. Before the train had pulled out, the head of Battery B was on the ramp. Before B had gone, C appeared and was ready to load.

At Charmes there was a similar precision of movement. We were surprised to learn how much we had profited by our one previous experience. The drivers made short work of refractory animals and the carriages seemed to roll into their places on the flats automatically.

These days were warm and working at such speed makes men thirsty. There was a little Y. M. C. A. hut on the ramp, and when the job was complete, the men were allowed to line up for a glass of raspberry syrup and water, and a limited quantity of chocolate, cakes, and tobacco.

Not until the trains had left did anyone know the projected destination of the regiment—that is, we had moved under sealed orders. Before the departure of his train, each battery commander had received an envelope with a typewritten command that it was not to be opened until he had passed a certain station. Inside each envelope was a rough engineer's map of the district north of Paris—a map covered with significant names—and a small typewritten slip of paper that said: "You will detrain at Nanteuil-le-Haudouin." We spotted it eagerly on our maps; its location indicated to us that we might either go in with the British, or swing more to the east through Soissons. There was another possibility: Were we going to lie in reserve behind the lines? Didn't the powers think us good enough for the big show, except in an emergency? Whatever the original intention, it was altered the next day, as everyone remembers.

Except for the customary struggles with a few unruly animals, the trip was tame enough, but there were plenty of reminders the next morning that we were close behind the busiest portion of the front. We saw many spreading nets of tracks, crowded with flat cars on which reposed battle-scarred cannon, camouflaged tanks, trucks, and automobiles. On everything could be seen the deep wounds of shell fire. We passed huge gun parks, ammunition dumps, airdromes, and dreary and interminable hospitals.

We gazed at such sights with a depressed interest and wondered if we would crawl through the outskirts of Paris. Then the trains halted shortly after noon at a small station, and an officer climbed aboard each one, presenting the commander with a new envelope. As soon as we saw it, everyone guessed that our destination had been altered.

There was a map inside—a 1–80,000, marked Meaux—and there was an order, brief and to the point. The division would detrain at Coulommiers and nearby stations and, on August 10, would commence a movement forward into the zone occupied by the First United States Army Corps. The infantry would be moved by motor buses to be furnished by the French; the artillery would go on its own wheels and legs.

Within half an hour after receiving that order the batteries were detraining. Battery A detrained in the yards at Coulommiers; Batteries B, C, and F at Chailly-Boissy; Batteries D and E, and the Headquarters and Supply Companies, at Saint-Siméon.

At that time, there were huge evacuation hospitals at Coulommiers through which thousands of Americans, gassed or wounded in the Château-Thierry salient, were passing. We listened, fascinated, to the gossip of hospital orderlies about the effects of big shell fire and concentrated phosgene and mustard gas.

The sky was full of airplanes constantly circling overhead. We tried to impress on each other that, although they were our own planes, discipline must be maintained as if we were at the front. As a consequence, the bugle blared alarm after alarm and our work was delayed. Still, we were willing that it should be, for in our ignorance we believed this great flock of airmen behind Château-Thierry meant that we had control of the air and that, therefore, our offensive and defensive dispositions would be made simpler and safer on this nasty front.

Battery A was billeted for that night and the next at the comfortable little village of Saint-Germain-sous-Doue. Battery B went to La Loge Farm, Battery C to Épieds. Regimental Headquarters, the Headquarters and Supply Companies, and the Second Battalion were at the comparatively large town of Doue.

Things had been fairly hectic on the Moselle, but that wasn't exceptional. Going into billets for a battery is always much the same problem, much the same mad struggle for a solution. And, when it's reached, the solution is always about the same. Yet invariably out of the confusion emerges a sort of order and comfort. Eventually, we became more than ever like a great traveling circus whose discipline automatically repairs the mistakes of a poor advance man. And that isn't intended as any reflection on our billeting officers and noncommissioned officers. They, as a rule, had too much to do, and they were restricted to too small an area by the advance agents of the division.

Some towns had a better welcome for infantry than for artillery, but that fact didn't seem always to be appreciated. Besides billets for officers and men, the artillery needed ground suitable for extensive picket lines and gun parks, and no matter how suitable the ground you couldn't establish either near the front without overhead cover.

Organizations, whether they arrived in the afternoon or during the dark hours, ran into much the same conditions in those billets north of Coulommiers. The billeting officers couldn't be all over the district so the noncommissioned aides, as a rule, faced the battery commanders alone. One always experienced a quick sympathy for these unfortunates. Invariably, they glowed with a naive pride. They always produced careful lists, showing the billets available and the number of men that could be housed in each. A battery commander going into billets, however, is only interested in two things initially: "The picket line and the gun park!" he cries as he meets his man.

Perhaps the glowing advance agent has let the battery slip past these vital points, and it may be necessary to turn the entire column around by sections in a narrow street. Battery commanders never take kindly to that, nor do tired drivers. It is seldom that the places chosen for the picket line and the gun park please commanders. The ground is swampy, or there isn't enough room, or the tree trunks are too small, or . . . Even the most indifferent commander can find something lacking in the most perfect park or line. The advance agent, of course, isn't to blame for these shortcomings. The town major, as a rule, has given him no choice. "That's it," he typically says. "Take it or leave it."

But a battery commander doesn't analyze causes when he is displeased by effects. He decides darkly to make the best of things and he considers his disappointed advance man. "All right," he says. "The thing's impossible. You ought to have done better than this, Smith, but it's getting dark. We'll make it do. Undoubtedly you've arranged to billet the drivers in a group close to the picket line and the cannoneers by the park. Explain your distribution to the first sergeant."

If the advance agent is a man of parts, he salutes, seeks the first sergeant, curses, and with him arranges some kind of a compromise. If, on the other hand, he flushes and stammers forward with facts about some of the billets being large and some small, and everything is scattered through the town and the surrounding country, he usually pushes the battery commander's patience too far. Then he sees himself as others see him.

As soon as the animals are arranged for, and he is certain his men will have some kind of a lodging, the battery commander turns his attention to the kitchen. The site of this, too, is more often than not an arbitrary selection made by the town's major, and it is nearly always in a farmyard, redolent and wet with manure, and thronged by an assortment of unclean animals.

It is at this point that the billeting noncommissioned officer generally goes back to his section a sadder and a wiser man. He mutters over his lists and his wrongs and tells everybody that he has done just what he was told to do by those above him in rank. He has, probably, but no one is sympathetic. The customary response of his friends is: "Can it, will yeh? It's a heluva billet yeh gave me."

That billeting detail became more unpopular than kitchen police, for it reduced corporals to the ranks and made officers lose faith in their men, and men in themselves. You see, we usually billet at the close of an exhausting day, when everything about us is strange, the black of night covers the world, and the whole battery is hungry.

No meal tastes as good as that first one in billets. You sit around on the grass or on a stone wall, eating with the comfortable assurance that for a few hours no violent effort will be demanded of you, that in a little while you will probably be able to go to sleep. Or, if it is in colder weather, maybe you carry your dinner or your supper to your new home, where a hospitable housewife gives you a corner by the fire and maybe crowns the Corn Willy with fried potatoes or a piece of cake.

Afterwards everyone, except for the guard and the few necessary details, seeks his bed. You climb a ladder into the loft of a house or a barn that is centuries old. You find straw there, sometimes clean, sometimes not, but always soft. You drop off to sleep with a healthy and abrupt unconcern scarcely known to civilian life. But there was enough going on in the Coulommiers area to keep any but the most-weary awake. As we strolled to bed that first night, we watched in the sky to the north vivid and endless flashes spreading and contracting with a variety of intensities. Someone chuckled self-consciously, and exclaimed, "Reckon the world has never seen such northern lights before." Judging by our experience of flashes in Lorraine, it was clear that the highest battle—even according to the standard of this war—was raging up there. In a very few days, we would be among those flashes.

As we watched that pallid, violent display we strained our ears for appropriate sounds. But the night was silent, except for a distant and amorous song and the rhythmical music of a breeze across the foliage. Then the song vibrated away, the breeze fell, and all night long, before that distorted sky, the silence was ironic.

Chapter Fourteen

ACROSS THE MARNE TO FORÊT DE NESLES

More detailed orders reached us the next day. We would take the road Saturday, on [August] 10, and march thirty odd kilometers before the next morning to Chézy-sur- Marne. The next night we would cover approximately twenty kilometers to a point to be chosen near Courpoil. The third night we would complete our journey to Forêt de Nesles, which had recently been cleared of the enemy.

We pored over our maps. The march would be a forced one, carrying us through the heart of the salient. Chézy was only a few miles from Château-Thierry. Courpoil probably smoked from the fierce fighting it had witnessed. Forêt de Nesles lay between Fère-en-Tardenois and Fismes.

We spent Friday getting ready; in our spare moments we wrote letters home. That afternoon we were summoned to Brigade Headquarters in Doue to meet the new brigade commander. He intimated the serious nature of our next step. Afterwards, Colonel Doyle gave the organization commanders an extended talk about aiming points and the identification of targets.

Since it was understood we couldn't safely start our march before four o'clock the next afternoon, everyone hoped for a good night of sleep Friday. The men needed it, but they didn't get it. About nine o'clock, Regimental Headquarters stirred itself and began sending orders to the battalions by bicycle messengers. The first order demanded that we be prepared to take the road by eight o'clock the next morning, which meant reveille around four o'clock. Other orders came to send teams and G. S. carts to various points to change and move equipment. It wasn't until two o'clock on Saturday morning that the excitement subsided. Bicycle messenger Montgomery came around then with a verbal order that we wouldn't move until the time we had been given originally: four o'clock in the afternoon. We took advantage, then, of what remained of the mutilated night.

The regiment was to rendezvous at Doue. It would take its place in the brigade column on the national highway beyond. So, at four o'clock, each organization mounted and pulled out of its comfortable billets. August smiled its best on us that afternoon. The cheerful countryside seemed reluctant to let us go. Natives watched us with emotionless faces; in their eyes, we saw the dull souvenirs of four years of departures.

In the old days of pitched battles, men walked from their bivouac directly into the obliterating shock of a fight that lasted a matter of hours. Maybe that was simpler than to move as we did, over three nights, into a battle apparently without end, with sights and sounds of a new and peculiar brutality crowding closer about us each moment.

We did get tired.

During our wait at the rendezvous, we drank hot coffee and munched cold rations. When we turned into the straight national highway, flanked by huge lime trees, we could see the entire brigade stretching before and behind us. French and American trucks snorted past without end.

The pleasant, warm sun sank lower. By twilight, on the outskirts of a town, we watched youths of the French 1920 class, freshened after their day's training, walking in groups and watching our dusty column out of curious eyes. Here and there, one strolled by the side of a pretty girl, shyly silent because of this undesired publicity. They waved hesitant farewells. In the village, little children shrieked after us: "Goodbye! Goodbye! Goodbye!"

The sun slipped away altogether, night closed around us, and by the last light we twisted through Épieds. The people gave us feeble cheers, but we paid little attention. We were already footsore, and even the mounted men, to save the animals, walked alternate hours. Our halts, because of the length of the column, had become extremely sketchy. Sometimes we missed them altogether, closing up a gap. And there were innumerable unexpected stops, when we dismounted, and were up and off again almost before our feet had touched the ground. Indeed, our feet weren't up too much. During the past month, we had been either in the line or changing station. We were soft. But songs brightened our worm-like progress along the dark, country roads.

The nighttime brought back the flashes to the sky ahead of us— they were not quite so pallid, they spread farther, they soared higher, they were streaked by ominous lines of ruddier flame. Always the traffic of the supply detail ground past us, forcing us to the side of the road, struggling desperately forward to feed the fires. A cheery voice flashed bravely back at the burning sky: "Gonna be some little fight, boys!" Another voice arose with a quavering, melancholy quality. Its song was something about a girl waiting at home, patiently and unselfishly, for a man to come back out of the fires. The ranks fell silent; the voice died away.

Somewhere ahead a rolling kitchen began to drop a trail of sparks. It wound, as the road twisted, like an unbelievably long and phosphorescent serpent. It kept pace with us. After a time, the odor of coffee floated back along the trail. Between midnight and the dawn, we knew there was a jewel of a cook up there. But was it safe, this red, serpentine trail? Are cooks ever safe near the front? Everybody saw the sparks and everybody caught the aroma. The fires were still distant. Nobody disturbed the cook, and the red serpent persisted until it was certain the coffee in the containers was hot and would stay so.

We drank our coffee between one and two o'clock, when we had halted on a high ridge. The flames seemed nearer and brighter than they had before. Or perhaps it was because the night was so dark up there. Then, for the first time, we distinguished star shells. They separated themselves from the flashes so slowly and disappeared so reluctantly that you couldn't be sure at first they weren't born of your imagination and your smarting eyes. You thought the first one, perhaps, was the Pleiades but less distinct than usual. They all looked exactly like that—tiny constellations blurred by the shifting glow ahead. But they were everywhere . . . so you knew what they were.

As we munched on a sandwich or a cracker and sipped the hot, fragrant coffee, everything impressed us as abnormally still. We missed the rumbling of the wheels, just silenced, and the rap of the horses' shoes on the road. In the beginning, there was only the slow shuffling of feet in the dirt as the forms—detached a trifle from the night by the flickering in the sky—formed a line by the rolling kitchen. And there was the occasional dull clashing of mess cups. Then a man spoke, and after a time another. It was usually only some banal remark, drowned and forgotten at once in this flickering still-ness. "You're spilling it on my wrist," says one soldier. "God bless you for the chow, sergeant," says another. Or, from the sergeant: "Move on! Do you want to delouse yourself in it?" Such aimless accents of the silence were forgotten at once.

Out of the subsequent, pallid calm stole the voice of battle and the men shifted their feet uneasily. It was the first of the cannon mutter to reach us from the flames. A quick activity thrust it back again. "Prepare to mount!" shouted a battalion commander. And then, "Mount! Forward yo-o-o-o . . ." These orders came down the line, growing apparently out of nothing as the cannon mutter had done, reaching a climax in one's own mouth, dying away on the long drawn vowel of the last command. We were moving forward again, drawing an odd and comfortable companionship from our rumbling, rapping progress.

At the scarcely perceptible birth of dawn, we were winding sleepily on the shoulder of another ridge that looked down on what might have been a long lake or a deep and gigantic river flowing between the hills. It was possible to guess, and here and there a man raised his head and stared. Someone spoke in a harsh whisper: "That's the valley of the Marne." (He whispered because we were somnolent and not alert.) The name he uttered possessed no dynamic power for us then. However, one fellow did manage to respond, "Didn't know it was so blamed wide." The first fellow offered to instruct him: "Oh, that's the mist." To which the second fellow exclaimed, "You don't say? Good-night!"

"Ha . . . a . . . a . . . lt!" sang the command down the line, like a savage chant. The regiment dismounted and, one by one, men dropped over against the bank and drifted into sleep, keeping a listless hand on bridles. The horses, though weary too, for the most part stood with drooped heads, not even troubling to nibble the lush grass. Now and then one would wander indifferently away from the feeble, restraining grasp of his master. An officer would rebuke the soldier sleepily, consigning the careless one to walk the rest of that stage. At such a time the world seemed drunk with sleep. A dim headlight pushed through the mist below—guiding one of the first trains, we guessed, to carry troops along the reopened Château-Thierry line.

As the dawn strengthened, it grasped the fringes of the mist and lifted it slowly from the valley. A stream like a ribbon, narrow and decorative, was strung across the fields. Tired eyes opened to gaze, with an expression of discovery, at the pleasant little river that twice had been wider than the ocean to Germany.

We resumed our crawling. There was no longer any reason in mounting and dismounting. We would go ahead for a few paces, then stop again. An anxiety grasped the command to get us somewhere beneath green trees before the light should grow much stronger. Then we saw the head of the column moving to the left where it was swallowed up by a large grove of trees. A sigh went up; we were nearly there. Each halt seemed longer than it actually was. We glanced upwards, we listened for airplanes. Then we, too, reached the fork and turned and entered beneath the friendly shrubbery. The chill of the night had disappeared before the mounting sun.

As we parked, an officer from headquarters ran about, whispering: "Keep everything covered up, and don't let anybody stand in open places. The Huns are watching these woods for bivouacs." Consequently, we draped the carriages parked in thinly roofed places with cut shrubbery. Then, we started the animals down a path behind a guide who knew where the water was, got our paperwork out of the way, and hurried the war diary to Regimental Headquarters, which had been established in the deserted town of Chézy-sur-Marne.

The rolling kitchens smoked; men forgot their weariness to form eager lines before them. Groups ate greedily among the trees—the forest was noisy with their talk. The colonel arrived and approached a party of officers on a tarpaulin, making a stupendous breakfast in celebration of having brought men, animals, and carriages through a stage that had worried everyone.

"Keep your seats, gentlemen," the colonel said. "I want to congratulate you on the way you handled your paperwork this morning." The group returned to its breakfast refreshed, for a word of praise after such effort was a tonic.

The illusion of a picnic, however, was never very convincing. The sunlight searched the woods and exposed souvenirs of the recent fighting. Half hidden by the underbrush were stained and eloquent garments: here lay a Hun helmet, a neat round hole through the front, and over there was the stock of a rifle. Men picked up such objects curiously and gathered them in little heaps, convenient for transportation.

The men prepared for sleep, but the sun seemed to laugh at this. That is the curse of night marches—you can't get a satisfactory sleep by day. There is a great deal to be done, robbing many men of the opportunity to sleep: guards must be posted; kitchens must go as hard as ever; animals must be more carefully cared for than when in billets or at an echelon; equipment must be cleaned; and the damage of the previous night's march repaired. All these operations manufacture a noise that disturbs those who do get a chance to rest. But it is the sun that irritates the weary more than anything else. No matter how shady the place you choose, the sun will find it out sooner or later, will grin in your eyes, will inform you that it is no time to be sleeping. Maybe you move to a better spot, but then a man shakes your shoulder, demanding information that he foolishly imagines you alone can furnish. If on such a march you can average three hours of sleep out of the twenty-four, you are lucky and inured to noise, disruptions, and the sun.

In Chézy woods there were other disturbing factors. The men's feet had suffered greatly, so it was necessary to treat them. But you had to stand in line for long periods waiting to get to the doctor. By the time you were treated it was probably dinnertime.

After dinner, nearly everyone that wasn't on duty strolled down the hill, through the grounds of a modern château, to the bank of the Marne. The water was dirty and, if one stopped to think, sinister. The afternoon, on the other hand, was warm and we didn't forecast many more opportunities in the near future to bathe. So, we filled the murky water with active, noisy bodies. On the shore, mature men mimicked the antics of schoolboys. Across the Marne frowned a land-scape stifled beneath the pestilential haze of war—a condition scarcely palpable, reminiscent of a land whose inhabitants have been swept down by some black plague. There weren't many ruins, but pervading everything—fields, farm houses, villages—was this sense of desertion and a morbid unhealthiness. It was like a picture from an artist whose melancholy and diseased brain has retained of the visible world no more than a sense of form.

All afternoon the activity about the banks mocked this oppressive landscape. From time to time, strings of animals were led down and watered. The antics of the bathers continued until dusk.

A few of our horses did not respond that day. We were already underhorsed. A new fair started: organizations swopped animals about so that no carriage should be left. That took time, so supper was a shadowy affair. We also policed the bivouac and lashed equip-ment to the carriages again, while souvenir hunters gazed at their stacks of trophies, shook their heads, and scattered the stuff about the woods. One man picked up a Hun helmet and thoughtfully beat it against a tree. "Seems tough enough," he mused, ". . . too darned tough." He flung it on the ground, thrust his hands in his pockets, and leaned against a tree. His attitude was, roughly, typical of the soldiers. The teams were harnessed, everything was ready, and we just waited for the word to move out.

The dusk had forced an unwelcome alteration of the woods. Instead of patches of sunlight, the grim souvenirs of battle scattered about determined the values of the picture. There was a chill in the air, too. One's sense of sleeplessness returned with the night. And the increasing darkness meant the resumption of those breathless pyrotechnics in the northern sky.

A little fellow, crouched on a stump with his hands clasped about his knees, gazed straight ahead. His face was immobile. You didn't like to look at it, because it seemed an expression of many more carefully guarded minds. You moved about, trying to throw the feeling off, this difficult conviction that the forest was crowded with homesickness. A man strolled up and put his arm around the little fellow's shoulders. With an understanding voice, he asked, "What's the matter, buddy?"

The little fellow sprang upright, like an animal startled by the appearance of a hunter, and answered fiercely: "Matter! Nothing the matter." He then burst into odd oaths, as if they might justify him. The other fellow gave him a cigarette.

Word came around that we were to be careful where we sat down that night, for there would always be the danger of mustard gas. Other messengers appeared with further updates. We were to cross the Marne on pontoon bridges at Château-Thierry. Carriages would cross on one bridge, with intervals of fifty meters, and individually mounted men would use another bridge, and dismounted men a third. An apprehension of shelling at the crossing from long-range guns saturated these orders.

With the last light, the word came to mount. Whips cracked and horses strained forward, as the carriages reeled drunkenly over roots and depressions. There were swaying escapes while men shouted warnings, put their shoulders to the wheels, and struck at the horses.

Where the woods trail turned into the main road, an officer sat upon his horse, repeating over and over again, like one reciting a piece: "Men may smoke, but must use automatic lighters. No matches will be struck tonight."

Brakes set, we slid down a long, curving hill into the valley. The column moved faster this evening. A soft moonlight gave an air of mystery to the few empty farmhouses we passed. Several groups of these suggested that we had reached the outskirts of Château-Thierry, but the road was longer than we thought. It was nearly midnight when we entered the city at last. Through a dark silence we became aware of multiple activity: the streets were full of half-seen figures that passed by us without words. The place might have been a rendezvous of criminals furtively intent on avoiding discovery. There were no lights, so we could scarcely distinguish the jagged remains of walls and buildings lined with fissures—which we knew were the graves of homes.

At the railroad, the column was cut by the passing of a train and the over anxious military police who closed the gates too soon. Beyond, teams tore through the dark to catch up and men rode back and forth keeping in touch with divided units. In a narrow street close to the river more military police were stationed. Their suppressed voices were scarcely audible above the rumbling of the wheels on cobblestones, as they repeated our instructions for crossing. Certainly the Hun wasn't so near!

We entered a wide place through the center of which the Marne flowed. More military police stood on each bridge nervously hurrying the crossing. But no shells fell. Our own progress on the planking drowned out the sound of guns and the hill ahead was a curtain against the northern sky. But once we were over, and had climbed the hill above the town, the voice and the gestures of battle became eloquent again. The passage of the river seemed to have brought us much, much closer. The sky was a wavering sheet of flame, no longer wan. It spread and contracted with a yellow intensity. Star shells stood out against it clearly enough, now. As the rumblings increased and diminished, one could almost guess the caliber of the guns engaged. An enormous mass of artillery was concentrated up there. It was folly to try to sing against that greater song. The column forged stolidly ahead.

We were in the heart of the salient now. Even by night the countryside and villages appeared haggard. The Hun's departure had been a matter of a few days, and he had not neglected his reputation in leaving. As we rode silently through village after village, we saw that they all shared a dreadful similarity: they consisted of clusters of homes—roofless and with gashed walls—that were filled with an odor that made the air reluctant in one's lungs. It was a compound of stale gas, of lime, of ancient plaster and woodwork, suddenly crumbled. It forced on us an impression of death, still warm; it suggested the proximity of departing souls. There seemed to be a connection between this sense and the ghastly light that flickered over everything. Between these dead villages the open country stank, too. At times we were sheltered by shell screens, raised by the Hun for his own safety.

Towards morning we munched on sandwiches and crackers, but there was no hot coffee. The fires in all rolling kitchens had been ordered drawn. Shortly after this meal we turned to the right at Courpoil—yet another slaughtered, empty, stinking town—and onto a rough road ascending a long hill. The halts became numerous and irritating and the road seemed interminable. In spite of the brief stage and our earlier speed, daylight would probably catch us again and the risk was greater here. And yet a little daylight might have offered a safeguard against this road, which degenerated with each meter. The *fourgons* and escort wagons lurched dangerously on it. Why the deuce, we wondered, were we struggling so far from the main road anyway? We'd have to come back by dark again over this risky trail. And our horses were tired. The only excuse that occurred to us was that we were going to a particularly safe and convenient bivouac.

As the east grew ruddy, the flashes faded. In the light of dawn, we saw a *fourgon* on its side by the road. The horses stood by gazing at it with rather a pleased air, while tired soldiers made unavailing efforts to get it up. "No sleep for those guys," we said pityingly. "They'll have to unpack everything, jack her up, and pack again." We decided it must be a peach of a bivouac that we were going to—it wasn't.

Just ahead two large masses of forest barely detached themselves from the slow dawn. There was an open field between. Some of the batteries were already strung out along the edge of the woods. The rest of the column halted, and a group of officers and men stood in the field, talking and gesticulating. One heard: "Who made the reconnaissance for this blasted thing?" There had been a reconnaissance the previous day, but something certainly had gone wrong.

We eagerly asked each other questions. The woods, in spite of their size, were for the most part choked with underbrush and the remainder was rough and honey-combed with infantry trenches. There wasn't room for the regiment to be under cover, and the Hun's planes might appear at any moment.

"And those woods," someone in the group of officers said plainly, "are full of dead things." Without calling attention to it, we had all noticed the thickening of the nauseating odor of wholesale animal decay. "It's bad for the men," he continued. Another officer chimed in, "The men have got to get used to it." To which the first officer said, "But it's better to see those things in the heat of action."

That, however, wasn't the point. We had to get covered up before the light grew stronger. The Headquarters Company and Regimental Headquarters got sketchily concealed in one piece of woods. The larger part of the Second Battalion got in the other. The First parked its pieces on the edge and cut foliage with which it covered everything. Opposite, the Supply Company employed the same makeshift.

The picket lines had to be placed inside. Those who entered the forest to locate these lines went softly, for it was still night in there. You didn't want to stumble over unseen obstacles. You imagined that the woods were still inhabited by an army, which for the moment slept. The trenches made angular scars between the trees—shallow, makeshift defenses of the retreating Hun. Their floors were littered with gray blouses, helmets, round Hun caps, Mausers, grenades, and belts of cartridges. Scattered between them were artillery ammunition dumps, the shells in wicker containers like wine baskets, or else in elaborate and expensive metal frames. As the light strengthened, we saw quantities of rations that had been thrown away, gasoline tanks, pioneer tools. If there wasn't an army in the woods, there was certainly the equipment for one. That day if we wanted anything— gasoline, for example, for an automobile or a sidecar—we went through no formalities. "Go in the woods and get it," we said. And the seeker obeyed and got what he wanted. But in there the odor was poisonous; everyone was warned not to prowl in the underbrush.

As soon as the picket lines were established, we went out, clinging to the edge of the woods, and almost at once the first Hun planes came over, but we were pretty well concealed, and they didn't trouble us.

The question of water became a primary concern. By taking the water carts all the way down the hill, water for the men could be drawn from a well in Courpoil. For the animals, the best that was offered was a pond a mile away. Because its banks were steep, the animals were watered individually from buckets. The process was tedious. Instead of watering three times that day, we were lucky to struggle into the mud and out again twice.

The lake wasn't any pleasanter than the woods. Scattered equipment littered its banks. When some of our men tried it for bathing, one or two of them cried out, and they all waded to shore, talking among themselves. When we asked what was the matter they looked sheepish, and said: "The lake is full of dead *bosche*."

There was a large farmhouse a few yards away, which had evidently been used for some kind of a headquarters. Its garden had been trampled and the fences broken down. In a corner was a new cemetery with rows of wooden crosses that were made, we guessed, from packing cases. They marked American graves; thus, we were glad there were so few.

One man said brutally: "There are a lot of things they didn't bury around here." He was right, and so we practiced making our lungs do with a minimum of air. On the higher ground, among the deeper shell holes, were many small and shallow ones that had been made by gas shells. Now and then you saw one whose bottom was yellow with the spewed mustard gas that had failed to volatilize. Everywhere was telephone wire, laid on the ground from position to position, but there had been no time to salvage it.

As we ate our late breakfast, we noticed that the flies were worse than they had been anywhere else, even at Souge. And there was a strange variety—a big, blue-nosed sort that fought to get at your food and, defeated, flew greedily back to the secrets of the underbrush. We ate, though, and we managed to even sleep in that woods. We failed to find in the Courpoil forest, however, the relaxation Chézy-sur-Marne had offered. There was more to be done: the animals required more attention, there were more airplane alarms, and there was more danger of men being caught in the open and not standing still.

That afternoon, Captains Ravenel and Delanoy rejoined the regiment. They had left Souge a while before, but had been unable to locate the regiment. Captain Ravenel, because he was senior, took command of the First Battalion in place of Captain Dana. Captain Delanoy assumed command of Battery F.

We had two lame men with us, who had failed to respond to treatment. A passing ambulance picked them up and carried them back to Château-Thierry. The surgeon in charge gave us some cheerful gossip. "Some of your infantry went in yesterday," he said in an offhand way, "and last night they sent out a lot of casualties. You won't want anything much hotter than you'll get up there." We thanked him, but we didn't press him to stay for supper. His gossip gave the persistent grumbling in the north a sharper threat. And yet, whatever the next day might have held for us, I don't think anyone regretted escaping from the Courpoil woods.

We didn't dare budge until the dusk was thick. Then we tore our improvised camouflage from the carriages and formed in the shell-ploughed field for the final stage of our march into the Oise-Aisne battle. The last of the sunset glow fought for a time against the violent and unnatural dawn in the north, for always the fighting intensified with the night. The gun chorus reached thicker, heavier notes. At one point a sheet of violet flame—supernatural in its vast luminosity—sprang from the earth and, while we watched speech-less and unbelieving, mounted to the very zenith and spread half the immeasurable circle of the horizon. During the several seconds it lasted, details of the landscape leered at us through a mauve daylight. The end of the world might come like that, we thought. We mocked our savage instinct to fall prostrate before a power greater than the power of man.

"Some flares those Huns have!" you said to your neighbor, but you weren't quite sure it was an ordinary flare. Was it some new device?

The violet sheet fell from the sky like a wind-swept curtain. The lesser fires resumed their flickering. Rockets and flares streaked always upwards, so that we lived in a chameleon twilight. It was as if a gigantic and undreamed of catastrophe had happened, could not be controlled, and threatened to sweep Europe. Men fought in its heart —that we would fight there, too, was a fantastic imagining. "Organizations ready?" a commander queried. Everyone reported ready, and so forward we went into the midst of this mad disaster!

The moment had obliterating demands. Our carriages were over-loaded. The *fourgons* were top heavy because of the horse covers, packs, and various paraphernalia lashed to the tops. Inside were our instruments of precision and communication. A picket line, perhaps, and heavy tools were slung from the axles in an attempt to lower the center of gravity. Sometimes a hand-reel cart flopped drunkenly along behind. A sensitive child would have wept at the sight of us. Of the attributes of vagabonds, we lacked only one thing—a fortune teller.

That long, rough road down the hill was damned as perfectly as stumps had been in our remote youth: horses were damned by drivers; drivers were damned by noncommissioned officers; noncommissioned officers were damned by officers; and officers were damned by other officers in order of rank, from bottom to top. Probably the language was quite polite, and it was only the intention that swore, but at any rate it got us on. We reached the foot of the hill at last and turned onto the main road amidst the ruins of Courpoil.

We halted at once in the shelter of broken walls. There was a block ahead and pretty soon motor lorries detached themselves from it and stormed petulantly past. Others, heavily loaded, wormed away from the other direction and demanded the right of way. Some of the trucks, we saw, belonged to our division ammunition train.

"What outfit, Buddy?" a chauffeur yelled at us.

"305th Field Artillery," a man answered thoughtlessly. Angry voices rebuked his indiscretion—at the front, no one knows what ears are about. The chauffeurs, however, recognizing us as of the same division, bandied about words. They said, "Believe me, you're going to some summer resort," or "Where there's a will there's a way, but don't forget your will." And, "Hey! You look as if you were moving from the Bronx to Brooklyn."

We didn't have much repartee. We were too anxious for the obstructing lorries to get by. An hour must have slipped away before the jam was broken. As we lurched ahead, a message came down the column, repeated from mouth to mouth: "Follow the carriage in front closely to avoid shell holes." That meant that the shells were falling on this road too fast for the pioneers. In spite of the advice, dodging such holes seemed an impossibility. We couldn't snake along from side to side in all that traffic. We couldn't stop until there was a chance to get past a hole. So we assigned dismounted men to walk ahead of the precious *fourgons*. We threatened dire penalties if they didn't give plenty of warning.

The forest of La Fère closed about us, shutting out the flames ahead and the wan light of the moon. We could see nothing. The man riding beside you was blurred by the heavy pall. We glanced to the right and to the left, trying to imagine the form of the forest and the things it hid. Our only clear sensation was that of the intolerable stench of death.

We halted. Would we never go on again? A double column of foot soldiers shuffled past. They, too, halted. We couldn't make out what service they belonged to, but it became clear something was wrong with them. They didn't seem to know where they were. They thought they may have gotten on the wrong road, but they weren't sure. They stood there beside us for a long time, growing more and more impatient. So there we were, hopelessly blocked, a rare target for a shell or an air bomb.

One of the scarcely seen men lamented, "Ah'd rather take my chawnces in the line than be walked ta death." Another soldier objected and said, "Not me. Ah can heah oy Mistah shell a singin' now. He says: 'gonta getcha, gonta getcha, gonta getcha. Bam! Done gotcha.'"

What appeared to be a huge light flashed out ahead and was immediately extinguished. It showed us that the foot soldiers near us were from a southern engineer outfit. Their lungs were good—they all burst into a huge and angry chorus: "Put that . . . match out."

Expressions of pity and disgust followed, such as: "Say, Bo! Put yo'sel on a plate an' hand yo'sel with a knife and fo'k to Mistah Jerry. But don't use me fo' the gravy." And, "Hey, Captain, take me away from these city fellahs that strike matches in the dark."

We all shared the shame of that one culprit. We tried to spot him to teach him a lesson, but the thing had been too quick and the night was too friendly a protection. We were from the city and, perhaps, the game of concealment came harder to us than to some others, but we thought we had learned it better. We had, as we found out later. And that particular crime wasn't repeated.

By this time the engineers had decided that they'd better try another road, so, without saying anything, they calmly countermarched, blocking the road more completely than before, and holding us up for another half hour, dividing our column at the same time. We got out of the ruck at last and onto a clear road, where we made fast progress, urged by the necessity of reaching Forêt de Nesles before daylight. Our hurrying hoofs and wheels echoed in the dead towns we passed through, and the walls shook from the reverberations of heavy guns just ahead.

We entered the outskirts of Fère-en-Tardenois, still under shell fire, and slipped through unmolested. Scarcely anything remained of the town—the largest in the district—now a heap of rubble with a few walls, like torn masks. It might have been the site of a prehistoric capital about which an archeologist has commenced to excavate.

Nearby batteries pounded away. Our horses, weary as they were, grew nervous and moved restlessly about during halts. On the other hand, the men, forgetting their surroundings and the warnings against gas, everything except their great weariness, sank on the banks of ditches and slept fitfully.

Daylight caught us again as we wound through the town of Nesles. It seemed impossible we should ever reach a bivouac at the time scheduled. Nesles was in ruins except for its storied medieval tower, which shells had only scarred. Beyond the town was a steep road, recently laid by the pioneers, which climbed to the forest. Even from there the forest appeared haggard and shell torn. The sloping fields between us and it were strewn with graves, dug where their occupants had fallen. Most of them had rough crosses from which German helmets hung.

The horses were unequal to the steep, hilly road. We manned the wheels and forced our way up. We entered between the broken trees, but felt we had arrived too late. There had been airplanes in the distant sky, and we had no doubt that the Hun knew there would be an artillery bivouac in Forêt de Nesles that night.

The place had been policed, after a fashion. The stench of death was less here than it had been at Courpoil. A regiment of pioneers was already in possession and had removed as much refuse of battle as they could. Everywhere about the forest floor were coffin-shaped holes. We guessed they were individual shelters from shell and bomb fragments. We learned to call them "funk holes," a term we later applied to far more ambitious refuges.

Anti-air guns opened all around us. "Tsching! Tsching!" we heard. Two shell cases whistled down in our woods. We put on our tin hats, but we knew they were no protection against shell cases. We recalled all the airplanes that had bothered us at Doue and asked the pioneers, with a perfect confidence, whether or not we had the control of the air up here. We felt that if the American air service was concentrated anywhere, it would be on this front.

The pioneers looked at us with pity. "The Huns," they answered, "own the air here and have a mortgage on the ether." They followed with accounts of American balloons brought down by Hun planes and unrestricted bombing attacks. Our hearts sank. We knew we had been seen coming in that morning. Yet we felt the pioneers must be wrong. The money spent, the men enlisted in the air service, and all those fellows flying about Coulommiers! Before many days, we accused the pioneers of uttering conservative statements.

A messenger found Captain Ravenel and took him to the colonel. The colonel introduced him to an officer who had just reported, as assigned, to the regiment—a Major George W. Easterday, who would take command of the First Battalion. And Captain Ravenel would return to the command of his battery. Major Easterday, we learned, had come originally from the Regular Army Coast Artillery, having entered the service from civil life. He had received word by telegram that he was removed from a few days of dalliance in Paris, after a lively share in the advance north from Château-Thierry. He then shot back into the show as a member of the 305th and was destined to remain with the regiment until it sailed from France.

So the forced march ended, and we were in the woods, which after disastrous experiments in other localities, was to become the regimental echelon. We breakfasted, unstrapped our packs, and stretched out to sleep. We were awakened almost immediately by the news that there would be a preliminary reconnaissance that afternoon. We studied our maps in preparation. The little party rode from the woods, and in an hour's time returned. There had been a blunder somewhere, for the rendezvous had not been made clear, and the various portions of the reconnaissance hadn't gotten together. So a real reconnaissance was set for early the next morning.

We dined to the accompaniment of mounting gun fire and crawled gratefully into our shelter tents, believing that no amount of noise could keep us awake. But that old metaphor of the "orchestra of the guns" is justified, and batteries and individual guns seem to have their own tones. When a great many are firing perpetually, as on this front, the tones blend into crashing chords. We fell asleep to this gargantuan lullaby.

After a few minutes, the hideous screech of the gas alarm had us up and snapping our respirators on. The screeching died away and, after a time, the gas officers went around singing out: "Masks may be removed." We went to sleep again. And again we were awakened by that unholy screeching. It happened three times. We told ourselves that the horns wouldn't awake us again, gas or no gas.

On the heels of the last alarm something else aroused us: we heard the throbbing whir of Hun airplanes. There were plenty of targets on the Vesle River, heaven knows, but we remembered our earlier fear that we had been seen coming into the bivouac that morning. And the throbbing grew. "Ba-room . . . ba-room . . . ba-room," we heard. It was as if the engines misfired rhythmically. Then, above the artillery we got the crunching detonations of large air bombs. Those airplanes were coming nearer. There was a squadron out and, if it wasn't after us, it would pass very close.

"Ba-room . . . ba-room . . . ba-room . . ." came the detonations, sounding always nearer and louder. They came in salvos of four, each burst half drowned by the next. And nearly overhead we heard the petulant rattle of a machine gun.

"That's the scout signaling to the bombers," someone said.

"Why, are our airplanes back in cheerful places while these fellows give us their droppings undisturbed?" we asked irritably. We thought of the other bivouacs, crowded with American soldiers, with the Hun birds merrily hopping from one to another.

The bombers responded to the scout, and their bombs fell on the edge of our woods with roars that made the artillery seem like childish fireworks. We smiled grimly as we thought of those fellows making us blow our bugles all day long back near Coulommiers. The Huns dropped several salvos, and throbbed away to other pastures.

Men were killed in the Forêt de Nesles that night, but our check showed us that the 305th had escaped. We crawled back to bed, and went to sleep, and didn't answer any more alarms until reveille dragged us out.

That was the first of many experiences on the Vesle with Hun airplanes working nearly undisturbed. Most men will agree there is no form of attack less pleasant. You approximate the sensations of an insect above which a giant foot wavers, waiting to descend to obliterate its target.

Chapter Fifteen

RECONNOITERING IN FRONT OF FISMES

The reconnaissance we made in the Fismes sector, on August 14, was about as much like our Lorraine ones as a pleasant day is like a period of violent storms. Nor was it as agreeable as a reconnaissance made during an advance, for here we faced a semi-stabilized battle. The Huns could see our little party, and they had registered everything. Still, all the reconnaissance efforts have one feature in common: they never work out exactly as one plans. They fail invariably to follow the pretty rules laid down by the books. At the front, you mold technique to the demands of the moment and to the necessity for quick results.

It is a matter of interest to preserve the field order that sent us into this, our costliest battle. The reconnaissance was made in pursuance to its provisions, as you will see below.

Headquarters 77th Division
 American E. F.
 14 August, 1918.
 FIELD ORDER NO. 23.
 MAPS: FÈRE-EN-TARDENOIS \
 FISMES [1:20'000]

1. The 4th Field Artillery Brigade will be relieved by the 152nd Field Artillery Brigade on the nights of August 15–16 and 16–17, 1918, in compliance with G-3 order no. 31, 3rd Army Corps, August 14, 1918.

2. The 305th Field Artillery will relieve the 16th Field Artillery, the 304th Field Artillery will relieve the 77th Field Artillery, the 306th Field Artillery will relieve the 13th Field Artillery assuming missions of organizations relieved. Necessary reconnaissance efforts on 14th and 15th August, as previously directed.

3. Relief will be completed as follows:

1st Night: (15–16) (a) Battery to be relieved in each position. \ battery 152nd Brigadewill be accompanied by an officer who will remain at the position. One chief of section of each J battery will remain at the position.

(b) Telephone operators, linemen, and observers of the 152nd Brigade will report to their posts and will remain in observation only.

2nd Night: (16–17) (a) Remaining \ of each battery relieved. One officer and 2 chiefs of section to remain at position until following noon.

(b) All specialists relieved, excepting one telephone operator and one observer of 4th Field Artillery Brigade in each post, will remain in place until noon following.

(c) Ammunition dumps will be turned over to 152nd Brigade.

(d) Battery combat train and other elements will stand relieved at 21:00.

(e) Ammunition train will stand relieved at 21:00.

4. Arrangements for exchange of wire, camouflage nets, etc. will be made between commanders concerned.

5. Elements of 16th Field Artillery, as relieved, will proceed to position in FORÊT DE FÈRE by roads to south through NESLES. Other

elements will use main road through FÈRE-EN-TARDENOIS. Elements of the 152nd Field Artillery Brigade will use the roads running north from the FORÊT DE NESLES.

6. The 302nd Trench Mortar Battery will remain for the present in the location where it is bivouacked. Reconnaissance efforts will be made to select suitable positions for this battery so that it may be put in position in the near future.

7. Command will pass to battery, battalion, and regimental Commanders of 152nd Brigade, as the relief of each unit is reported complete.

8. Command of the artillery of the sector will pass to Commanding Officer 152nd Brigade at 8 A.M. August 17, 1918; P.C. 152nd Field Artillery Brigade will open at FRERE CHATEAU at the same time, same date.

By command of Major General Duncan, J. R. R. Hannah, Chief of Staff

So, we set out to study the ground. The regimental, battalion, and battery parties left the Forêt de Nesles together and trotted down the hill to Mareuil, where the 4th Field Artillery Brigade had its headquarters. It was a warm, brilliant day. Therefore, we knew we would see and be seen. We dashed past parties of pioneers repairing roads that had been damaged by shell fire the previous night. In the stricken village, ambulances stood outside a distributing station and on the ground were many stretchers bearing forms—some still, some restless, each covered with a secretive issue blanket on which the wounded man's tin hat and gas mask rested. Ether and iodine cut the pervading chlorine odor.

Brigade Headquarters was a one-story building, originally a café or a rural hostelry, that was now dilapidated. The dusty square in front of it was white with chloride sprinklings. Opposite, an arched gateway admitted to a large courtyard surrounded by stables and dwellings. Our party was herded in there and commanded to keep out of sight, because Hun planes were constantly passing overhead, expressing an impudent curiosity. So we got as many horses as we could in the sheds and kept the rest close to the walls. Then officers and enlisted men made themselves inconspicuous and awaited the result of the conference of field officers that continued in the café across the street.

Every soldier, I think, has noticed that daylight acquires false qualities because of one's own perceptions. To all of us there was an unnatural tone to that brilliant sun, streaked occasionally by enemy planes. Perhaps another planet might have light like that. You heard men commenting about it with little laughs.

Restlessness grew upon us. Would the conference never end? A group of field officers came from headquarters. With serious faces, they glanced about uneasily. Some of them appeared a trifle undecided. They paused, forming little groups, to which representatives from our party attached themselves. Gossip drifted into the hot, restless courtyard. We heard, for instance, that one of the batteries that the 305th was going to relieve had had forty casualties during a burst of harassing fire the afternoon before. There was always harassing fire, it seemed, where we were going. We would have to take up new positions, we said confidently. From the gossiping groups the depressing word slipped that there were no positions much better than the ones already occupied.

The colonel came in and said, at first, we would have to go forward from that point on foot. Those of us who had studied the maps groaned, for the road went diagonally toward the frontline and our positions were many miles away. The colonel reconsidered and talked again to some of the officers of the 4th Field Artillery Brigade. Doubtfully, he decided, we might ride as far as Regimental Headquarters with an interval of two hundred meters between pairs.

No officer or man that took that ride cared much for it. We curved up the hill past the half destroyed Romanesque church and turned onto a main road on the crest. There were, of course, no shell screens and, to the left, we could look all the way to Jerry's temporary home.

One of the men expressed our general emotion: "I feel all undressed up here," he grinned.

Everywhere along that road were nice, fresh signs left by the enemy, pointing the way to dressing stations, to ration and ammunition dumps, and to shortcuts for the various villages. And there were newer French signs, regulating traffic and repeatedly calling attention to the exposed nature of the highway. In the vicinity of a small group of buildings ahead, large high explosive shells were vomiting black smoke We guessed that the group was Chartreuve Farm, the Regimental Headquarters of the 16th Field Artillery.

We waited in a lane, behind the shelter of a wall, until the rest of the party had come up, then hurried across a courtyard into the farm. Two or three habitable rooms downstairs were packed. The colonel and the majors conferred behind a closed door with the field officers of the 16th. Less important but quite intelligent young men gave us the sector gossip while the Hun continually reminded us he knew where we were.

The sector gossip was simple: it was a rotten place we were going to and there wasn't much we could do about it. Jerry had a big concentration of artillery opposite and he was using it with an admirable and murderous skill. We listened mutely to recitations of casualties. We sensed some joy on the part of these young men that they were going out; and a brotherly sympathy that we were going in. This conference, too, ended at last, and the 16th gave us a bite from their field kitchen set beneath great trees in the pretty grounds of the place.

The Colonel and his party went no farther just then. The two battalion parties continued on foot, out of the friendly trees, across stripped fields, and into the ravaged village of Chéry-Chartreuve. Even on that busy day the 305th rendered even more bibulous the name of this dissipated appearing town. It was known ever after among us as "Sherry Chartreuse."

At the first corner, a military policeman stood between wrecked buildings. He reminded the more careless of us to carry our gas masks in the alert position. Another policeman, a hundred meters beyond, advised: "Walk farther apart, sirs. They're giving the road hally-lool-yah right now." And they were—the louder whistling of shells preceded explosions close at hand. A bank on the side of the road towards the enemy was pitted with funk holes gouged out by infantrymen. Into these we ducked when the whistling warned us of a dangerously close explosion. We must have resembled animals of absurd habit that hopped aimlessly from place to place.

This erratic progress brought us to our first view of Les Pres Farm —place of unbeloved memories.

A huge hangar rose where a country road crossed the main highway. The number of shell holes testified to the enemy's interest in that crossroads. The country road climbed a bare slope to a cluster of buildings, a third of a mile from the hangar. Our first impression was of a large and dignified stone dwelling house with half a dozen outpost trees and wings of sheds and stables reaching behind it around a large courtyard. To the right were two small stone dwellings, with a horse shed and one or two outhouses just below them. The bare slope stretched upward for another half mile beyond the farm, blatantly broken by three battery positions, whose only protection was flat tops. Wherever you glanced you saw the mortal and redolent remains of horses in grotesque attitudes.

Jerry saluted us. He commenced raking those exposed battery positions. From beneath the flat tops soldiers scurried like an indignant party of ants whose hill has been disturbed. As we climbed the slope, we couldn't help admiring the nicety of the Hun fire. Their volleys walked through the positions then walked back again until there was so much jetty smoke you couldn't be quite sure where the shells were falling.

"Hundred and fifties," we muttered.

"Battery positions!" someone sneered. "Targets! That's all!"

The farm at first appeared deserted. Then we saw a red-headed soldier peering at us curiously from a funk hole dug close to the wall of one of the smaller buildings. This one, nearest the enemy, we had been told would be the First Battalion command post. The other would be used for a similar purpose by the Second Battalion. The large farmhouse and the courtyard were occupied by the infantry for a dressing station and a reserve position; and the 306th, it was understood, would establish a battalion command post there. The farm, it was clear, was already crowded. From its exposed position it was obvious it would give Jerry plenty of practice.

Each of the two battalion parties went into its future little home. We were greeted by the following: walls decorated with coarse cartoons made by the Huns, very recently departed; logs piled in the rooms above the cellars in an insufficient effort to hurry the burst of a direct hit; bedding rolls tumbled about; and the remains of a meal and a few gay copies of *La Vie Parisienne*—incredibly out of place— strewn across the top of a greasy deal table.

We were also greeted by some men, with sleeves rolled up and a tendency to scratch, and innumerable flies—on the walls, on the men, obliterating the neglected food. The men welcomed us. When, they wanted to know, would they get out?

We examined the cellars next. There was one under each building—stuffy, fly-choked places with rough bunks improvised and, inevitably, the switchboards in the places of honor.

Gossip was unnecessary here, for the place spoke for itself. Still the men did tell us some things. This was the 16th's first trip to the front; they hadn't expected to stay here long. "We used," a major said, "the observatory for a bleachers. I'm not joking."

Decidedly he wasn't. There were casualties in that observatory. We had to move it. As long as we stayed there, the ridge was raked periodically by high explosives, gas, and air bombs.

We fought the flies away from a map and studied the dispositions. It was proposed to place batteries D, E, and F in the three positions the Huns were harassing on the hillside. From a rear window we could see a grove of trees just across the road, a few hundred meters from the farm. Battery C would go there—on a forward slope. We would have to walk some distance to inspect the possibilities for the other two batteries.

We set out after waiting for what we thought was a quiet moment. It might as well be said now that there were no quiet moments in or near Les Pres Farm, until the Hun moved back to the Aisne early in September. There was never a time you could go about your work there with a feeling of comparative security. Always, shells were bursting near you or whistling unpleasantly close. To give the devil his due, it was great artillery work, and it was devilishly uncomfortable. We learned afterwards that we had made Jerry dodge rather more than he had us.

Now the Hun opened up, as we walked across the fields to the southeast, but we managed to reach the Battery A and B positions and express a decided disapproval. We stood on the edge of a deep valley, where B seemed fairly well off with a little natural foliage to break the angles of its camouflage. Battery A was a hundred meters forward in the open with only its flat tops to make a futile attempt to deceive the Hun airmen.

The valley—called Fond de Mézières on the map and later renamed by the soldiers as "Death Valley"—was full of artillerymen and infantry, bathing in a narrow stream, washing clothes, playing ball, or dreamily watching their horses as they grazed. "It's doomed," we said to each other. And, it was. A night or two later, the Huns filled it with gas and high explosives, collecting a heavy toll. We decided at first glance to have nothing to do with it, even as a location for our kitchens or first aid stations.

We learned a lot that afternoon about the radius of Hun shell fragments. They seemed to follow us wherever we went. They disturbed our consultations and they hurried our walks. Even so, it was nearly six o'clock before we got through and took the road home, dodging along the line of funk holes to Chéry-Chartreuve.

As we walked hot, dusty, and tired through the town, we noticed a Y. M. C. A. canteen in a half-ruined building. We greeted it with joy that day, but later on the place would impress us less pleasantly. Chalked across the door by some German was the legend: "hier wasser" (hey water). A big, cool-looking pump stood inside, and the next room held a counter with chocolate, cakes, cigars, and cigarettes.

We wandered on, refreshed, to Chartreuve Farm, where our horses waited for us.

Regimental Headquarters, we learned, would not remain there. There was a farmhouse a mile or so farther back—considerably safer to all appearances—named La Tuilerie.

Forêt de Nesles impressed us as exceedingly peaceful and remote from danger, when we trotted in just before dusk. We smiled. Clearly the lesson of the previous night had not been wasted on those who had stayed in the woods that day. Let the Hun airmen come! The forest floor was fairly honeycombed with elaborate funk holes; some were even covered with sheets of elephant iron.

The 305th learned early on the wisdom of taking every precaution possible, and undoubtedly it is due to that habit that our casualty list is no greater. But that night we faced the uncomfortable facts of Les Pres Farm. We had to go there and it was clear that because of the amount of artillery already in, and the nature of the terrain, there were no really good positions to be had. Those on the slope above the farm, however, probably could be improved on, and it was decided not to use more than two of them, and only temporarily. Batteries A, B, and C, however, would start at least in the 16th emplacements. The communication experts were as troubled as the battery commanders. It was going to be a job to keep those lines working; and lack of equipment would have to be combatted as well as shell fire. "We've got to take our losses," everyone admitted, "but we can try to hold them down."

Those who had made the reconnaissance had brought back to Forêt de Nesles some stirring descriptions. In our bivouac, no illusions remained and each man went about the work of preparation with an extreme care, with a thorough understanding.

That day Major Miller became regimental surgeon, replacing Captain Parramore, who had been invalided to a hospital.

At dusk on August 15, the two pieces prescribed from each battery were ready to start. We had hoped that by leaving early we would dodge some of the night congestion on the roads. And those roads would be shelled. "Keep your platoons moving," officers said with an effect of prayer. Whips cracked, the horses strained forward, and our sections jolted out of the friendly and haggard forest.

Chapter Sixteen

LES PRES FARM AND MUCH SHELL FIRE

Early as we were, the roads were crowded from the first. The two other regiments of the brigade had had the same idea of an early start. Quads bearing ammunition and ration trucks bumped along, their drivers sarcastic and anxious. There was a great deal of infantry out—some *fantassins* [French foot soldiers] and very many of our own doughboys. A lot of heavy firing made the dusk noisy. The darkness came down nearly impenetrable and ominous.

Frequently, now, the column had to halt. These halts showed there's plenty of chance in war. Battery B's platoon had its captain; A's was in command of a lieutenant. During one of these halts, B slipped past A, and a little later got what might have been A's share.

But it was all rather confusing and conditions got worse on the main road above Mareuil. Shells came down perpetually, like unseen fingers tearing the black pall of night. One knew that they wouldn't all fall over or short. The halts were continual and, because of the congestion, we couldn't keep the carriages separated.

Battery E got it first. Shrapnel popped overhead, but nobody bothered much about that. Then a high explosive shell burst on the road in the midst of the platoon, and horses reared and tried to pull free, making queerly human-like sounds. It was impossible to tell at first how much damage had been done. Officers and noncommissioned officers rode up and down the line, shouting and exhorting, but they might as well have saved their breath. There was no panic among the men. Nor, miraculously, had a man been hit. Two horses had been killed, and their teammates were dangerously active.

"Cut 'em out," came the quick command. "Haul 'em over to the ditch, if you can. But let's go on." The flashes from bursting shells helped the drivers, as the dead animals were cut out and drawn to one side, and the platoon moved ahead.

It wasn't all shrapnel and high explosives. As the column approached Chartreuve Farm, gas shells came over in a dangerous concentration. Reluctantly, men put on their respirators, shutting out what little light there was. They struggled with frightened horses and got the awkward masks over their muzzles. They went on through a suffocating blackness. The few commands given were choked and had to be mumbled from mouth to mouth.

It was under these uncomfortable circumstances that B suffered. The column was blocked again near the Chartreuve crossroads. B was just short of the junction, clearly a registered point and consequently a dangerous one. Yet there was nothing to do about it. Some outfit has to be caught at or near crossroads in these blocks. You can ride ahead if you like, and try your hand at straightening out the tangle, but in the majority of cases you come back with nothing accomplished and you stand still, or sit your horse, and pray for the movement of the units ahead of you.

The Hun came down on the crossroads and some of the shells fell among the waiting cannoneers and drivers of B. Even in the blinding respirators, it was easy to see that men and horses were down. The horses screamed, and there came a whimpering cry from some hurt fellow for his mother. Nor was there any panic here. An amateur of the National Army cried out cheerily: "It would be a hell of a war, boys, if nobody got killed."

But then someone asked, "Where's the captain?" The captain's horse stood riderless near the head of the platoon. Lieutenant Montgomery found his orderly and all the anxiety was removed. The captain had gone ahead on foot to try to break the jam. Lieutenant Montgomery sent a messenger to report what had happened and, with his own hands, attended as best he could to the wounded.

There was nothing to be done for Private John W. Whetstone, as he was killed instantly. It was clear that Private Harry E. Kronfield hadn't long to live. By rare good luck an ambulance was struggling through the jam at this point. It picked Kronfield up and hurried him to a first aid station, but he died before morning. This ambulance also took Private Douglas Tredendall, who was so severely hurt that he was evacuated and never returned to the regiment, and Private Joseph Horowitz, whose injury was particularly unfortunate as he was the platoon's medical orderly. His task of mercy was very brief— with one arm blown away, he was evacuated and we didn't see him again. First Class Private George A. Thomas was wounded less seriously. By the time these men had been cared for and the horses cut out, the jam broke and the column pounded on towards Les Pres Farm.

On the way up, D battery had no casualties. Its first platoon went, as did E's, temporarily on to the hill above the farm. There was a lot of gas there and several bursts of heavy shelling. By choosing quieter moments, however, Captain Starbuck got his guns in and his limbers and caissons started for home.

Corporal Connie F. Geer was in charge of the second-piece caisson. On the way back, the traffic had thinned out a good deal so that the column moved rapidly. Corporal Geer had been particularly cheery and helpful during the trying moments when the caissons had dumped their ammunition at the position. On the return journey, he was at the rear of the column and went back often to make sure there was no straggling. The train must have been half way home when one of his men reported Geer missing. A search of the road was unsuccessful. The shelling was still heavy, and it was necessary to get men, horses, and carriages back to the echelon. A report was made there and then Lieutenant Hoadley set out with a party. They found Corporal Geer's body at the lip of a fresh crater close to the side of the road. His death had probably been instantaneous. He was buried that day in a quiet corner of the Forêt de Nesles.

Even at the echelon, the night didn't wear itself away very comfortably. Regimental Headquarters had moved to La Tuilerie Farm that afternoon. At midnight a messenger arrived with a note from Colonel Doyle for the battalion commanders, explaining the arrangements for going in. As a few of the dispositions were altered, there were excited conferences. One such conference, some of us will recall, was held in a *fourgon*, heavily blanketed with horse covers. Even so, the light of the single candle within escaped wanly here and there. Outraged cries roared through the forest: "Put out that light, you fool!" In response, someone said, "If you want to croak, go and do it by yourself."

It was impossible to heed these compliments. If important dispatches arrive, they must be read. What to do about the present one was a problem. The solution gave Captain Henry Reed a pleasant automobile ride through quarrelsome firing to headquarters. Once there, he found out that the document hadn't been intended to change anything, so we went ahead on the basis we had agreed upon the day before.

So, the details went up on the morning of August 16. The movement of a detail was never a very dignified proceeding, as details went in for efficiency rather than appearance. The surrey—there was never anything less military in appearance—was always an absurdity on a shell-torn road, but it carried a lot of stuff. Doughboys used to grin at the group of very military appearing horsemen followed by a couple rambling cobs, which drew this vehicle with its fringes flapping from a bent top. Underneath were piled switchboards, telephones, instruments of precision, and spare wire.

Everybody got to the farm and pitched in. In spite of the fire, officers and men of the battalion details got an idea of where the lines ran and how they were laid. They also appraised the task that lay ahead. These lines were continually shelled out. Some improvement could be made by relaying here and there, but at best it was going to be nasty work. For the Huns had so much artillery and ammunition that they didn't hesitate to snipe with 77s [anti-aircraft guns] or Austrian 88s [small caliber shells] at a single man at work in the barren fields.

The detail men in such warfare have rather the worst of it. They work, as a rule, in pairs—on the lines, or in an exposed observatory, or on the edge of woods—doing the careful work of a surveyor under the most distracting conditions. And, it is always simpler to be brave in a crowd.

The yellow intelligence sheet for that day, too, informed us that the enemy was taking an increasing interest in Les Pres and its neighboring positions. Things were noisy while we settled ourselves there. The B position, which we had thought the best of the lot, got a pounding during the morning. The B men escaped, but the 16th had a number of casualties. Captain Ravenel reconnoitered a fresh position, and Major Easterday decided that he should move his first platoon there that night and bring his second into action alongside of it.

Major Wanvig had put F directly into a new position near the Les Pres crossroads, and he settled on positions for D and E on the slope of the valley beyond Chéry-Chartreuve, so that none of the sections took many chances with the emplacements on the hill.

The observatories looked nastier than their reputation, but we had to use the ridge above the farm. The regimental and the two battalion observatories were there, so close together that they were really one. Besides, the ridge was sprinkled with the observatories of other organizations, with division and corps stations; and the infantry had a reserve line near. All this activity added to the discomforts of that exposed place. Lieutenant Thornton Thayer had spent the previous night there and had got the lay of the land. We sent our observers and operators up and, although an officer of the 16th remained for several hours afterwards, practically took over at noon.

Lieutenants MacNair and Graham were already down with the infantry, and we sent eight enlisted men to them to act as runners. It was found advisable at the start to alternate this work between the two battalions, so that after the first day only one officer and one group of men were with the infantry at one time. Such liaison was particularly dangerous in this sector. The infantry received a lot of high explosives and, because of the low ground near the Vesle, suffered from gas more than the artillery. Yet it was really the only liaison we had, beyond rocket signals. It was difficult to maintain a telephone line between infantry headquarters, in spite of the division liaison order that gave the infantry the task of laying and maintaining such a line. We put a wire through to a forward observatory at Mont-Saint-Martin, very close to the frontline, but because of the constant movement of battalion headquarters and the shortage of men, the infantry never hooked up with it. We connected with the infantry net through one of their switchboards, and when they had wire communication with their front-line troops, we did too. But in such a type of warfare runners furnish the only dependable communication, and our men were on the road day and night.

The sun set hot and red that first night in—and with the sun's going, Jerry awakened to a new interest in us. There were no dugouts, so the men not on duty crawled into such funk holes as existed, or into the stifling cellars at battalion headquarters.

Privates Shackman and Silber had already been sent to the observatory to act as operators. Lieutenant Thayer, after leaving the shelter of the cellar, joined up with Corporal Tucker and dodged up the hill to relieve the officer and the men of the 16th.

At Boston, as the observatory was called, the only protection was two narrow trenches, five or six yards apart—one for the operators, the other for the observers. They were less than six feet deep and had no overhead cover.

A few minutes after the arrival of our party, a thick curtain of high explosives descended on the ridge. The ugly little volcanoes bracketed Boston while our men crouched in the trenches. The curtain lifted. Perhaps it was just an evening hymn of hate and the rest of the night would pass without music. But . . . in five minutes the curtain was down again. The bracket narrowed and fragments of shell shrieked over the trenches. Sand stung the faces of the little party. "O.K. . . . . O.K." said the soldiers.

Lieutenant Thayer and the 16th officer decided to take their men to a flank until the show ended. "Jump out and run for it after the next shell," they directed. Right after one shell burst closer than before, the little party clambered from the trenches. Some were quicker than others. A following shell hit directly on the lip of the smaller trench, causing the 16th officer to fall back, his rain coat drilled full of jagged holes. Private Martin W. Silber slipped in on top of him, and the rest turned back without hesitation to see what could be done. They lifted Silber out; he was dead. The 16th officer had not been injured.

Those that remained dashed to the left and fell in shell holes, where they waited for the curtain to lift again. But gas came in for a time with the high explosives. They put on their respirators and worked from shell hole to shell hole until they were out of range.

In the command posts at the farm, everyone knew the ridge and the crossroads were getting it. Our men were in the observatory and our platoons before long would have to pass the crossroads.

A drop on the switchboard fell. "Silber's dead," the operator commented. He commenced to test: "O.K. . . . O.K." He paused, then whirred the magneto of his home telephone, and said, "Red line out, sir." A moment later he reported two other lines out. That's the way they went at Les Pres. So, the linesmen left through the noisy darkness with coils of wire and testing telephones over their shoulders.

In the First Battalion cellar, the operator called to Major Easterday: "Second Battalion wants you, sir."

The major lifted the handset. "Tucker. Which one is he?" he asked. You see he had only been with the regiment a few days then.

"What's the matter with Tucker?" Reed asked.

"First Battalion says they've just heard lie's been killed," responded the major.

There were close personal friends of Tucker's among the detail in that cellar, who swore softly as they went about their jobs.

As the major replaced the telephone handset on the table, the blanket that hung as a curtain at the cellar entrance waved. A hand drew it aside and in stepped Corporal Tucker. Our men didn't believe in ghosts; they grasped his hand and a laugh burst out. Tucker denied the Second Battalion's story and made his report. Thayer had sent word by him that he was going to establish a new observatory. We had gone over the ground with a fine-tooth comb. The change in the location of the observatory would only be a matter of a few yards. A digging detail was ordered up to him with a guide. Lieutenant Klots, with a number of bandsmen bearing picks and shovels, arrived about the same time and started to dig in a regimental observatory. Corporal Caen ran up to stand by the telephone in the old observatory until the change could be made to the new ones. And all night the Hun remembered the ridge with high explosives and gas, while stray airplanes swooped low there, to let a bomb or two fall.

The curtain in the cellar swung in again. "For the Lord's sake keep that curtain down," somebody grumbled. "If an airplane sees this candle, we'll be bombed out in a jiffy." But it was a battery commander, who had halted his platoon at the crossroads. He took off his helmet and the perspiration poured from his hair. What, he asked the major, should he do about his platoon? He didn't want to lose his men or his pieces if he could help it, and the shelling down there was particularly vicious. Nor was there any way around.

"Watch your chance and take them through one at a time," the major said shortly.

The battery commander nodded, replaced his helmet, and backed cautiously out.

"Somebody on the line for the major of the 16th," the switchboard man called.

The 16th officers had sat there for some time, waiting only to hear that the relief was complete before striking out for quieter parts. The 16th major answered the call and looked annoyed. We gathered that an ammunition dump at the C position had been hit and was burning. His officer in command there evidently wanted to know what he should do. "Go in and put it out," the 16th major said, and lowered the handset.

Almost at once, it seemed to those in the cellar, the same drop rattled again and the operator asked for the same officer. The 16th major picked up the handset with a frown. Then his expression altered, and when he spoke his voice had changed, too. "Wolff is dead," he said to Major Easterday, and everyone knew he spoke of the officer in command of the C position, who he had just ordered into the burning dump. "Wolff is dead," he repeated, "and Dean, the only other officer I have there, is wounded. You don't take over until the relief is complete. I'll have to get one of the 16th's officers. Who is Robinson?"

"One of our battery C officers," Major Easterday answered.

No one asked about Robinson for a moment, because it seemed certain that he had been struck, too. But the 16th major shook his head when at last the question had been asked. "No, the major said, "he's taken command. He seems to know what he is about."

Robinson did. It was for that affair that he and Corporal Johnson were awarded the Distinguished Service Cross. Wolff and he had been sitting together in a funk hole, and Wolff had just said to him, expecting to leave with the last of his battery in a few moments: "You know, Robinson, I'm not so sure I'm going to get out of this place alive after all." He had laughed a little, and just then the shell had tumbled into the dump, and he had telephoned battalion headquarters to ask what he should do about it. Wolff and his assistant Dean and Robinson had all gone in, carrying dirt to throw on the popping shells.

Robinson had just gone out for more dirt and Dean was starting to when the explosion occurred. There had been shrapnel there, and while it was still bursting, Robinson had dashed in. Corporal Johnson, without any command, without any request, had followed him, and they dragged out Wolff's body and the wounded lieutenant. It was then that Robinson had reported.

The major of the 16th looked very tired. At last he shrugged his shoulders and called up his colonel, and explained: "Wolff's dead. Dean's hurt. Burning dump. What? One officer of the 305th, but I'm getting an officer over to stay until the relief's complete."

It seemed at times as if that formality—the relief— would never be accomplished. We got reports from A and, at last, from C. But B hadn't reported its second platoon in yet, or that its command post moved to the new position. So we sat and waited.

We had a number of gas alarms during the evening. Time after time, our gas guard had wound his klaxon horn and, time after time, we had struggled into respirators and the switchboard operators learned how difficult it was to talk intelligibly through a mask. But we suspected nothing worse than mustard gas. While we sat impatiently there, an officer of the 16th stumbled down the cellar steps and through the curtain. He seemed to be in a hurry and his face was white.

Then, from a corner a quiet voice spoke: "There's phosgene in this cellar." Sure enough, the penetrating, sickly odor was apparent to everyone. Masks went on with a rush. The newcomer, however, didn't disturb his. He waved his hand deprecatingly; it trembled a trifle. "Don't bother," he told us. "I think I've brought it in on my clothes. Those shells are all over the hillside. Good Lord! I tell you one of them fell at my feet. Don't know why the rotten thing didn't hit me. When are we getting out, major?"

The major shook his head and told him that nobody knew. It was B that held us up, and we tried them again. This time there was no answer to our call. We tried them through A and C. They were out of touch with the world.

Over there on the edge of Death Valley, the B signal men worked frantically with a coil of twisted pair that had been snarled half a mile from the new battery position. We established runners from that point to the battery, so that the relief could be reported and communication of a sort maintained until daylight, when the battalion detail ran a new line in.

At midnight, then, the 16th was through and it went out of Les Pres Farm, leaving us our own masters. We gazed upon our new kingdom. In the stifling cellars, men not on duty tried to sleep. They lay sprawled on the dirt floor, endeavoring in their restlessness to keep out of each other's way, with their respirators conveniently at hand.

At the positions, men crouched in funk holes, sleeping by turn. One soldier lay moaning softly with a bad touch of shell shock. Now and then a soldier paused and spoke to him sympathetically, for the hardiest realized that this was illness, not cowardice. You had only to feel his weak and rapid pulse. The surgeon was on his way.

Details struggled with the flat tops, softening angles against the daylight. Nearly motionless, the rocket guards gazed in the direction of Boston. Nestling against the lip of the hill was a wan patch, like a dying bit of fox fire. It was a blanketed shelter tent, with flaps down where two officers worked over the intricate figures of new barrages.

Even in that unrevealing starlight each man you saw projected an expression of extreme weariness. Already many were ill with the dysentery that got us all sooner or later. And there was no prospect ahead of real sleep, as long as we stayed in that place.

There seemed no diminution in the fire even when the stars paled. The details took advantage of the first light and went over the lines while Hun airplanes loafed about the ridge and the positions. Instead of the brisk freshness of early morning, we breathed the warning odor of animal decay.

The last officer of the 16th walked through Les Pres Farm, asking about his horse, reminiscing disjointedly about his escapes. We watched him go without saying anything, wondering when we would follow him and how.

Chapter Seventeen

THE COST OF BATTLE

Somebody said they called our observatory at the Mont-Saint-Martin crossroads, Pittsburg, because it was so smoky. We inherited the name from the 16th, but there's probably something in it. Yet it is extremely doubtful if the Huns ever knew we had an observatory there. The instruments were behind a ruined garden wall, with a little foliage to protect them from airmen. The personnel, always limited in number, was taught to keep itself out of sight when airplanes appeared overhead. Against the heavier firing they sought refuge in an old wine cellar.

The crossroads were the reason Pittsburg received so much shelling. Our ambulances used the main road through Mont-Saint-Martin, tearing past Pittsburg and through the dismantled village. Always the Huns let them have it at the crossroads and through the town. That's really the reason the town was destroyed, for one fails to recall a definite bombardment of any of the buildings. After a time, they tried carrying the wounded through on stretchers, but these seemed to draw as much fire as the ambulances.

The first aid station at the farm was a busy place. The ambulances would scurry up the road and into the courtyard, unload, and hurry back again. Others would arrive empty and load up with men for the evacuation hospitals. At that point, we had a good many cases on the spot, for the shelling didn't let up until the enemy had retreated to the Aisne. Because the courtyard was the center of the farm, the Hun tried to put his center of impact there, so it received a large proportion of the hits. Certainly an average of five or six men a day must have been killed at the farm, while our division was in on the Vesle.

The Second Battalion had established its first aid station under Dr. Moore in its command post. The First Battalion had located Dr. Cronin in a draw between A and B Batteries, planning to use Dr. Moore for its command post and C Battery casualties.

The Second, however, decided on the second day to leave Les Pres Farm and move back to Chéry-Chartreuve, which seemed less exposed and less attractive to the Hun gunners. So the First arranged with the infantry to use its first aid station for local casualties.

Gas cases came in large numbers; the infantry sent out hundreds of men daily from the Vesle bottom. When Lieutenant Graham was relieved on the 17th, his eyes were seriously inflamed from mustard gas, and he was sent back to the echelon for two days. If he had asked, he might have been evacuated and so have escaped his fate of a few days later. But Graham knew how short of officers the regiment was, and he insisted on carrying on with his duty in spite of his painful condition.

Consolidation here differed radically from similar tasks in Lorraine. We were so busy that we straightened things out as we went along, and we were often surprised to learn how efficient our makeshifts were. For it must be remembered we were fighting under new conditions. Until this time, there had been very little of this semi-stabilized warfare. The 305th faced new problems, solved them, and gave instructors and secret pamphlet writers something to pass back to newer outfits. And, always, the digging on the ridge went on until Lieutenant Thayer was comparatively comfortable.

The firing, meantime, increased. On August 18, the day selected by Major Wanvig for the removal of his command post from the farm, Jerry commenced to take the most flattering care with us. A number of the Second Battalion telephone men, under Sergeant

Point, were salvaging wire in preparation for the move. Point, I recall, was spinning a reel, calling out good-natured encouragement to his workers. A group of the First Battalion men stood behind the farm wall, commenting on what appeared to be a relief from the heavy fire. Just then, the rising shriek of a shell made itself heard. For a moment we gauged the sound: that shell was going to fall mighty close, the shriek was right on us. Then it ceased and no explosion followed.

"Guess I'm getting jumpy," a man said. "Heaven knows where that bird burst."

"A dud," another warned. "Watch out for the next one." Still, one doesn't worry enough about the shell that doesn't burst.

We went about our business and within two minutes a shell exploded outside the window of the First Battalion command post, filling the room with smoke and knocking the adjutant, the telephone officer, and the sergeant-major across the room. Outside, Point and his men had dropped to the ground, while a fragment flew across to a shed against the farm wall, killing one of our horses. Everyone picked himself up, grinning, and sought shelter. The prospect was uncomfortably clear: that was the commencement of precision fire on the farm and it was no time to salvage wire.

Point, with Lieutenant Fenn and Sergeant-Major Applegate, stood close to the wall of the Second Battalion command post, apparently safe from the burst of any projectile coming over the building. We hadn't learned to appreciate yet the sharp angle of fall of some of the German howitzers. That cost Point his life. A shell whistled over the building and burst in the garden a few feet in front of the wall. The three fell to the ground, but a fragment flying towards the house caught Point in the back. He got first aid and was hurried away; he was as optimistic as ever, talking of a quick return to the detail. But Dr. Moore was doubtful, because his lung had been torn. In a few days, word came back that the sergeant was dead, and the Second Battalion now had a loss difficult to repair.

The Second Battalion left Les Pres Farm that afternoon, just when an order came down that each battalion should put forward four pirate pieces. Major Wanvig decided his Battery F, near the crossroads, accounted for his four. A and B were designated to each send out two pieces, and Captains Dana and Ravenel went on reconnaissance and chose the best positions available at some distance from each other, in a draw to the west of Saint-Gilles. They were on a forward slope, and it was necessary to lay lines to them approximately three kilometers long. There was always difficulty getting ammunition up, and it is probable that the pieces would have been more valuable in their battery emplacements. But, as has been suggested, the higher command as well as the lower was often experimenting, and at the time the move seemed useful. It proved a decidedly uncomfortable one.

That night the limbers were brought from the Forêt de Nesles. Details were already at work on the emplacements and the lines, when Lieutenant Brassell started up with the A guns, and Lieutenant Montgomery with the B. One of the telephone men met the party with the cheerful news that the Huns had been strafing the positions all evening. With the usual optimism of the American soldier, the cannoneers grunted and said that in that case things would probably be quiet for a while. So they sent the limbers away and manhandled the pieces into their emplacements. Shell by shell, they carried the ammunition in and piled it in the least exposed spots. Then, they started digging funk holes, for they saw they would need good ones. By midnight the Hun shells were dropping again, and the men drew off to a flank until the show seemed over. As soon as they had returned to their digging, Jerry popped at them again, appearing to follow them with an uncanny malice as they scurried for safety.

It was always more or less like that in those pirate positions. There were two regular programs that the men could foresee and guard against: one at 12:30 p.m., the other at 6:00 p.m. But in between came impromptu concerts that couldn't always be avoided.

Both A and B got plenty of attention. Both had the same difficulty bringing up ammunition, and both suffered from a similar lack of officers and men. Sergeant Buchbinder was put in charge of the A pieces, and Sergeant Martin got B's. There were no extra cannoneers. That meant, from ten soldiers at each position, men had to be found to serve the guns, to post guard, to dig shelters, to carry cooked rations from the kitchens three kilometers away, to lug in ammunition, scrape, and polish it, and to attend to odd jobs of sanitation and getting back the wounded. It must be remembered, too, that at that time nearly everybody was suffering from the weakening effects of dysentery.

No one knows how we got through such labor without sufficient sleep and with unsatisfactory food. Still, after a heavy shelling, even the digging went on with a strong rapidity.

The second night in, the *bosche* took a particular dislike to the B positions. Sergeant Martin ordered his men to the flank, but one of the early shells killed Private George J. Lucking and wounded Private Fred Scheuner. Two men volunteered to carry the wounded man three miles to the first aid station. He was heavy so occasionally they had to rest, pausing in a dugout. While one of them remained with Scheuner, the other hurried to the battery position and got a detail with a stretcher. Corporal Kelsey and Privates Terry and Elliot went back. The little party put Scheuner on the stretcher and started in. The Hun seemed to have a special sense for such missions—this time he opened up with gas. While the shells fell around them, the stretcher bearers put on their respirators remarking: "At that, gas is a darned sight better than H. E. [high explosive shells]"

On August 20, Sergeant Bernhardt's section relieved Sergeant Martin's. That same night a sergeant, who was an extremely good churchman, went up with two G. S. carts and an escort, bearing ammunition. The uncomfortable main road was his only practical route, and Jerry showed him that he had a better trick than high explosives for ammunition escorts. An airplane swooped low against the moonlight and began pumping machine-gun bullets at the sergeant and his horses. It seemed that both divine intervention and more speed was needed, as perhaps a combination of the two might avail. So the sergeant, thorough in all things, prayed devoutly: "Lord God, help us now!" And to the horses, with a different sort of fervor, he cried: "Get up, you ___s." The entire party lived to tell the story.

The difficult and disturbed routine of the pirate positions continued. Sergeant Buchbinder was carried out, wounded, on August 24, and the affair ended when the next day the *bosche* informed B that it had the pieces bracketed to a meter.

The early morning had been particularly quiet. The crews sat comfortably about one of the pieces, smoking after an early luncheon of cold chow and coffee. Not a bad looking place, they agreed, when Jerry let it alone. Evidently Jerry had had enough of them, and what an afternoon it would be to make up sleep! We heard a sharp "whiz!" —and the racket started with no more warning than that. An avalanche of metal then descended.

At the first whistle each man scurried for his funk hole. These little shelters, which looked like deep graves, had grown during the week. Each cannoneer crouched in his, listening to the angry shrieks of the fragments, to the splintering of trees, imagining always that he was the sole survivor of the party. It was fire for destruction of the most intense sort.

At the end, each crawled out and looked for the mangled bodies of his comrades. All that digging hadn't been wasted. The entire group stood there, half-dazed, but unhurt. The position, however, was in ruins: the trees lay in a twisted mass, Sergeant Bernhardt's gun was out of action (a huge fragment had passed through the recoil mechanism), and the telephone lines had been torn to pieces. That was the end of those pirate positions, and orders came to salvage what was left. The limbers appeared that night and drew the guns out and two G. S. carts arrived and loaded the ammunition.

Out on the open road one of the G. S. carts broke down. There was a good deal of shelling, and another airplane took a hand, dropping a bomb very close to the party. The limbers had gone on and the guide was evidently with them. The drivers of the carts, who had never been on the road before, were at a loss. Private Margid, who had been at the position from the first day, volunteered to stay behind until the cart was fixed and then he would guide both in. After another breakdown he got them to the position, and once more the firing batteries of A and B were united.

During these days, the men at the regular positions hadn't had any too pleasant a time. Private George L. Forman was killed on August 16, while walking from the Battery A position to the edge of Death Valley. On August 18, Captain Douglas Delanoy was wounded at an improvised observatory near Boston. He had an old German dugout for protection, and at the first shell he started to slide into this. A small fragment caught him on the knee, making a trivial wound—but his leg stiffened, he was evacuated, and he did not return to the regiment until the last of October. This left Lieutenant Derby in command of Battery F.

On August 21, while firing a normal barrage, Battery D's No. 2 piece was destroyed by a premature burst, as B's had been on the range at Souge. Fortunately, the full gun crew was not in the pit. Private Walter Rubino was killed and gunner Corporal Arthur Roos escaped with a bad fracture of the skull, probably because he was for the moment adding the duties of No. 2 to his own and was not on his seat when the lanyard was pulled. He was in the hospital for more than two months, but was eventually returned to his battery. Sergeant Jacob Metzger and Private Joseph Cohen were also seriously wounded and evacuated to America.

On the next day, August 22, the regiment lost its first officer at the front. Although he was still suffering from his gassing of a few days before, Lieutenant Graham had returned to duty, relieving Lieutenant MacNair at the infantry battalion command post. During the evening, he walked with Captain Belvedere Brooks of the 308th Infantry to a shelter near Savoie, known as Cemenocal Cave. The Huns had not, apparently, fired on this point before. Standing nearby were a number of other infantry officers and a large group of enlisted men. This congregation seemed unsafe, and Lieutenant Graham spoke up about it.

When a shell (that turned out to be a dud) came over and fell near the party, Captain (later Major) Breckenridge cried: "Look out!" There was a rush for the entrance of the cave; Graham and Brooks with the other infantry officers stood back to let the men in first. A second shell burst in the midst of the little group and Graham, Brooks, and a second infantry officer were killed. Lieutenant Bruce

Brooks, Captain Brooks' brother, was at that time assigned to our regiment. Captain Breckenridge got word to him, and telephoned Major Easterday of Lieutenant Graham's death.

Lieutenant MacNair happened to be in the Second Battalion's command post. He was hurried down to the infantry, while Lieutenant Ellsworth O. Strong was summoned from the echelon to replace Lieutenant Graham. Corporals Hickey and Rice and Privates Golden and Aasgard, who were on duty with the infantry, carried Lieutenant Graham's body over heavily shelled roads to Les Pres Farm. Chaplain Sheridan was summoned and the lieutenant was buried in the little cemetery on the Chartreuve Road, where so many of our men lie.

Three days later, Lieutenant Strong, who had relieved Lieutenant MacNair, was killed with a number of infantrymen near the same spot while going about his work with the quiet and confident ability that characterized everything he did. After that Lieutenants Klots and Brassel alternated on liaison. Lieutenant Klots was touched by a machine-gun bullet in the arm, but fortunately the wound was not serious, and he was back at work within a few days.

The question of officers was growing more serious daily. An order came through requiring the regiment to send one captain, three first lieutenants, and five second lieutenants to America to serve with new organizations. The colonel chose the following: Captain Fox; First Lieutenants Brooks, Dodworth, and Stryker; and Second Lieutenants Beck, Sawin, Schutt, Walsh, and Wemken. These officers left the Forêt de Nesles on August 26.

It was about this time, too, that the chief of artillery reminded Lieutenants Camp, Church, and Fenn of their recommendations at Souge. The first was sent as instructor to the Field Artillery School at Meucon, the second to Valdahon, and the third to La Corneau. As one can understand, the officers that remained didn't get much rest; an organization with two officers for duty was lucky.

One is reminded of the battery commander who was summoned to Division Headquarters to testify about some alleged short firing.

"On the day in question," he was asked, "did you have an officer with all your guns?"

He answered promptly: "I did not, sir."

And oh the disapproval of those Olympians whose lot in war lets them ask such questions! "And why not?" this Olympian demanded with an air of, "Young man you shall be tried."

There was a map present and the battery commander put his finger on it. "Because," he answered, "one of my guns is here, another is here, a kilometer from the first, and the other two are here, three kilometers away. I am the only officer on duty with my battery."

The telephone details were at it day and night, but communication on the Vesle was kept open. Working on the lines, as they did, the telephone men became experts in judging the probable point of impact of a shell. They knew when to duck, and they did it—under orders, some of them, at first. Without this ability and this touch of commonsense, a telephone man wouldn't have lasted long at Les Pres. It wasn't, however, always possible to duck. Sometimes there were too many shells in the air. Sometimes, too, the Huns used an Austrian 88 with a flat trajectory that was on you before you could really hear it coming.

On August 26, Corporal Schweitzer and Private Fred Isler were on the line from La Tuilerie to the First Battalion command post. A portion of this line was strung from old telegraph poles, and the pair carried a ladder as well as their testing instrument and spare wire. They had tested as far as the Chéry crossroads when they heard a big shell coming. They didn't have time to get rid of their impedimenta and duck. The shell burst too close to them; Isler was hit in the temple. Schweitzer carried him to a first aid station in Chéry, but he died without regaining consciousness.

Two days later another telephone man went. Regimental Headquarters had desired all along to establish an observatory forward with the infantry, although observation of any sort down there was difficult. A point had been located, and it was desired to run a line to it. Such a line would have to cross the open ground in front of Boston from the woods to the left, which were full of bodies and under constant fire. It was practically the same ground that so many infantrymen and artillerymen had attempted before to lay with wire —that was shot out almost as soon as it was laid.

Captain Gammell, Lieutenant Willis, and Private Frank Tiffany believed the importance of such an observatory made an attempt necessary and, as you never get anything in war without trying against odds, they set out towards Mont-Saint-Martin, paying out the line as they went. It was a brave effort that should have succeeded. But the Huns sniped at the trio, probably with an Austrian 88, and Private Tiffany was hit in the leg and back. The two officers carried him to Les Pres Farm; he died shortly after.

Such sniping was always to be looked out for, as it was particularly dangerous. So was the intermittent dropping of single shells about the farm, at intervals of a few minutes, all day and night. On the other hand, the concentrated firing of the Germans, although irritating, it was by no means so risky, because you could tell after a fashion what to expect and when.

The Him introduced an appreciable amount of system into his shelling of the farm and its neighboring positions. Let us say it is 8:30 in the evening, when the last light softens the shattered buildings. Here and there groups of men stand close to the walls—several are coiling wire on an improvised hand reel, one glances at his watch. "Must be time for the evening shower," he says. Several men yawn in response. The groups scatter, some slip into the cellar, while others seek the shelter of the walls where a few funk holes have been dug. In a moment there is no sign of life about the place, except for a delayed ambulance plodding up the hill and a curious head that projects cautiously from the cellarway.

"Whiz-z-z . . . bang!" breaks the silence. The ambulance scurries into the courtyard, the curious head disappears, and shells follow one another with a relentless rapidity. It is like the cracking of several whips with long lashes—the crack of one is lost in the swish of another. These are "105s" [105mm shells]. In the cellars and behind the walls, the men are safe enough from these shells, except from a direct hit, and their chances are fairly good although all the shells are certain to fall within a limited radius.

The switchboard operator turns his crank and gets Regimental Headquarters for the major. "Raining hard," the major reports to the adjutant.

"How hard?" the adjutant inquires.

"Pouring," says the major bluntly.

It's not altogether pleasant to be asked such questions when you're in the midst of the storm. Somebody's got to stick his head out to verify the size of the shells and somebody's got to count them. The first time we had this particular drubbing the major asked us for estimates of the rate of fire, so that he could tell them back at Regimental Headquarters. One officer, in an honest effort to be conservative, put his reply as low as a hundred shells a minute. Another said seventy-five. A third objected, saying, "That's all nonsense. It can't be more than fifty a minute—a little less than one a second." The noise made him seem like a poseur. We got out our stop watches. The rate of fire averaged just eight shells a minute.

After a few days, "pouring" was enough to indicate that particular strafing. At the end of twenty minutes, everyone yawns and prepares to go about his affairs. The racket suddenly ceases, the curtains are thrust back, the men slip out, clinging close to the walls because that intermittent firing will continue all night, more dangerous than the expensive burst we just experienced.

It was amusing after one of these noisy, shrieking concentrations to watch men ducking at the whistle of a projectile that would probably burst a kilometer away. They did have that effect; they put one's nerves on edge. You never got accustomed to many fragments flying past you with a sound like the crying out of mad witches. Always after these exhibitions there were fresh holes in the roofs and walls of the farm, and usually another piece of the cellar steps would be knocked away.

These strafings annoyed the cooks. Even here, they clung to their fires as they had done in Lorraine. After one of the first concentrations, we rushed out and checked up on the men. A cook was missing. "Who saw him last?" we asked.

"I saw him in the kitchen just before the shells came in," a kitchen police answered.

Hesitatingly, we stepped to the open door of the kitchen. It was quite dark in there, except for a red glow from the stove against the end wall. In that red glow, we saw a body outstretched. We tiptoed in, the body stirred, and the head, we could see, was hidden in the hot oven. We drew the man gently away. It was our cook, his face the color of a well-broiled lobster. For nearly twenty minutes, he had lain there with that pandemonium raging outside, his head at least protected, if rather painfully so.

Another day the surrey came up with rations. Manzo, the First Battalion cook, looked them over and forgot all about the war for a time. He told everybody what Sergeant Bayer and Ramstad had turned over to him: fresh beef, potatoes, rice. "No Corn Willy for dinner today, boys!" he cried. Manzo was the most popular man in the army. We had lived on Corn Willy, goldfish [canned salmon], and beans for so long that the thought of fresh food was a little heady. Manzo and his assistants set to work, and extraordinarily pleasant odors slipped from the kitchen. "Bet the flies don't get any of my dinner," one man boasted. It is doubtful if any Christmas feast was ever looked forward to as eagerly as that meal.

Then, the tragedy happened. The Huns began to shell, and off their scheduled time. Manzo and his assistants were forced reluctantly from the kitchen. They left the dinner cooking on the stove; fifteen or twenty minutes of absence wouldn't hurt it. But all the time that Manzo was inhabiting a shelter he was uneasy. When the bombardment lifted, Manzo and his assistants crawled out and hurried to the kitchen. A moment later Manzo came rushing out. He saw Major Easterday, flung up his hands, and burst out: "Maje! The Hun! He shoota da hell outa da kitch!"

The major was as interested as everyone else in the feast. The entire detail crowded into Manzo's temple. Few had the courage to gaze for more than a moment on the scene of sacrilege. A shell had come through the end wall, had landed on the stove, and burst there. The remains of dinner were on the floor, the walls, and the ceiling. Strong men wept, while Manzo went sadly back to his tins of hardtack and Corn Willy. For soldiers must eat. It is such outrages that breed hate.

While some of these escapes held a touch of humor, they were rather too close for comfort. The affair of the dud at Regimental Headquarters, for example, might have had a very different ending. On the morning of August 22, a 105mm shell ripped into the building, through a room in which Captain Fox and Lieutenants Klots and Willis were standing. It then tore through the wall into the next room and passed through Colonel Doyle's cot, which fortunately was not occupied. The shell failed to explode, but had it exploded, Regimental Headquarters would have needed some new officers.

Chapter Eighteen

SPIES AND THE ADVANCE

Chéry-Chartreuve did not prove to be the ideal command post the Second Battalion had hoped for. The Huns undoubtedly knew the town was thick with headquarters and therefore, logically, shelled it a good deal. So Major Wanvig decided to move to a cave in dead space in the steep hillside to the east of Chéry.

The move was originally planned for August 24. On the morning of the 23rd, Regimental Headquarters called for a number of barrages, then abruptly shortened the lines. This meant to everyone a strong enemy attack; perhaps that vast effort we had sometimes looked for to recapture the lost ground in another drive for Paris. As a matter of fact, the enemy did get La Tannerie and portions of the south bank of the river that morning, but they were unable to hold their gains for very long.

In the midst of the confusion born of this rapid and unexpected work, Major Wanvig telephoned from Regimental Headquarters to move the P. C. (Post Command) at once. At that time, the battalion staff was really too small for its routine work. Lieutenant Fenn gave the difficult task of wiring the new P. C. to Sergeant Froede and tried to keep things going from the old headquarters.

All afternoon and evening the batteries continued their firing. At midnight, a complete program came in from Regimental Headquarters for a rolling barrage to accompany a counterattack by our infantry. It was hurriedly figured, and rapid firing went on until 5:00 a.m., when word came to cease firing. It was explained that there had been a misunderstanding and the infantry had not counterattacked. So much ammunition was expended that night that stray dumps were scoured for serviceable shells. Still, before many hours, a counterattack was staged that reached its objectives. Without interfering with its program, the Second Battalion got into its cave where it was never once shelled.

That night was exceptional, but every day and every night an enormous quantity of ammunition was fired. Under such conditions there were inevitably charges of short firing. The Germans had a number of guns in the vicinity of Rheims that occasionally treated infantry and artillery to a few shells. These seemed to drop from behind us, although what we suffered was really only enfilade fire. It is not extraordinary that the infantry should have thought some of these puzzling shells were shorts from their own artillery.

One day, Captain Whelpley was sent from Regimental Headquarters to investigate such a charge, which had been advanced by Captain C. W. Harrington of the 308th Infantry. Captain Whelpley lost some time at Les Pres Farm waiting for a guide, so that it was dark when, after a hazardous walk, he reached Captain Harrington's command post to the north of the Vesle. It seemed impractical to return that night, and so Captain Whelpley intended to start at daybreak. With the first light, however, the Huns put down an intensive barrage that lasted for an hour, which made a shell hole a pleasanter place to be than in the open. This was followed by an infantry attack in strength. Captain Whelpley picked up a rifle and told Captain Harrington he would help. With a party of men, he moved to the edge of a patch of woods to observe and cover Harrington's left flank. He also maintained liaison with neighboring units. His party killed ten Germans and captured three. For this voluntary assistance to the infantry at a critical time, he was mentioned in division orders after the Armistice. If it had not been for the colonel, who asked for an explanation of Whelpley's absence, the story of his courage might not have been made public.

Charges of short firing were always investigated but never amounted to anything on the Vesle. For the regiment, short-officered as it was, had developed a facility with figures and execution that left small room for mistakes. The lessons learned here made the problems of the Argonne for the 305th comparatively simple. Such experience is not gained without a continued cost.

The enemy got First Class Private Frederick J. Weeber of Battery E on August 25. He was in his gun emplacement with another cannoneer when an "over," intended for the Chéry crossroads, fell just outside. "Look out!" Weeber called to his companion. He didn't duck low enough himself. The other man escaped, but Weeber was carried to join that great silent army that lies in the shallow graves of Champagne.

The Huns favorite type of warfare seemed now and then to be aided by a brutal sort of luck. As I said some time back, we were taught not to care as much as we had for the Y. M. C. A. in Chéry-Chartreuve. That lesson came on August 28. Even if the passage was risky, it was a relief to get permission to leave one's position and dodge to the pleasant odors and companionships of that little store.

On this day, there was a long line of infantrymen and artillerymen waiting in the street to get to the counter. But a shell, which seemed guided by an evil genius, fell in the middle of the line, burst, and harvested eighteen casualties. Of our regiment, Private Charles C. Rosalia, Battery E, was killed; and Privates Rasmus Hanson, Battery E, Dona J. Monette, Battery E, and Corporal Alexander Landsman, Battery D, were wounded. On the whole, though, one wonders that we didn't have more casualties in that heavily shelled, unprotected sector. We suffered a good many more than we liked, but the regiment felt that its intelligent discipline kept the number down.

Naturally, there were some duties that had to be done blindly, as it were, without using brains or anything else to protect yourself. For instance, barrages had to be fired whether your position was being shelled or not. And rocket guards, when their comrades scattered for the funk holes at the first warning shell, had to stand their ground and take whatever came.

Private Hackett of Battery B was caught like that one night. He remained sitting on an empty ammunition box, his glance always on Boston Ridge, while his more fortunate friends got out of the way. He was pathetically reminiscent of the well-sung young man who stood upon the burning deck when he very well knew he ought to have been nearly anywhere else. A shell burst at Private Hackett's feet, and when the smoke and dust cleared away, he still sat upon the box, and his gaze was still on the ridge, but now his feet were in a new crater. So he lived to become known admiringly as "The Salvage King." His own description of the moment was: "Think? When the thing went off I expected to see myself in little pieces."

On the Vesle, spies were more dreaded than in Lorraine. The bitter nature of the fighting placed in a spy's hands the lives of more men. Over several nights, we noticed the unequal flashing of a lamp on Boston Ridge. The infantry there had seen it, too. Many efforts were made to catch the operator, yet none met with success. If he was a spy, he was an amazingly clever one. If he was a telephone lines-man, carelessly using, against all orders, a light as he worked on a wire, he was lucky far beyond his due. At any rate, after a few nights the flashing ceased.

The order from General Bullard, which follows, tells its own story.

P. C. Third Army Corps, August 31, 1918—21:30 Hr.

G-3 Order

No. 56

1. During the attack of the enemy against Fismette, August 27, someone in American uniform ran among our troops shouting that further resistance was useless and that one of our officers advised everybody to surrender. These statements were absolutely incorrect because further resistance was not useless and no officer had advised surrender. Nevertheless, because of lack of training and

understanding, the results were as follows: Out of 190 of our troops engaged in this fight, a few were killed or wounded, about 30 retreated fighting and escaped, and perhaps 140 surrendered or were captured.

2. A person who spreads such an alarm is either an enemy in our uniform, or one of our own troops who is disloyal and a traitor, or one of our own troops who has become a panic-stricken coward. WHOEVER HE IS, HE SHOULD BE KILLED ON THE SPOT.

3. In a battle there is no time to inquire into the identity or motives of persons who create panic, disorganization, or surrender. It is the duty of every officer and soldier to kill on the spot any person who in a fight urges or advises anyone to surrender or to stop fighting. It makes no difference whether the person is a stranger or a friend, or whether he is an officer or a private.

4. The day before the attack on Fismette, a German soldier was seen and then mortally wounded by our men in Fismes, far inside our lines. He was well stocked with food. Having lived for many years in America, it was possible for him to get an American uniform. And, because of his knowledge of our language and customs, he would have been able to create doubt and disorganization among our men.

5. Division commanders will cause this order to be read to each company or platoon in such manner as will insure that every member of the command thoroughly understands its contents.

By Command of Major General Bullard:

F. W. Clark Lieutenant Colonel, G. S.,

A. C. of S., G–3

The attack against Fismette, mentioned in the foregoing order, was one of the last determined offensive efforts of the enemy on this front. It became clear about the same time that a vast German retrograde movement was in contemplation. Any change from Les Pres Farm would be a welcome one.

The intensity of our firing increased, while Jerry's waned. Undoubtedly we were making his plans difficult to carry through. On the night of September 3, the observatories reported many fires in Perles and its vicinity. A huge sheet of flame advertised the explosion of a big ammunition dump. Towards morning of September 4, the Hun-made fires thickened. Evidently great quantities of stores and the buildings that had housed them were being destroyed as an alternative to leaving them for the Americans. The Hun fire nearly ceased. Anyone who was there will recall the blessed relief of being able to stroll about those positions at last with a feeling of comparative safety.

Word came that the infantry was already moving forward. The artillery would follow in support. Strong combat patrols were already in contact with the enemy. It was understood that if a battalion of infantry were sent as an advance party across the Vesle, Battery D of our regiment would cross, too. But the Hun went faster than the most optimistic had prophesied, and the entire regiment started forward on September 5.

The old positions were policed and equipment made ready on the night of the 4th. Early the next morning, the limbers came down from the echelon, whips cracked, and, after those unpleasant weeks about Les Pres and Chéry, the regiment was on the road again.

Since they had been widely scattered, the batteries followed the most convenient routes while agents kept them in touch with battalion headquarters. Regimental Headquarters went forward to the desolate ruins of Fismes and established itself in a cellar. Opposite the cellar steps an alley ran between tumbled walls. The horses, motorcycles, and bicycles were placed here as the safest place in the vicinity.

Shortly after the party had arrived, Private Wallace Fisher, of the Headquarters Company and motorcycle driver for the Second Battalion, entered this alley and started to make some repairs on his machine. He was the only man there, so no one else saw what happened. In the cellar, they heard a dud fall and then another shell come over and detonate across the street. Corporal Tucker ran from the cellar to see if the horses had been struck: two were down and the third, which curiously had been the center one of the trio, was unhurt. Tucker saw that both motorcycles had been smashed. He saw Fisher lying beside one and called to him. Fisher didn't answer, and the scout went closer, and discovered that Fisher had been killed. Tucker reported the news back across the street, and a party buried Fisher in the garden behind headquarters, making a rough cross from the wood of a splintered door for his grave.

Battalion commanders, along with their captains or reconnaissance officers, started forward early to select new positions in the vicinity of Savoie and Saint-Gilles. It rained hard, and the complaints were bitter and many—at first. A little later, the men realized what a blessing the bad weather had been, for the Huns still held control of the air and with better visibility he would have dropped more bombs and directed better fire on our columns, which crawled by daylight along crowded roads. He would have interfered more disagreeably with the taking up of the new positions. One fellow did appear, flying low to get beneath the mist. The battery machine gunners greeted him with shouts, sending such well-directed streams of machine-gun bullets at his plane that he left the cannoneers to settle their guns in peace.

While it was perfectly obvious these positions would be occupied only a short time, they were consolidated, after the habit of the regiment, as if they were intended for the duration of the war. The cannoneers dug in, and officers and details figured firing data, and ran long, difficult lines for only a few hours of use. First Battalion Headquarters had moved out of Les Pres Farm to a house near Mont-Saint-Martin. It was necessary for its batteries to be in telephonic liaison with it.

After only a little firing, the order came to move again at midnight. The limbers had been echeloned in the neighborhood, so that there was no delay in starting. Everyone knew the next stop would be nearer the enemy and that the guns must be in position and hidden before daylight.

The batteries rendezvoused near the crossroads between Fismes and La Tannerie. Battalion Headquarters went ahead to the crossroads, which threatened to be an unhealthy place. The Huns did commence to shell it, but most of their projectiles fell to the right in low ground. Here again the rain proved its friendliness, for in the wet soil the majority of the shells buried themselves without exploding.

Nevertheless, such waiting was nervous business, for there was always the prospect that the Hun would sweep, or at least shift, his deflection. He seemed, however, to have lost some of his skill, or else he imagined himself directly on his target. The column grew restless, asking: "What's slowing us up?" and "Where are we going anyway?" Whispers filtered back, telling us: "We're going across the Vesle. It's the bridges that are slowing us up." There was a dramatic quality about this realization—for we were going across the Vesle, to those very heights from which Jerry had pounded the regiment for so long! Everyone was curious, too, as to the kind of bridge we would find, and about the cost of building a bridge in such a place. The news, moreover, brought some apprehension. If the crossing of the Marne had caused misgivings, the passage of the Vesle created graver ones. The Hun artillery must surely have it registered. It was inconceivable that we could get over without a shelling. Perhaps that explained the delay. The bridge might be down, or it might be blocked by dead animals and broken carriages.

Long drawn, the command to "get ahead" ran down the line. Horses stumbled forward. The luminous faces of wristwatches appeared like fireflies here and there, as the men took a check on the time. Almost immediately, the rumbling of wheels on planks could be heard. Word was passed along that there would be two streams to cross. At each, men would dismount and lead their animals over most carefully, for there were no side guards or rails, and the column wasn't using any flares to guide its feet.

The carriages rumbled on the planking. Down below, between steep banks, rushed a narrow and black stream—the Ardre, about a kilometer from its junction with the Vesle. There was no disturbance there, and the column was swallowed by the crumbling outskirts of Fismes. Just beyond, the road swept to the right onto the main highway to Braine, and so came upon the Vesle.

The only light—and it wasn't much—was from Jerry's distant

flares and star shells. It became clear to the men that the enemy was after this second bridge. The rustle or shriek of arriving shells was perpetual, but there was an odd scarcity of detonations, and there was no halting. At the river, the reason became apparent. Again, the enemy had failed to register quite perfectly and the low ground and the rain were friendly. Most of the German projectiles were duds.

The Vesle River was scarcely wider than the Ardre, but the bridge seemed narrower and riskier than the other. Drivers led their horses and cannoneers manned the wheels. There was only one casualty, which aroused a laugh that made itself audible above the shells: musician Scharf, acting as messenger, was crowded over the side and splashed into the deep, unpleasant current. They pulled him out, and he went on his way, laughing, too.

The column hurried through Fismette, into which the regiment had sent so many shells, and scattered into the positions selected during the reconnaissance efforts of the day before. The First Battalion commenced to dig in a kilometer south of Blanzy, near a confluent of the Vesle. The Second Battalion, which had come from its Chéry home without taking up intermediate positions, swung more to the west, and with its batteries side by side established itself on the slope of a deep ravine across from Perles. By daylight every battery was in place.

The First Battalion settled its command post in a road-repair-man's house on the Fismette to Blanzy road. There was no cellar; the only protection was the stone walls of the building. The Second Battalion chose a German dugout in the ravine between Perles and its guns. Regimental Headquarters, meanwhile, moved forward from Fismes on September 8, and came upon what proved to be about its nastiest experience of the war.

It was the custom for our headquarters to remain with Infantry Brigade Headquarters. Near Blanzy was the cave of La Petite Logette, a huge hole that the Hun had long occupied, digging from it many galleries. It was a perfect shelter except for one thing: its very appearance proclaimed it a gas trap. With no opportunity to judge, Regimental Headquarters established its command post with Brigade

Headquarters in the cave. Engineer and medical officers worked nearly all day to purify the air of this formidable hole. They declared the main portion was safe when Colonel Doyle arrived later that afternoon, but even then the place retained an unhealthy and ominous atmosphere. The doctors had boarded up the more suspicious of the galleries, and they warned the men against invading the remainder. The men, however, were very tired. The mere fact that such a place had been chosen as command post was a recommendation to them that it was safe. And some of the galleries were a good deal quieter than the main portion of the cave.

Regimental Headquarters set to work at La Petite Logette, quite a different affair from Regimental Headquarters on the table in the mess hall of Jl at Camp Upton. There were about forty men attached to it at that time. After dark, when all the soldiers not on missions should have been in the large cave or near the entrance, a check was taken and a number reported as missing. The searchers entered the forbidden galleries and found a number asleep or resting, quite unaware of the risk they ran. All of them were gassed to some degree. They were removed and treated, and the night's work went on.

About midnight, a new condition stealthily disclosed itself. Men sniffed the air of the main cave and discovered it was clearly poisoned. So much gas could not have escaped from the galleries. The Huns, beyond question, must have buried gas shells in the floor of the cave, surrounding them with an acid, perhaps, to eat through the casings and so release the fumes when the occupants were without suspicion. Consequently, most of those who had spent the evening in the cave were unfit for duty. There was no other shelter nearby, but the colonel ordered everyone out of the cave.

"The entire medical staff (officers and men)," to quote Colonel Doyle's account of the evening, "had been gassed and were unable to give any assistance. Colonel Doyle alone remained in the cave, giving aid to a constant stream of gassed men."

As is usual with slight cases of mustard-gas poisoning, eyes suffered most of all, and so many soldiers were temporarily blinded. After their eyes had been bathed with a weak alkaline solution, the victims were hurriedly evacuated. A few were more seriously affected.

Colonel Doyle, meanwhile, worked until 4:30 in the morning, when he was forced to leave the cave. A medical officer of the Engineer Corps, who had been summoned, took his place. The effects of the gas on the colonel were slow. He stayed by the telephone all day, and it was only after a hard day's work, towards ten o'clock, that he lost the use of his eyes. As long as he could talk, however, he insisted on staying with his regiment, and he was not evacuated until midnight. The regiment did not lose him for long, but he suffered from his experience for many months afterward.

The list of officers and men who were more or less gassed in this extraordinary incident includes: Colonel Doyle, Captain Gammell, Captain Mitchell, Lieutenant Klots (his second wound stripe), Sergeant Bromm, Sergeant Mamluck, Sergeant Major Miller, and Gillette, Hoffman, Kurash, Palmer, Pullen, Saloman, and Wallach.

The regiment had struggled through its most difficult days with an insufficient number of officers. When the word came that it was to receive replacements, officers and men took the news skeptically. Only two or three had come in before the crossing of the Vesle, but now the rush commenced. First Lieutenant H. J. Svenson had arrived on September 1, but he was invalided away on the 14th. Second Lieutenants George E. Putnam and Jesse W. Stribling had reported on September 3, but the real influx came when the batteries were in their new positions across the Vesle.

On September 8, Second Lieutenants Stedman B. Hoar, and David J. Macleod, a veterinarian, reported. On the 9th, came Second Lieutenants Osbon W. Bullen, Johnston Copelin, Raymond E. Dockery, Leon H. Hattemer, and Harold Holcomb. On the 10th, the arriving stream of subalterns seemed a beautiful dream. That day brought Second Lieutenants Roy H. Camp, Thadeus R. Geisert, Edward W. Hart, Albert B. Hill, Waldo E. McKee, Thomas M. Norton, Reuben T. Taylor, John G. Teichmoeller, Philip A. Wilhite, and Charles L. Graham.

These were practically all young men from the Artillery School at

Saumur. They were distributed among the three headquarters and the batteries, and made the fighting between the Vesle and the Aisne far simpler than it had been in the shorthanded days of Les Pres Farm.

For self-sacrificing work in the Vesle-Aisne fighting, Lieutenant Thayer, Corporal Ramsdell, and Privates Shackman and McCune received divisional citations.

This campaign was in many ways far less exacting than the preceding one. The regiment, to be sure, was opposite the pivotal point of the Hun line between Soissons and Rheims but, although there was plenty of artillery opposite, the shooting seemed poorer and there were fewer casualties.

The weather played its share, too. The brilliant, warm days of Les Pres Farm were replaced by much mist and rain. The nights, too, were colder. The men, therefore, did not need much urging to dig them-selves in. Very few German dugouts could be used, because their openings were in the direction of hostile fire, but German straw could be carried from its old home to the new hillside apartments of the Americans. Tiny, living souvenirs may have come with that straw, but one acquired those anyway and it seemed a small price to pay for warmth (that is, before the S. O. S. inspectors got at the regiment). There's no point in wasting words on cooties, for practically every man and officer knows all there is to say about them.

Observation brought its difficulties here also. There was no satis-factory observatory near the First Battalion command post, so Lieu-tenant Thayer pushed forward to the very frontline of the infantry. On the edge of the ruins of Serval, he found a deserted house that stood on high ground in a salient of the American frontline, so that it was exposed to fire from three sides. Yet, while nearly everything else in Serval had been destroyed, this building was comparatively whole.

Lieutenant Thayer didn't attempt to get his men in or run a tele-phone line until after dark. The line was long and difficult to keep open, but for the most part communication was maintained. By using extreme care, the presence of observers in the house was kept from the Germans. Only once, while the regiment was in that position, did the place get a direct hit. Yet it was necessary to make reliefs, to carry in food, to bring water from a well in Serval, and to have telephone men coming along the line whenever it went out.

Here is an example of a conversation you might here in the lower room, after a telephone man has crawled in and lies on his back, catching his breath.

"You fixed the line all right," says one of the observers gratefully. "What kind of a trip did you have?"

"As per usual, kid," the telephone man explains as he rests. "All the way across, Jerry threw G. I. cans at me as if they didn't cost a cent. When I gets to the foot of the hill here a machine gun goes pop-pop-pop-pop. I plays possum, but for a long time, every time I lifts my head, pop-pop-pop-pop he goes again. Honest, George, I've never felt very harsh towards the *bosche* but, George, when they turn a machine gun loose on one poor linesman every time he moves his little finger, I say they ain't right-minded folks. Can't tell me any atrocity stories I won't swallow now, George."

The interior of the stone house was given over to perpetual watchfulness. Old clothing was hung across the front windows, so that no one would be silhouetted for the benefit of the Germans, and behind these the instruments were placed. Day and night Lieutenant Thayer and his scouts watched the Germans and the effect of our fire, within calling distance, practically, of his victims.

Positions very much less exposed didn't fare so well. When the regiment crossed the river, the Supply Company moved forward from the Forêt de Nesles to the grove behind Les Pres Farm, in which Battery C had been stationed until September 5. By all the rules of the game that should have been a safer place than Forêt de Nesles. The Supply Company had two men killed during the war and both were lost in this place.

This tragedy recalled the earlier charges of short firing. With all of the batteries far forward, no such explanation could be advanced here. Evidently the Hun's guns near Rheims were at work again. The Supply Company men indulged in the wildest hazards to account for this strange shelling. There was talk of supernaturally concealed guns left by the Germans when they had retreated. There were whispers of an extraordinary underground railway on which the *bosche* moved big guns to convenient trapdoors within our lines. For, until the Rheims explanation was generally passed around, this fire did look like magic.

It was on September 11 that these shells got Wagoners Jackob E. Jackson and Fiori Fillici. There had been some firing, but at three o'clock it lifted, and the men poured from their funk holes and returned to work. Jackson was cleaning harness at one of the wagons when the company clerk came up and spoke to him. The wagoner was very happy, for he had just that day received a letter from home, telling him that his wife had presented him with a son. He displayed the letter to the clerk, and they chatted cheerfully about the future. With the Huns falling back all along the line, they agreed it might be only a few months before Jackson would be on his way home to this new arrival. The clerk promised to look after the additional government allowance that was due to Jackson's wife following the baby's birth.

"My wife," Jackson said, "needs the money very much, because things are so high in the States." He said nothing more after that. The clerk climbed into the wagon to search for something the captain had left there, and at once the Huns resumed their odd shelling. The third shell, the clerk said, seemed to burst directly beside the wagon. A piece hit him in the leg, inflicting only a slight wound. When he climbed down, he saw Jackson lying on the ground and a medical orderly bending over him. A piece of the shell had struck him in the back of the head; he died on the way to the hospital.

Fillici was killed during the same bombardment, although he was a short distance from the echelon. He had started on a horse, without saddle or bridle, to get some medicine from the Veterinary Detachment. Fillici had volunteered for this service because the company veterinarian was occupied at the moment. He had been advised to take a shortcut, but instead he chose the main road.

The news of his death was brought by French soldiers, who had been working on the road. The shell, they said, had burst very close to Fillici, knocking him from his horse. Fillici had been killed, but the horse had not been scratched. The Frenchmen said that the same shell had killed a captain and a lieutenant of the 305th Infantry.

When one considers the number of shells that fall idly, it is astonishing to count up the amount of damage a single shell, better aimed or carried by chance, will accomplish. The First Battalion got one of these at its command post near Blanzy on September 15. For days, shells of all calibers had fallen about the place without accomplishing any more damage than tearing up the soil. Then, this one arrived: it fell at the picket line, where the horses stood in a row. Private Aimer M. Aasgard groomed a horse near the end of the line. Near him sat a group of telephone men, winding wire on makeshift reels—a necessary diversion of the telephone detail when there was nothing else to do. The men heard the whine of the approaching shell and realized from their acquired judgment that it would fall very near. They called out a warning and ducked. Aasgard wasn't quick enough—a tiny fragment cut into his neck, severing the jugular vein. Dr. Cronin hurried to the doomed man, who died within a few minutes.

The same shell caught Corporal Leonard Cook, of the telephone detail, in the knee, disabling him and putting him out of the war. An ambitious telephone man, he was evacuated grumblingly and was never returned to the regiment. Other fragments cost the detail eight more of its vanishing horses. But these serious moments were the exception; life north of the Vesle was far less complicated than it had been around Les Pres. There were, of course, minor casualties.

First Class Private McGranaghan gave Sergeant Hickey an opportunity to distinguish himself. McGranaghan was hit while working on the Serval line. Hickey, who had been on duty in the observatory, picked him up and carried him over a crest to the first aid station, all while exposed to machine-gun fire. Individual instances of courage like this were innumerable. These men, however, don't say much about what they do themselves. Unless someone happened to see their bravery, their acts drifted into that vast blurred background of devotion and sacrifice against which the American soldier fought.

Between the Vesle and the Aisne the Second Battalion was even more fortunate than the First. Major Wanvig's command didn't have a single casualty in the Perles positions. Hun airmen gave the Second Battalion one bad night and might have done a lot of damage. A bomber created the impression that he had located the emplacements, for he dropped a number of flares over them, and followed with two bombs in the ravine, which missed Battalion Headquarters, and he dropped one on the slope close to the guns, which splintered a number of trees.

A group of men from Battery D had a close run of it. They had made themselves comfortable in a large German dugout that had only a sheet of elephant iron for overhead cover. At the first flare they decided there might be safer places, and sought one. When they returned a few moments later, after the plane had throbbed away, they found their pleasant home had become a mass of twisted elephant iron, ploughed up dirt, and ruined equipment. The third bomb made a direct hit on the dugout, in which they had just before been crowded into for warmth.

The regiment fired as persistently here as it had done in the Les Pres and Chéry positions. Barrage after barrage was thrown ahead of our infantry on La Petite Montagne, which because of its pivotal situation, was of great strategic importance. Before it was captured the order came for the regiment to move to other pastures.

Chapter Nineteen

THE ARGONNE

The first intimation that the 305th had that it would be relieved was brought by advance parties from General Garibaldi's Italian division. The sight of these strange faces and uniforms indicated to everyone that the regiment was going out for a well-earned rest. How deceitful that opinion was, everyone remembers, but the occasion was important and exciting. All our men of Italian parentage greeted the newcomers with joy and hospitality. There was much excited conversation, and there were more interpreters than could possibly be used.

While the Italians reconnoitered, the Americans packed—joyously, too. The prospect of billets, baths, and cooked food was alluring after more than two months in the line. The thought of quiet, after a month of such fighting as the Vesle had developed, was frankly welcome.

The movement commenced on the night of September 15–16. No one had any idea where we were going, except that it was to the rear. And the belief in billets was touchingly firm. The columns wound down roads on which they had advanced under shell fire, through the fragmentary and odorous remains of Fismette and Fismes, past Les Pres Farm, at which some fists were shaken, through Chéry-Chartreuve for the last time, and to the crossroads just beyond where the two battalions rendezvoused.

When the last man was up, the regiment took the road to the left through Dravegny, where our infantry was regrouping, and then Coulognes-Cohan to the Bois Meunière, which was selected for the first bivouac. Between eight o'clock in the evening and three o'clock in the morning, the column covered twenty-three kilometers. After the exhausting work of the past two months, it was a tired, nearly voiceless column that rode away from the flares, the flashes, and the star shells. Many drivers slept on their horses and the cannoneers, doomed to walk, stumbled forward, only half awake.

There was a delay of nearly half an hour just before reaching the bivouac. The column halted as if automatically. The men rested where they were, deciding it was quite like old times. Impatience seized a group of officers, and they rode forward to learn, if they could, why the halt continued. Ahead the road was open save for one obstacle: a machine-gun cart rested in the middle. On the seat was a dozing driver and attached to the cart was a mule, supremely indifferent and content. The group awakened the driver hurriedly. After yawning the driver explained: "Reckon Jinny's decided she's gone far enough tonight." Jinny and her master then suffered the application of united brute force and watched the column go by.

It was on this first stage that Battery F wandered astray. In the dark it mistook the 306th column for our own and followed it for some time, until scouts located it, explained the situation, and led it back to the fold.

During the day, men fought against the light and the noise again for a little sleep, and at eight o'clock moved out once more. In the early morning, the carriages rumbled across the Marne on an engineers' bridge at Verneuil. The average man's sensations were very different from those aroused by his previous crossing at Château-Thierry. Again, the river was a dividing line, with the country seeming immeasurably less disturbed to the south, and thus the march lost its sense of being conducted under the menace of airplanes.

At Mareuil-le-Port, where that twenty-kilometer stage ended, an officer brought us joy with several motor trucks assigned to the regiment for the transportation of a certain number of dismounted men. Sixty were chosen from each organization and put under the charge of Lieutenants Brassel, Putnam, and Copelin. Although it wasn't generally known at the time, the destination of these trucks was La Grange, three kilometers northwest of Sainte-Menehould. The rest of the regiment, which was condemned to the long hike, continued to foresee a glorious rest ahead. The rumor was that the billets were a four-days march away.

Mareuil-le-Port had other cheering features. The weather still held fair, for the country was not yet scourged by autumn, and the area was pleasant to men fresh from the gashed slopes and devastated forests of battlefields. The gun park, the picket lines, the straight rows of shelter tents were arranged in pleasant fields. In the village, where the civilian population went about its business, there were shops for the first time since Doue, specializing in a fresh cheese that nearly everyone added to his rations. Best of all, the column didn't form again until 10 'clock on the morning of September 18, so we had all one day and a large part of the night for rest.

The roads were now not particularly congested, so the regiment traveled rapidly, which is far less fatiguing than a snail's march with many halts. It was generally known by this time that the French were routing the column, and were keeping it off the congested main lines of supplies. Therefore, twenty kilometers were covered by eleven o'clock on the morning of September 18, reaching the summit of a high hill at Grauves, near Épernay.

The weather threatened here, but the place had matters of interest. It was in the heart of the champagne country, and the wine was plentiful, cheap, and harmless, as far as one could judge. Thirst was excusable after the last two miles of that stage—the horses would have given up the grade, if the men hadn't encouraged them and put shoulders to the wheels.

At four o'clock the next morning, the regiment was on the road again. Its route lay through the plains of the Marne, a rich country sheltering farms and vineyards that had not experienced the harsher touches of war. There was an added spur to muscles and spirits this day, for wasn't it the fourth stage? Wouldn't night see everyone in the paradise of rest billets? But, the march closed towards noon at La Ferme Notre-Dame, twenty kilometers southeast of Châlons. "That's all right," men said wisely. "They're putting another day on the march to make it easier for us. We'll sleep tonight and get there tomorrow."

Yet certainly no one would have chosen to stop at Ferme Notre-Dame "to make things easier." It was a place at once beautiful and abominable. There was only one well, for instance, at some distance from the main buildings, so that it took five hectic hours to water the animals just once. Word passed around that the start wouldn't be made until late the next morning. So we fitted in. A short march, then rest, baseball, baths, delousing!

In fact, the regiment didn't move out until 6:30 on September 20, but the stage lengthened into twenty kilometers, and ended during the middle of the afternoon in meadows near Cheppes-la-Prairie on the bank of the little river Guenelle. For the first time, doubt appeared in men's faces: "What does it mean?" they asked one another. "Ah," someone answered carelessly, "we'll get there tomorrow or, if not, the day after. This isn't so bad." Nor was it bad for men or animals: the one bathed and washed clothing in the river; the other grazed contentedly in the lush meadows.

Suspicions, too, were lulled when Captain Reed was ordered by Brigade Headquarters to reconnoiter to the south in the vicinity of Bassu for the next night's bivouac. Swinging further to the south, of course, meant rest. But the next morning that hope died. A change was announced: the regiment wasn't going south, for French officers appeared and warned commanders of the necessity of seeking concealment most carefully from now on. At 5:30 on the afternoon of September 21, the regiment moved out—to the northeast, and everybody knew it meant the front again.

The attitude of the men in face of this abrupt change was stimulating. No matter how brave or bloodthirsty he may be, a soldier who expects rest and is suddenly shot back into the line must experience a vivid disappointment. The 305th had the air of having foreseen such a fate. They talked cheerfully of a huge, new offensive that couldn't possibly be successful without the presence of our regiment. If there was any grumbling, it was done under the breath.

The march was quick. After twenty-five kilometers, the column halted at 11 p.m. in Bussy-le-Repos, where we found a confusion already suggestive of the front. The 304th had bivouacked in and around the town, few billets were available for headquarters, and the nearby fields were crowded. Our regiment settled itself where it could.

If there had remained any doubts, they were dispelled here. Captain Olney, from Brigade Headquarters; Captain Reed, from the First Battalion; Lieutenant Wilhite, from the Second Battalion; Lieutenant Klots, from Regimental Headquarters; and officers from the 304th and 306th were ordered forty kilometers forward by motor truck to Les Islettes to make a reconnaissance, locate positions, and figure data. This party left on the morning of September 22—the advance guard of the brigade into the Argonne.

At Les Islettes, they were met by French Corps Artillery officers assigned to support the Americans. These Frenchmen had foreseen everything, which was fortunate in view of the difficult and tricky Argonne terrain. They took our officers to the point near Florent, which they had selected for the regimental echelon. Then, they led them carefully forward almost to the frontlines and pointed out positions for the First Battalion that were a kilometer due east of Lachalade, and other positions for the Second Battalion that were a kilometer and a half northwest of the First. These choices were clearly the best available, so the reconnaissance party set to work checking up targets and data.

While they figured in the forest, the regiment resumed its march, leaving Bussy-le-Repos on the night of the 22nd, to bivouac for a few hours at Verrières the next day. The column went on that night to the vicinity of Sainte-Menehould. And, for the moment, Regimental

Headquarters established itself at the Florent echelon, from where it superintended the regrouping of the command and made arrangements for its entry into position at the earliest possible moment.

The men who had come by truck from Mareuil-le-Port had had a good rest. Moreover, they were full of the gossip of the sector and possessed rumors without end about what was going to happen. In many respects, the situation was fruitful for rumors.

Positive orders, for instance, came from the highest command that no American soldier was to risk exposure to enemy observation unless he wore a French uniform. That made scouts and observers near the frontline masquerade. It also meant that a surprise attack on a gigantic scale was in the wind. Yet no one suspected then how big the scheme really was. Indeed, the terrain seemed badly suited to anything of the sort. War here had practically paused for more than four years. The reason lay before everyone's eyes—the woods and the hills of the Argonne.

Here, one of the few points where "position warfare" had persisted, both the French and the Huns had developed deep and elaborate trench systems. A large proportion of the work was in cement and there was an elaborate net of barbed wire. The prospect of attacking such defenses head-on was not a cheerful one. It was whispered, however, that our doughboys were only waiting for our support to go over.

The situation, meantime, remained placid. There was very little firing and as far as we knew there were no raids. Either the *bosche* had been fooled and didn't know what was gathering, or else he was waiting with a little surprise of his own. A day or two now would show which was to be true.

Both battalions moved into the positions selected near Lachalade, during the early morning hours of September 24. At the same time, Regimental Headquarters went forward to Ferme Ferdinand. Those positions were trying on both officers and men, not because of enemy harassing, but because of their exhausting natural difficulties. Out in front in No Man's Land, and for a considerable distance back, the forest survived only as a ghostly collection of stripped tree trunks. Two thousand meters to the rear, however, where our guns were placed, it had suffered less, and there was a dense underbrush with practically no tracks. As a consequence, the cannoneers had to chop a way in. The pieces were unlimbered on the road, then manhandled a half a kilometer through the brush to their emplacements. That would have been hard enough by daylight and before the dawn it was a task for a Hercules with the vision of a cat. Still it was done before sunrise and the work of consolidation got under way.

These positions were in a piece of forest known as the Bois des Hauts-Bâtis. They were near some old French reserve trenches, in which our infantry waited for the great moment. The doughboys didn't seem to know exactly what was going to happen to them, or seem to care particularly. The difficulties of the terrain failed to appall them. With curiosity, they watched the artillerymen as they went about their labor.

Ammunition was the chief difficulty. The firing would be intense; as a consequence, vast quantities of shells would be required at the emplacements. Time was short—word to commence firing might come at any minute. Yet a point on the road about four hundred meters from the guns was the nearest place to which projectiles could be transported on wheels. The G. S. carts dropped them there, and the battery men carried them one by one through the tangled underbrush. This work went on during September 24–25, while everyone wondered if the *bosche* wouldn't observe such diligence and compliment it with a little heavy fire.

An odd incident happened on the 25th. There hadn't been a single high-explosive burst near these positions, nor were there any later, yet on that day six gas shells fell among the pieces of the First Battalion, or in the road nearby. One of these shells cost the regiment a valuable messenger: Private Carlos Montgomery, who was thrown from his bicycle by the explosion. Pieces of the casing had struck him in the knee, and before he could get his mask on, the gas had burned his eyes severely. He was evacuated and invalided to the States. Within a few yards of where he was injured, another gas shell fell beneath a G. S. cart, which five men were manhandling, but it failed to injure or gas even one of the five.

From the start, it was clear that in the Argonne we would face new difficulties of observation. Here and there were observatories cleverly concealed in trees or on the heights above the Biesme River, which ran through the French trench system. Officers and men disguised as *poilus* climbed into these, but found the outlook from all was unsatisfactory. Communication, on the other hand, was comparatively simple in the first Argonne position. Regimental Headquarters, the two battalions, the observatories, and the infantry were closely grouped. Later, when the advance commenced, those in liaison with the frontline had a good deal of difficulty keeping headquarters informed as to the details of a changing and hazardous situation.

At last the orders came down: the regiment would open fire at 2:30 a.m. on the morning of September 26. The volume of noise that burst forth at that moment was greater than the Argonne had ever known. The terrific uproar came as a surprise to the men on the receiving end of the gun fire, for they had not suspected such a mass of artillery would be collected for the drive. Whatever they had learned, the Germans were stunned by this merciless fire. It continued until the infantry went over shortly after daybreak and then shifted to a rolling barrage. Finally, because of the rapid advance of the infantry and shortage of ammunition, it ceased altogether for a time.

Runners brought back word of what was happening out in front. Over the cement trenches and strong points, through the mazes of barbed wire, and the natural barriers of the forest, the infantry made an advance that first day of three kilometers. The artillery would have to move forward at once. The limbers were hurried down and the pieces went over difficult roads through the old French trench system for three kilometers to the vicinity of La Harazée. Regimental Headquarters established itself in the remains of the town, and the two battalions went into position side by side within two thousand meters of the new frontline. There were dugouts here—large, luxurious, and fairly safe. So the personnel of the three headquarters and the batteries made themselves comfortable.

It soon developed that there would be no letup in the drive: it would proceed at once and new missions were assigned. It was during those days that citizen officers and soldiers displayed an exceptional cleverness and adaptability. They located their guns and their targets on the map and—frequently without registration, and as frequently without observation even—blazed merrily away. It was like firing a revolver in the dark, and yet when the regiment moved forward, it could check up on its accuracy. Then dead *bosche*, destroyed shelters, machine-gun emplacements, and a torn forest offered their mute and terrible praise.

The second day the infantry made an advance of two kilometers. After that it slowed down for a time and, by lengthening the range, the entire regiment remained in these emplacements until September 30. On that morning the First Battalion decided to get farther forward. Major Easterday left at seven o'clock to reconnoiter for new positions and Captain Reed was to follow with the battalion at ten o'clock, to a point near the Abri du Crochet. The infantry had captured this important and pleasant place a day or two before. On the map it appeared as a crossroads; it was, one estimated, scarcely one thousand meters from the frontline.

We expected that distance would soon be decidedly widened. And it was, to some extent, but for a time now the progress of our infantry was reduced to nearly nothing. There were a number of reasons including: the effect of the first rush was over; the men were tired; every battalion had had serious losses; and, while the Germans gathered themselves for a stand, several divisions—probably nearly two hundred thousand men—were rushed to their support. In addition to these fresh odds, the country had become if anything more difficult than at first. Then, before the advance could get fairly started once more, the affair of the Lost Battalion helped hold things up. But on this day of Major Easterday's reconnaissance, the advance continued if slowly.

The battalion halted short of the crossroads while Captains Reed,

Dana, and Ravenel, and Lieutenant Kane rode forward to find the major. After they had joined him, the major said he had chosen positions a kilometer and a half to the rear. Coming up, the battery commanders had seen these positions and were by no means enthusiastic. As usual, Major Easterday was ready to weigh the opinions of his battery commanders. In the meantime, Captain Reed had pushed through a fringe of trees and had seen positions on a slope to the right that he believed had possibilities, if a small amount of cutting could be done. Major Easterday approved and studied the ground more closely with the battery commanders. Together they located positions in which no cutting at all was necessary, in the altogether delightful Abri du Crochet.

Delightful is really the word, for here, in a sort of amphitheater, the *bosche* had developed the rarest refinements of position-warfare life over four years. The place possessed enormous and intricate dugouts, some of them bored into the rock for nearly a hundred feet. They were furnished and some had food, even, left by the hurried Germans. Chlorinated water was forgotten for a time, as the dugouts were well stocked with mineral water, and some had stronger liquids. Shower baths invited, firewood was cut and piled, and tramways ran here and there for convenience in bringing up supplies. The dugout network extended so far that battery command posts fared as well as the battalion.

The Battery A commander had an experience the first day that illustrates as well as anything else the elaborate scheme of the system. The A commander and B commander had their eyes on the same dugout: Captain Ravenel got to it first, so Captain Dana chose another one some distance away. Everyone had long since learned to examine such places for traps. Therefore, Captain Dana and Lieutenant Stribling went in at once with flash lamps and searched through the galleries. When they came to a door they halted, for something with a slow stealth moved beyond the panels. In whispers, the two officers discussed the situation. A German spy might have been left behind to wait in this comparatively safe retreat until he could slip through the lines with a plan of the American artillery dispositions. There was only one thing to do—the door had to be opened. The two loosened the pistols in their holsters. Captain Dana raised the lamp and flung the door wide with a sudden gesture, prepared for emergencies. Across the threshold stood, in much the same attitude and with much the same suspicions, Captain Ravenel.

Chapter Twenty

ALWAYS THROUGH THE FOREST

The First Battalion remained at the Abri du Crochet for a week, while the Second stayed in the position at La Harazée, both supporting, after October 2, the famous Lost Battalion. The ring of fire with which the 305th circled Major Whittlesey's command was credited with a measurable share in his salvation. The Second Battalion, moreover, had an officer with him during those black days. Lieutenant Teichmoeller, of Battery D, had been in liaison with Major Whittlesey when the jaws had closed.

The story of the Lost Battalion has been told often enough. A word here will suffice to explain the artillery's perpetual support of the trapped men.

On October 2, the infantry was forging ahead, scarcely able to maintain flank and rear liaison because of the broken and overgrown terrain. Suddenly, the enemy appeared on both flanks and to the rear of the First Battalion and Companies E and H of the Second Battalion of the 308th Infantry. This party of about six hundred men and officers had made a quick forward thrust of half a kilometer or so. It then became clear that neighboring units had failed to keep pace. Major Whittlesey's party, therefore, was in a trap from which the Huns were evidently determined it should never escape. For six days, they pounded the little command while the Americans did everything possible to relieve it. For six days, the party went almost entirely without food. Airplanes were sent over to drop rations and ammunition, but in the thick woods they had difficulty locating the suffering group. For six days, its personnel accomplished Homeric deed, endeavoring to guide the airplanes and to get messengers through. Its losses were great. Of those who came back after the relief, few had not been wounded. Lieutenant Teichmoeller, although completely exhausted and ill from lack of food, was one of the fortunate ones who had not been hit.

During this period, the strain was felt almost as thoroughly by the artillery as by the supporting infantry. Our batteries fired constantly, and our agents and observers forward did what they could to locate the command and to report on the result of the fire.

On the second day—that is, October 3—Lieutenants Burden and Klots, with Private Cox of the Headquarters Company, were forward with the infantry, trying to get some light on the situation. They were crossing an open space when they spied an observation tower in the woods ahead. "It's an observatory," Lieutenant Klots said, "and if there's anybody in it he'll snipe at us."

Someone was in it, and he began sniping with a 77. The party took shelter in a crater. After a time, when the sniping had ceased, the three made a dash for some trees on the flank. They reached the shelter, but the grove itself was getting a good many shells. Lieutenant Klots pointed out a low bank. "Looks like dead space under that bank," he said. "Why not wait there and have a smoke?" The others agreed, but it turned out that the bank did not furnish dead space for the men. A number of shells fell nearby, then one dropped directly in front of the party, and the backlash got all three. Lieutenant Burden was badly hurt in the thigh; Private Cox got a painful and disabling wound in the leg; and Lieutenant Klots was struck by a fragment on the instep. Other shells would come. There was only one thing to do —make a run for it.

Lieutenant Burden started first and reached the thicker woods out of the line of fire. Cox tried it, but went down after a step or two, realizing for the first time that his leg had been fractured. Lieutenant Klots carried him back to such shelter as the bank afforded and remained with him until some infantrymen came along with a stretcher and took him to the first aid station. Cox did not return to the regiment. Lieutenant Burden had had a narrow escape and was not discharged from the hospital until sometime after the Armistice, when he was assigned to work in Paris in connection with the peace parleys. Lieutenant Klots's wound was slight, and he returned to duty within a few days.

On October 4, while firing in support of the Lost Battalion, C Battery lost a man in a premature burst. A piece of the tube struck Private Edgar A. Blethen. Lieutenant Robinson was the first to reach him, but the man had been killed instantly.

With the relief of the Lost Battalion, the infantry resumed its advance, and it became clear that the artillery would be better off in new positions. Regimental Headquarters had left La Harazée on September 27, for Fontaine aux Charmes. On October 9, it went forward two kilometers to the Dépôt des Machines. The First Battalion moved considerably further into the Bois de la Naza, but remained only a few hours. After taking position, it found that the infantry had gone so far ahead it would not be profitable to fire. It continued, then, to a point a kilometer west of Chatel-Chéhéry, where it remained for one day, firing semi-steel shells on German works near Grandpré.

The Second Battalion, on October 8, left La Harazée for the Stolzenfels dugout system in the rear of Binarville, two hundred meters to the left of the Binarville to Vienne-le-Château road. The battalion did not fire from these positions, which it left on October 9, for positions further forward, about half a kilometer to the right of the village of Langon.

On October 10, the entire regiment moved forward through the Bois de Langon to the vicinity of Grandham. Regimental Headquarters was established at Malassise Farm on the Aisne. The First Battalion took up positions to the east of Grandham, while the Second went a trifle further to the north to Hill 208. The regiment remained in these positions by shifting its ranges, always within reach of its targets, until the 77th Division was relieved on October 17.

The regiment experienced a real mystery on October 10, and it was not a pleasant one. Sergeant Orville C. Cooper of Battery B was the victim. He had served as First Sergeant of the battery since the early days at Camp Upton, and had been much appreciated by Captain Ravenel. According to the report made by the battery clerk, on the night of October 10, Sergeant Cooper was called from his quarters at the battery echelon near Lachalade by a soldier unknown to anyone in the battery. The soldier said that the sergeant was wanted by Captain McKenna at the supply dump, about five hundred meters away. Sergeant Cooper took a shortcut, and about half an hour later he was brought back to the echelon by an infantry guard detachment. He had been badly slashed in the throat and about the body, evidently by barbed wire, in falling from a narrow footbridge after his assailants had beaten him on the head with a club. In a semi-conscious condition, he was evacuated by Major Miller. Captain McKenna had not sent for the sergeant, and a searching examination of property, and a careful questioning of the personnel in the vicinity of the echelon, failed to yield the slightest clue to the assault. Sergeant Cooper was so badly hurt that he was invalided to America.

On the next day, Battery D had an unusual casualty. Private Rodney J. Lecours, who had been guarding ammunition, lay down in hay on the side of a road near Binarville and fell asleep. His head was towards the road; other men were asleep nearby. A motor truck, not seeing these men in the dark, drew up at the side of the road and one of the wheels passed over Private Lecours, killing him instantly.

During this last stage of the operations, which had commenced on September 26, the regiment suffered a particularly depressing loss: First Lieutenant Sheldon E. Hoadley was killed on Sunday, October 13. He had left his battery position and was riding along a road to the rear when a shell burst near him and a fragment struck him. He received immediate attention, but there was no chance. He died a few minutes after he was hit, while on his way to a dressing station in an ambulance.

Rumors of a relief became persistent. Horses and men were worn out. Since entering the war in Lorraine, the regiment had left the front only to change position. In other words, it had had no rest at all. The supply of ammunition was uncertain, too, and the materiel needed attention. Both Grandpré and Saint-Juvin, divisional objectives, had been taken. Then these rumors crystallized into fact: the 78th Division relieved the 77th on October 17 and 18. Regimental Headquarters moved back to Lachalade, resting for a few hours at Langon. The batteries followed out, with baths, delousing, and rest waiting for them at Lachalade.

We soon learned there was to be more than simple rest. The regiment was allowed a number of passes for three days, exclusive of travel. The fortunate departed gaily for the vicinity of Paris or the Riviera. As it will be sadly remembered, in many cases they were met as they descended from the trains at their destinations by military policemen who presented them with telegrams. These missives recalled them at once to the regiment. The reason was obvious: the division was returning at once to its place on the frontline. There would be a new and vigorous offensive.

Chapter Twenty-One

THE LAST PHASE

The clans gathered again at Lachalade, and made ready to hurry back to the line. During the period of rest, everyone had found time to read the papers. It was known that the Germans had asked for peace, that notes had passed back and forth, but at the front no one took the news very seriously. There was too much to be done. The men had become so absorbed by the war that at last they had borrowed something of the French attitude. The thing appeared eternal.

The question of transportation caused worry and wonderment. The regiment had received replacements of men but not of horses. How was it going to be possible to move guns and ammunition with the few animals left? The answer came a little later, in an unexpected form.

On October 27, the echelon was established at Chatel-Chéhéry, and Regimental Headquarters and First Battalion Headquarters settled themselves in a house at Cornay. The First Battalion guns were a kilometer to the north; the Second Battalion guns were in the same valley as the First, but to the left, near the town of Marcq. Major Easterday and Captain Starbuck made a careful reconnaissance of the front, and then firing opened on November 1.

Again, the advance was large and, on November 2, the regiment moved forward to the vicinity of Verpel. At midnight on that day, the transportation problem was solved in a radical fashion. The orders for the move had evidently come from high up. Colonel Doyle summoned Majors Easterday and Wanvig to Champigneulle and told them that the regiment would be split. The First Battalion would continue as a combat battalion and the Second would act as its combat train, turning over to the First its horses, wire, telephones, and other equipment. Such a move was inevitable, for more than once in heavy weather the horses had been unable to draw the pieces without the aid of cannoneers. The weather could be counted on, now, for much rain and the consequent mud.

The battery commanders, in pursuance to this order, met Major Easterday in Verpel and the dispositions were settled upon. The echelon was established there, with Captain Derby, who had recently been promoted with Captain Pike, placed in charge. Captain Storer was given command of the combat train and instructed to always keep two thousand rounds in the train. Major Wanvig and his staff, of course, were responsible for both the echelon and the train.

Under these new conditions the regiment moved forward towards the Meuse River. On November 3, the firing batteries passed through Buzancy, a town that had been fired on by the Germans and was in flames. Civilians, who had been under the German yoke for four years, hurried to the rear with what belongings they could save. They were clearly grateful to see the Americans, but such emotion as theirs did not express itself demonstratively.

That night and the next morning pirate guns were sent out. Lieutenant Robinson took one, Lieutenant Mitchell took another, and Lieutenant Warren W. Nissley took a third. These officers, with a piece each and a cart full of ammunition, went forward to the infantry and fired on whatever targets the infantry commander chose. It was dangerous work. Our officers went into position, practically in the open, and fired at German machine-gun nests, later receiving a gratifying amount of praise from the infantry.

On the evening of November 3, the First Battalion moved forward to the village of Fontenoy, where the Regimental Headquarters also located its command post. First Class Private William Kuttler, one of the regimental messengers, was killed on the road near Fontenoy that day. He was walking behind an escort wagon and was close to a party of infantry when a shell burst in the bank at the side of the road. Kuttler was the only man of our regiment hit, but seven infantrymen were killed and a number wounded. On the same day, Lieutenant Charles Graham was wounded by a shell fragment and evacuated.

The regiment remained in Fontenoy on November 3 and 4, then moved into Stonne, placing the three firing batteries in position in the valley to the southwest. The civilian population of Stonne welcomed the Americans as saviors. Men and women said the Germans before fleeing had instructed them to take refuge in the church, promising not to shell the town for twenty-four hours. However, scarcely had the Germans gone than the place was drenched with gas shells and, of course, the civilians had no gas masks.

The next day another forward move was made to Flaba. Rations were scarce; often the men had given of their issue to the civilians. Here, the civilians gave the soldiers black German bread, which the hungry men had not experienced before. The result was a sad amount of indigestion and a heightened sympathy for those who had been compelled to live for so long under the Hun's food regulations.

There was no firing from these positions and, on November 6, Batteries A, B, and C moved a half a kilometer to the east of Haraucourt, into range of the heights across the Meuse. The Second Battalion, acting as combat train, had kept pace with all these changes and had assured the supply of ammunition. Here the regiment remained until the signing of the Armistice, five days later.

On the day the pieces moved into the final positions, the regiment had its last casualty in action: Second Lieutenant Leon H. Hattemer, who had come to the 305th on the Vesle, was killed by a machine-gun bullet, while in liaison with the infantry. The nearness of the war's end made his death seem all the more unfortunate.

Lieutenants Burden and Bullen, and Private Gormley were

mentioned in division orders for their work during the Argonne fighting. A few new officers were assigned during this last offensive. Second Lieutenant Augus R. Allmond had come on October 10, and three other officers were with the regiment for a few weeks but were transferred away again to other branches of the service. On November 10, the day before the Armistice, Major Edwin A. Zundel was assigned to the command of the First Battalion to replace Major Easterday, whose promotion to the rank of Lieutenant Colonel had just come through. Easterday, it will be remembered, had commanded the battalion from the Forêt de Nesles to the Meuse Heights—that is, during the battalion's most active combat experience. For his aggressiveness and his daring in reconnaissance, he was cited afterwards in division orders. He was a familiar figure near the frontlines on foot, on his horse, or dashing about in a motorcycle. One time he and his driver wandered past the pickets and into a village filled with German soldiers who were preparing to depart. Easterday told the driver to turn around, and before the Huns had recovered from their astonishment, he was rushing back to his command. By virtue of his new rank, he went to Regimental Headquarters as second in command. Along with Major Zundel came Second Lieutenants Solomon Abelow and Horace Heyday. The next day the war was over.

The Armistice had been announced to the regiment during the morning, but the regiment was skeptical and went about its business. When the firing stopped, the men attended to their routine duties and grinned wisely whenever anyone tried to tell them the show was at an end. The silence at last made an impression and, once a band appeared playing victoriously at the head of a regiment of Moroccans, the majority conceded that there might be something in the rumor.

There were, however, cases of chronic doubt. Sergeant Joseph of the band, for example, had been left some distance in the rear to guard a reel cart. He picked up what he could to eat from neighboring units, but on the whole he was a hungry sentry. On November

14, a doughboy passed him in his isolated retreat, came up, and burst into a laugh. "Hay, Buddy! What you wearing your gas mask in the alert for?" he asked.

"Orders," said the sergeant.

A guffaw from the visitor, who said, "The war's been over three days."

"I've heard that before," replied the sergeant drily. But somehow this fellow managed to persuade him.

The minute the great fact of the Armistice was absorbed, the talk was of home. Originally, the word was that the 77th would go into the Army of Occupation. But that was altered and, except for a few officers and men who were transferred to units ordered up, the division moved out of the line.

The 305th was billeted in Verpel for a time, and then the period of leaves commenced. After one or two stops on the way, the regiment detrained at Latrecey and marched to Arc-en-Barrois, a charming and hospitable village in the Haute-Marne, where rumors of our departure remained until February 9. Here, an elaborate schedule of training went into effect based on ancient methods of firing, so that some had a good time talking wisely and extensively about aiming points, designation of targets, and "P" minus "T." Also, scandals of ammunition and equipment were laid bare at leisure. And, everybody was brought into close personal touch with the High Cost of Living.

But there was a difference about us. Officers and men followed out the appointed schedules, but their spirits were at home. There was no desperate and necessary future to which this training led. It had the air of killing time and keeping men occupied. And many soldiers wanted to learn things that would be useful to them on their return to America and work. The days slipped away beneath heavy skies and a downpour seemed nearly perpetual. Athletics started up with soccer football on New Year's Day.

In the midst of rumors of our early departure came the epidemic of Spanish influenza. We had had a number of cases, and Lieutenant

Danforth Montague had gone in December, and there was an uncomfortable feeling that the dread disease was always with us. In the latter part of January, men began to report sick by the score. On One day thirty were evacuated and on another we said goodbye to forty. The evacuations climbed to fifty or more, and we knew each day that some of the men that climbed—feverish and ill—into the ambulance, would not come back. During this emergency, Major Miller worked day and night, but sporadic cases of typhoid complicated his labor. His ultimate success, however, permitted the regiment to leave for the embarkation center on February 9.

The bitter cold, the snow covering the ground, and the prospect of cattle cars didn't dampen the joy the men took in this move towards home. The train, composed of ancient cars, just crawled. A journey that one might take in a regular train in eight or nine hours consumed for us—cramped, cold, and uncomfortable—about sixty hours. We recalled the days before the Armistice when we had been of more value to people generally, and when we had been rushed long distances into action at an express-rate speed.

That trip will be eternally colored in our minds by Lieutenant Arthur Robinson's death. After accepting all the chances of the front with a cheerful and inspiring indifference, which had won for him the Distinguished Service Cross, Robinson was accidentally killed on the night of February 10, at the little station of Châtillon-sur-Cher. He had stepped from our train, which was standing on a siding. The fastest train on the road—an American special—tore by at a terrific rate of speed striking the open door of a compartment. Robinson was struck by this door. He was buried with full military honors in the American Cemetery at Angers. It was not like a death in action and everyone, officers and men, had liked and admired Robinson. His death cast a persistent shadow over the regiment.

On the evening of February 11, the 305th entered Malicorne, a pottery town on the beautiful Sarthe River. The people were rather different from those at Arc-en-Boisson, but after a time they learned to like the Americans. We stayed there until the 17th of April, drilling, getting reviewed and inspected, and chasing the elusive cootie, so that we should be rushed through Brest. The weather was sufficiently warm to permit us to develop a baseball team, which remained undefeated at the close of an extended divisional season.

When we reached Camp Pontanezen on April 18, we realized that we were, indeed, veterans. And, we realized we had really been pioneers in the American Expeditionary Forces, for Pontanezen had grown out of all recognition since our visit the year before. Back then it had been nearly as the French had turned it over: a group of old barracks and a few tents. Now, it covered many acres and the original camp was lost in the midst of countless huts and tents. Whatever horrors the place may have contained, we failed to experience. We were there only two days, and the weather was clear and warm. On Sunday, April 20, we marched into Brest, survived the mad confusion of loading baggage and men from pier to tender and from tender to ship and, by nightfall, we were packed on the transport USS *Agamemnon*—formerly the German liner *Kaiser Wilhelm II*. We sailed at noon on the 21st, out of the harbor of Brest, "a good deal wiser," as one man put it, "than when we had landed." The boat was uncomfortably crowded, but no one cared. We were going home. The weather, moreover, was good, so that scarcely anyone was ill. And the *Agamemnon* was fast: at nine o'clock on Tuesday morning, April 29, we saw the low shore of Long Island and picked up our pilot at *Ambrose Channel Lightship.*

The story of that day of homecoming is still in everyone's heart—a trifle vague, perhaps, because it was difficult to realize that we were, after more than a year, again in New York Harbor and that, where twelve months before we had slipped out hidden between decks, we were now steaming noisily in surrounded by cutters and ferryboats decorated with banners and filled with shouting friends. Everything, indeed, was reversed. But on the pier there were still men and women who gave us things to eat and smoke. We piled on to the same ferryboats and went around the welcoming town to Long Island City.

That night we reached Camp Mills, and the next morning, after a final delousing, half the regiment went home for forty-eight hours, while the other 50 percent followed two days later. Even then you had a feeling that you were through. You could already count the hours that separated you from a return to a normal life, a final rupture from the service to which everyone had given himself wholeheartedly, but with which nearly everybody wanted to be done now that the emergency was over.

The parade alone held us. We went to New York Monday morning for that, left our equipment in the 9th Regiment Armory, spent Monday and Tuesday night home, and on Tuesday morning marched up Fifth Avenue from Washington Square to 110th Street, where we saw the last of our division and brigade commanders.

The return to Camp Upton the next day was the commencement of the final phase. It was there that our regiment had been born, and it was there it would end its career. Upton had altered little, yet it seemed oddly different—but that was because it was ourselves who had changed.

At Upton the machinery of demobilization seemed to be out of repair by day and to grind only during the dark hours. After three nearly sleepless nights, the last formalities had been complied with and organizations gathered in a pouring rain for their final pay and their railroad tickets home. Men glanced proudly at the red chevrons on their left arms signifying discharge. They walked, in formation for the last time, to the familiar railroad station, where organization commanders and officers gave them their discharges and shook hands as they passed through the gates—civilians after one of the best jobs soldiers ever did. And with this breaking up of the 305th Field Artillery died a good deal that was fine, a good deal that you couldn't see vanish without regret. Yet, although it may seem paradoxical, few would care to watch its full resurrection, because that would mean also the rebirth of the conditions on which it was built.

No more that great communal chorus: "When do we eat?"

No more the revolt in one's heart at the best cursed music in the world: "First Call!"

No more tearing one's hair at Paper Work!

No more elaborate language or strong-arm competitions with the Red Hats!

Even the first sergeant got a sympathetic thought that last morning. His piercing whistle at reveille held a special significance. And so did his loud, uncompromising, and final: "Outside!"

Afterword

THE ANGUISH LEFT BEHIND: CHARLES WADSWORTH CAMP'S INFLUENCE ON HIS DAUGHTER MADELEINE L'ENGLE

Charlotte Jones Voiklis

The box is round, and fits in the palm of my hand. It is deep red, almost oxblood. Covered in leather, with a delicate leaf pattern in gold around the rim, the box's top comes off smoothly. The interior is lined with velvet and contains a small package wrapped in paper with my great-grandmother's writing on it: "glass from the cathedral of Rheims, picked up by C. W. Camp while under fire." Inside the paper are five pieces of glass, thick and gray and sharp—they hold hard history. Also inside is a pin made up of two long guns in the shape of an "X", of the type worn by soldiers in the field artillery units of the US Army in World War I.

The first time I saw the box I was in a Manhattan Mini Storage, sorting through cartons of manuscripts and journals, and letters and photographs, all of which had come from my grandmother's nearby apartment and from her house in Connecticut. She had recently died and I was responsible for figuring out what to do with all of these things. That is a difficult task in any family, but when the person who dies is famous, it adds another layer of difficulty and responsibility: which personal effects and records are worth keeping for the family, and which are important to keep for posterity and scholarship? Where does everything go?

My grandmother was Madeleine L'Engle, most famous for her children's classic *A Wrinkle in Time*, and author of more than sixty books. Her father, Charles Wadsworth Camp, a journalist, critic, and novelist, was rather less famous than his daughter. His manuscripts, books, and letters were also in that storage unit, kept by his widow after his death in 1936, and then moved from Florida to Connecticut to New York over the next eighty years.

I had mixed emotions when I discovered the box: I was excited that such an artifact existed, curious about its story, and a little in awe of holding it. But, I was also overwhelmed by the weight of history in all of the cartons, and not just my grandmother's things, but also her father's. What to do with all of it?

I had a kind of reverence for and clarity about my grandmother's things: I would inventory, preserve, and be a good steward of them. About Charles's things, on the other hand, I held a kind of resentment. Were his belongings going to be my responsibility, too? Could neither my grandmother nor great-grandmother (Madeleine Camp) have figured out what to do with his things themselves?

I noted the round, red box on my spreadsheet and put it back in the carton with the other letters and writings. Copies of Charles's published books were in Connecticut; over the years, I had opened them to see if they were interesting, but I was never compelled to venture into them very far. They were mostly horror or thriller novels, some of which did well at the time of publication, and were adapted into films and plays. His two nonfiction books: *War's Dark Frame* (1917) and *A History of the 305th Field Artillery* (1919) were different because they weren't novels, but also didn't hold much fascination for me when I was growing up. I liked made-up stories, not war histories.

My grandmother was a storyteller, and not just in written word but also in conversation. It is how she liked to engage in dialogue with people, offering narratives that invited the listener to imagine they were part of the same story. When I was a child, she told me stories while cooking, gardening, riding the bus, and during quieter moments sitting in her lap or, when I was older, drinking tea. She told stories about her parents and grandparents, as well as more distant relatives, particularly those who settled in north Florida when it still "belonged" to Spain. She left the South for college and only went back for short visits. She always considered herself a New Yorker and always felt that she had escaped the South. While her relationship to her southern roots was full of contradictions and ambivalence, she took pleasure in the stories she chose to tell (and that had been chosen for her, for there were many that had been passed down to her). She continued to tell those stories, not just to her grandchildren, but to her readers, too, keeping their memories alive but also fixed like a doll on a shelf that could be taken down and petted but not played with too roughly. I know there are other stories, too, that are not quite so pretty or proud, more like the shards of glass in the leather box—sharp and full of pain. Not all stories about her own parents and childhood were doll-like. Some of them were alive —shimmering, shifting, escaping her deft narrative skills—and created in me a curiosity and skepticism.

As a child, I imagined I knew what Charles smelled like from her descriptions: Egyptian tobacco, whiskey, and starch. My own grandfather smoked cigarettes (though I was sure Egyptian tobacco was lovelier than unfiltered Camels), drank whiskey, and there was a spray can of starch in the laundry room I would sniff to try to get the full picture.

After many years of listening to and reading stories about my grandmother's parents, here is what I know for sure. Charles and Madeleine's mother—also called Madeleine, and I'll refer to her as Mado here to avoid confusion—met at a wedding in Jacksonville, Florida. Mado's family was from there and much of Charles' family had recently moved there from Crosswicks, New Jersey. Twenty-eight-year-old Charles had a reputation for being handsome and outgoing and charming. A Princeton graduate, Charles was working in New York City as a writer—a theater critic, playwright, and journalist—when he met Madeleine Barnett, daughter of a prominent local banking family. At twenty-six, Mado, who was an accomplished pianist and not considered beautiful, was thought to be past the marriageable age. Still, they were a "love match." After they married she moved with Charles to New York, and together they travelled all over the world for his work. He had a portable typewriter and a camera, and while I am not sure that photography was part of his journalism assignments, I have found negatives of landscapes and portraits of Europe—such as the Cathedral in Milan, the canals of Venice, a family in traditional Dutch dress, and the deserts of Egypt—all of which were taken sometime between their wedding and the start of World War I. When he went to Europe in 1916 to cover the war (some of his articles were published in *Collier's Weekly* in 1917, and he may have been published elsewhere, earlier), Mado stayed behind. His experiences as a correspondent and later as a soldier impacted him deeply and had reverberating effects on his wife and daughter (who was conceived while he was on leave from training). After he returned from the army, his *bon vivant* attitude slid more into alcoholism and depression. He had a hard time finding work and was physically weakened by the effects of gas—either mustard, phosgene, or chlorine, the record is unclear and the use of gas was ubiquitous—and never fully recovered. The 1920s and 1930s were tumultuous globally—the flu pandemic, prohibition, the depression, the rise of fascism—and for the family as well, as they moved from New York to France to Florida. He developed pneumonia after attending a Princeton football game and died in 1936, when his daughter Madeleine was seventeen.

From the stories my grandmother told about them, Charles and Mado, despite all of their challenges, seemed both exciting to me and happily married. One of her favorites about them appears in her first work of nonfiction, *A Circle of Quiet*:

> In the hall, between my parents' room and mine, hung an old etching of Castle Conway, in Wales. It's a charming picture, and I have always loved it because of the story my mother told me about it. . . One hot summer evening, long before I was born, she walked through the hall and glanced at the etching of Castle Conway and said, "Oh, Charles, it's so hot. I wish we could go to Castle Conway."

"Come on!" he cried, and swept her out of the house without tooth-
brush or change of clothes, and into a taxi, and by midnight they
were on a ship sailing across the Atlantic. In those days a trip could
be as spontaneous as that. My parents were not poor, but neither
were they, by today's standards, affluent. Father was a playwright
and journalist, and their pocketbook waned and swelled like the
moon; this must have been one of the full-moon moments.

The story always felt implausible to me, and does so even more
now: Could a woman at that time travel without petticoats and hats?
How did they pay for things so spontaneously in those days before
ATMs or credit cards? But I enjoyed this tale and was intrigued by it.
Its implausibility, however, did create in me a bit of distrust in both
the story and the teller. The same etching hangs in my house, but I
have never been to Castle Conway.

My grandmother would hint at other, less kind truths in stories
that were meant to be amusing. Once, as Charles and Mado were
coming home from a party, he looked at her and said, "You know
dear, I really was proud of you tonight. You looked almost pretty."
Perhaps he was making a distinction between superficial prettiness
and more enduring beauty, but I can't be sure. Another story recalls
how he took his daughter to a restaurant that he frequented and the
waiter, who knew him, gave him very disapproving looks until
Charles made a point of introducing Madeleine as his daughter (as
opposed to his mistress, one assumes). This makes the family rumor
(that was only shared with me when I was well into adulthood) that
Charles gave Mado gonorrhea (for which she had to have a hysterec-
tomy) more believable. But again, I can't be sure.

Madeleine said that her father had the ability to "make a party
go," and to make people feel interesting and important. Charles could
also be a tyrant at home, and he and his wife argued about
Madeleine's education. Charles always won. Madeleine switched
primary schools in New York three times. I remember being told that
the first switch was because a friend of the family had encouraged it,
because of the school's reputation. Mado was enthusiastic about the

new school, but Charles had his reservations. Madeleine was very unhappy there, but Charles did not allow her to leave it, apparently in order to teach his wife a lesson of some sort. Later, it was Charles who insisted that Madeleine be sent to an English boarding school in Switzerland, while they were living in France, even though her mother wanted to send her to the local school. It is hard for me to reconcile the gay Charles with the stern Charles, and I don't know how my grandmother forgave him—but I believe she did.

A portrait of Charles hangs in our living room. Painted by Gilbert White in Brittany, sometime before Madeleine was born, its soft grays and greens convey a whiff of French Impressionism. Charles looks young and vital and mischievous, Edwardian in his confidence and entitlement. My grandmother told me that her mother didn't like the painting because "you could see the martinis in his eyes." It was always said with a laugh, but I knew the comment carried with it a terrible sadness. She also wrote about the painting in her memoir, *The Summer of the Great-Grandmother*: "I never knew the man in the portrait; I knew only the man whose world had been demolished in that war, and who took eighteen years to die." In her earlier nonfiction, she was clear about his faults and talked about his moods and alcohol abuse, and she was compassionate about his professional disappointments. But as time passed, in both family conversation and the nonfiction she continued to publish, she talked about him differently, becoming less willing to discuss the difficult parts of his personality, or the ways in which she struggled with his influence and legacy.

Shortly after finding the box, I visited my mother in Connecticut at Crosswicks (named after the town Charles had grown up in), the house she grew up in, which my grandparents bought in 1949, and began living in fulltime around 1952—and where I now live. On that visit, I looked through my great-grandfather's *A History of the 305th Field Artillery*, about the battalion he joined in 1918. I wondered if he would describe finding those pieces of glass "while under fire," but I found the book was a general history, not a memoir. I looked up Reims Cathedral (spelled as "Rheims" during his lifetime) and learned that it had been bombed early on in the war. That is to say, 1916—well before Charles's deployment. Therefore, my great-grand-mother's note on the paper holding the glass was evidence (at least to me) of a conspiracy to make Charles a tragic figure. I wasn't buying it.

By the time I was an adult, I did not have a good impression of Charles. There were things about my grandmother's relationship with her father that made me uncomfortable: she always seemed to be insisting that he was a good and kind man, when evidence appeared to contradict that. I felt that she did not tell the truth about him. Over the years, I felt her stories about him began to leave out the less flattering portions—his alcoholism, his violent temper, his auto-cratic manner, his inability to support his family. And, because I knew she wasn't telling the whole truth about *everything* about him, I began to doubt she was telling the truth about *anything* about him. Had he really been gassed in the war? Was he a good writer? Had there been true affection and respect among their family of three?

I held onto this ungenerous portrait of him until more than a decade after Madeleine's death, when my sister and I began working on *Becoming Madeleine*, a biography of our grandmother intended for middle grade readers. As I read through family letters and journals, Charles began to take shape for me in a different way. There is still much about his life and personality that I don't know or understand, but could certainly discover through further research, but, as with my grandmother, I prefer not to make his life an object of scholarly study. I became more interested in and forgiving of my great-grandfa-ther when I read the letters he wrote to his daughter, particularly while she was in high school. He seemed sincerely interested in her writing abilities. He also encouraged and challenged her in ways that shaped her, ways she never forgot and were forever missed when he died a month before her eighteenth birthday.

April 21, 1935

　　Dear Daughter,

　　. . . On our way back we stopped at the post-office, and found your letter with the poems, "Night Cry" and "Night." I like them both. In the first: "The moon shivers behind a thin cloud" seems a very suggestive and rememberable [sic] line, but of course all appreciations and criticisms are personal. Mother thinks that "Night" has too much use of the word light, but my feeling is that your intention was to make your effect with just that repetition, and I believe with a little smoothing it will be very successful. Of course "The night with its thousands of eyes—" is in itself very lovely, and you are perfectly free to phrase it so; but that old war-horse, "The night has a thousand eyes, the day but one," I think detracts from the dignity of your line, because in the public mind it has much the same appeal as a crooner's song; but you must decide for yourself. With a little smoothing you have a lovely thing there.

She wrote letters to him that were equally personal, specific, and intimate, even when she was very young. This one was written around 1926 when she was eight, when she and her mother were in France visiting Mado's father:

Dear father,

I am very sorry I scolded you, but you see, I had not gotten any letters from you then.

Just think, I went to a real circus, the seals were lovely, and the Elefhants were sweet, the Lions were tortured, the man was so crual; The clowns was very funny, (there were too clowns) and there was a very strong nothing but mussles, to men I mean; Just Imagine it was a three ring circus. There was one woman who did tricks on a trapise and when she had finished she took a doll she had and made her do tricks on the trapise. There was a man who jugled his hats and clothes and his cigar and every thing, and another who jugled plates and everything. There were to men who did stunts. There were horses tame, in other other parts horses wild a man riding a horse and all sorts of animals. There to very interesting things, there they are, 1. There were to men one man balanced a long pole on his face while the other man climedd up to the top and did stunts on it. 2. There a lot of men and a seesaw and a VERY high thing behind the seesaw. One man got on the high thing and the other on the lose side of the seesaw, then the one on the high thing jumped of and landed on the high part of the see saw and sent the other man whizing through the air to land on a chair wich another man held up for him. I have not told you half the interesting things that they were, but I am tired now so love and kisses

    Madeleine

Madeleine's note was written on the back of her mother's letter, in which she details the hassle and expense of travel. In it, Mado used her favorite phrase: "I'm as blue as indigo," and, she explained, "This certainly has been a disastrous trip for me—I feel as if I never want to come abroad again. Madeleine is delighted at the prospect of going home."

The photos, journals, and letters my sister and I examined during our research softened my mental portrait of Charles, and he became more like Gilbert White's painting: a little blurry, but pleasing to look at. I wanted to know more, and so I became determined to read his published works.

I was sorry to find that the fiction still did not compel me. I have not been able to find copies of the film adaptations of his work (at least one film was produced before the war, and another after his death; Mado was a careful executor of his estate and tried to keep his books in print). It took me longer to crack open his nonfiction. Up until then, I had rarely read nonfiction for pleasure, and I figured if his novels hadn't grabbed me, his war stories weren't going to either.

I was wrong.

When I began to read *War's Dark Frame*, I was struck by the fact that it was both reportorial and engaging—a kind of journalism without the pretended (and impossible) objectivity, but nonetheless trustworthy and factual. This was a reporter with a point of view, who used dialogue and "man on the street" examples, displaying a dry humor and literary wit. These qualities can be seen in a passage in which he describes his arrival in Nancy, in Lorraine, a region in

France that saw heavy fighting during World War I and was very close to the front.

> He drove us into Nancy whose chief beauties, in spite of the bombardment, remained intact. There were ruins, however, in the vicinity of the station, which the Germans had been unable to hit directly. An apartment house in the middle of a block had recently been struck, and all that survived was a heap of rubbish in a yawning hole. More pitiful, more anger-producing, was the rubble and charred beams that marked the site of a children's school. If it was the purpose of the Germans to make the innocent suffer in Nancy, they have achieved an admirable success. [. . .] Close to this circle of devastation lay the hotel, so far practically untouched, in which we were to spend the night. "Perhaps," our officer grinned at me, "a shell will fall through *monsieur*'s bedroom and furnish America with a *casus belli*." I patiently explained to him that I entered the war zone at my own risk, but his wit intrigued him, and each time he repeated his joke we tried to laugh.

Charles begins his narrative on the ship, traveling from New York to Liverpool, a dangerous crossing in those years because of German submarines. After setting the scene, he relates the story of a young woman, traveling with her mother, who catches his eye because she is not in mourning. "She wanted, moreover, to talk about her experience," he says. "That, too, was in her eyes. Because of the past, possibly because of something she approached, she desired to tell her story." And Charles listened.

The young woman's story hinges in part on the consequences of missing a letter, at a time when transatlantic communication was slow and unreliable, even for those in the armed forces who had easier access to cables. Her story reminded me of the difficulty Madeleine's parents had when communicating about her birth: Charles wasn't able to assure his wife that he was well, or that he had heard the good news and was thinking of them, until a month later.

From Liverpool he makes his way to London, where he witnesses the first Anzac Day parade, commemorating the losses sustained by the Australian and New Zealand Army Corps. He continues on to France, through another submarine zone, and onto Paris, where he connects with his government tour guide or handler, an elderly Quaker doing his conscientious objector service. Because he has a press visa, Charles is given supervised access to munitions factories and certain towns and villages—places that the war authorities found useful to have foreign journalists report on, in order to shore up support for a war that was having such a devastating effect. (Charles was aware that he was serving that purpose and, as a supporter of the Allies, he thought the United States had a moral calling to also enter the conflict, despite how horrible war is.)

From Paris, he makes trips to various parts of the front, and to reclaimed towns and areas, documenting the devastation and the resilience of the people. He doesn't miss an opportunity to note the brutality of the German forces against civilians and even their own soldiers. He visits the Cathedral at Reims, about ninety miles north-west of Paris, which was famously shelled in 1914, and was still subject to random fire from the German Army not far away. In *War's Dark Frame*, Charles recounts the visit:

The cathedral of Rheims proves how absurdly conservative photography is. A picture of the twin towers and the rose window won't give you a sense of unbelievable tragedy, or an instinct to speak not at all or in whispers. [. . .] There were with me officers and soldiers hardened to the filth and corruption of war. Some of us had seen devastation more complete and no less excusable than this. Yet no one failed to respond to that sense of suffering which seemed to have survived its physical source. It is, of course, impossible to say how far our knowledge of what had happened here gave birth to such thoughts. It is merely significant that we all experienced them. One visualized rows of bandaged and groaning men, stretched on the straw or crawling about with awkward, incoherent motions, like mutilated insects. The vaulting seemed to retain the echoes of cries and curses. Openings showed where the Germans had sent incendiary shells to burn their own wounded.

Such anguish leaves something behind it.

In reading that passage, I suddenly realized that during that visit Charles must have carefully taken pieces of glass from the site and wrapped them in paper. That they had indeed been collected "while under fire," though he was not yet in uniform. I had done a disservice to Charles, and to Mado. Such anguish does indeed leave something behind it, and also carries itself into the future, as Charles's biography shows.

One of the striking things about Charles's discussion of the war is his continual notice of its democratizing influence. He remarks, for instance, in *War's Dark Frame*, on the loosening of social mores that govern both gender and class relations: "A little more than two years ago this woman would not have spoken to a stranger" and "So the perpetual strain, its general distribution, draws people to each other for comfort, because so many of them can say: 'I understand.'" He boldly praises "the social justice of compulsory military service" and states that the social effects are "the best thing that can be said for this war, the finest thing that can survive it."

I admit that I looked for those moments in the text that would be consistent with my own opinions, and hoped I wouldn't find any rank racism or other offensive content. However, to be fair, I must note that class chauvinism does persist in his account. He tells of meeting a soldier on the front who confesses to being an author himself. Charles identifies him by his pen name—G. A. Birmingham—to evade the censor (who would have been reading and approving Charles's newspaper dispatches). The Irish novelist's given name was James Owen Hannay, about whom Charles writes: "The thought of those rollicking Irish stories and plays made his presence here seem an injustice. It gave the lie again to the so-called British apathy. It was one more example of how every social and intellectual class is feeding this monster of war." And later, his own class prejudice becomes visible when he writes, "Those who man these filthy craft are largely of the naval reserve class—men out of comfortable homes and convenient clubs. Consequently, they bring to their work an exceptional intelligence."

While reading *War's Dark Frame*, I also looked for how Charles chose to address his American readers. This was in the spring of 2022, and as reporting came in about the Russian invasion of Ukraine, I wanted to understand if he thought, or indeed hoped, his journalism would sway public opinion in favor of the United States entering World War I. He lectures his readers a bit, taking them to task for their complacency, but he is also always sympathetic to the distance that made that easy and the difficulty of committing to something that they knew would bring death and sorrow.

America, with its lights and its careless pleasure seeking, attained a visionary quality. Was it possible such a place actually existed? At first one was happy at the prospect of that refuge, but the bugles continued, blaring the truth of this war, and one became ashamed, reading in such a state a vital wrong which sooner or later would have to be paid for.

Charles was given access to certain parts of war-torn France by the war powers who considered he and other journalists as part of their propaganda efforts. Although he wasn't a war enthusiast, he was sure that the United States had a moral obligation to assist the Allies, and his journalism tried to convince others. From reading accounts like Charles's, how could America not be moved to help? In *War's Dark Frame*, he reports:

> Luncheon commenced. There was tactful talk of America, our position in the submarine controversy, our political conventions, the possibility of our entering the war. There was—as always at such gatherings—an undercurrent of wonder, never quite reaching the surface, that we should have found it to our best interests to have held aloof.

Moral considerations were joined by practical ones: while the United States had a standing army, to answer the call in Europe would require a draft. That hadn't been done since the Civil War, and America had changed and grown a great deal since then. There was real concern, both at home and abroad, as to whether a draft army could be effective. Again, Charles is direct about this:

> I gathered, not particularly from this conversation, rather every-where in England and France, that a belief had grown since the beginning of the war in our lack of homogeneity. We were, it was suspected, incapable of direct and concerted action. In those days the men who were actually treading the exhausting mill frequently placed upon us—whether justly, who can tell?—the taint of many races, the incoherence of too vast a variety of creeds and desires and antipathies.

In his next book, *A History of the 305th Field Artillery*, he will tell the story of the creation of this new armed force. In the meantime, he ends his first work of war reporting, *War's Dark Frame*, with the following words:

For a question had survived through my months in Europe: "How long before we, too, will be at war?" As I drove on, the question drifted inevitably into a statement, brutal and unescapable: "We, too, will be at war. It will not be long."

And he was right. I estimate his reporting trip took place between April and September 1916. The United States entered the war in April 1917, and shortly thereafter Charles volunteered. As a former member of the New York National Guard, he was an officer and served from August 1917 to May 1919. He was assigned to write the history of his battalion by his colonel, in the interim between the Armistice (November 11, 1918) and their return home in late April 1919. The history includes drawings, as well as short, firsthand accounts of notable incidents written by several soldiers.

If *War's Dark Frame* has the agonized but clear purpose of rallying support for US involvement, and a narrative engine that pulls the reader through, *A History of the 305th Field Artillery* has a little less urgency. Readers (as did the author, Charles) already know that the

Allied Forces triumph. Still, this firsthand account is compelling. Charles doesn't give a plodding chronological account, rather he writes about military bureaucracy, personified by what he calls "Paper Work," and the pull of gossip and rumor in the ranks. There is still his signature narrative skill in deploying humor, dialogue, and scene. He takes the reader through the first days of the first US volunteer army (at a time when everyone was aware they were creating something new) and its growing pains (which he likens to a newborn baby) at Camp Upton on Long Island; then through to its deployment to France and its movements along the front; and finally, describes the four long months between Armistice and homecoming, when intramural sporting events abounded to fill the days.

I looked for Charles in his own history but he kept himself hidden, except for a few instances when he lists assignments and refers to himself in the third person. I can, however, infer a few things —such as when my grandmother must have been conceived, while he was on leave in February 1918. Rumors had been circulating for days that the battalion would soon set sail for France and people were anxious:

In some measure, Washington's Birthday cleared the air. It fell on a Friday. We began to speculate our future when we were informed that on the holiday there would be a parade and a monster Division ball that night in the armory of the Seventh Regiment. And, as many of us as possible would be given passes between Thursday evening and the following Monday's reveille. "Looks like a farewell show and a last chance for a good visit home," was the common interpretation of the events. This view was strengthened when we struggled to town Thursday night and word passed through the train that the absent battery commanders had been recalled.

Two months later, Charles's battalion sailed to France. And Mado, who had suffered numerous miscarriages during their ten years of marriage, was put on bed rest. Madeleine writes about her mother's pregnancies in *The Summer of the Great-Grandmother*:

I was a much longed-for baby. It wasn't for want of trying that my parents were childless for so long. But Mother could not hold a baby past three months. "All I needed to get pregnant," she once remarked, "was for your father to hang his pajamas over the bedpost." She had miscarriages all over the world—Paris, Berlin, Cairo, and—I think—one in China. Sometime toward the end of "the war," my parents' war, World War I, Father came home on leave from Plattsburg, and I was conceived. Because Father was sent immediately overseas, Mother was able to spend most of the nine months in bed. Even so, I am a witness to her determination. The first doctor she went to told her that she could not possibly carry a baby to term, and that if she did not have a therapeutic abortion, both she and the baby would die. Then she went to a Roman Catholic doctor: that was in 1918. So I am here to tell the story.

There is nothing in Charles's history of his battalion about his daughter's (my grandmother's) difficult birth, or the difficulties the family had communicating about it. Letters were written and cables were sent, but Mado still had not heard from Charles by mid-December, and she was worried. On December 13 she writes him:

My dearest husband—

If you could see your little flower of a daughter, I am sure you would forgive her for not being a boy. Oh my dear, I am so thankful that she is here and healthy and perfect and I wouldn't exchange her now for all the sons in the world.

She is considered a perfect miracle in the hospital and everyone is interested in us, and so if you were only here to share my happiness. It is worth all the long months of waiting and the hours of agony at the end. Dear one, I have been pretty sick and am hoisted up in bed for the first time this afternoon. Baby is two weeks old today. [A cousin] phoned yesterday that your mother had sent you a cable yesterday after noon. I wonder if you have had mine sent November 30 and if you know what a proud father you should be? It seems so strange not to have heard from you yet. I got your letter of

Nov 21 a day or two ago. There was nothing between that and the one of Nov 13 I received Thanksgiving Day, so I must have lost one at least.

I do hope you have had a nice time on your leave and that your cold is gone. I have worried over your ears, dear. Never mind about the promotion dear. It is hard luck but lots of others have been treated the same way and I am so happy that you are safe and whole. Nothing else matters. Have you any idea yet when you are coming home? It can't be so long, yet every week will seem an eternity now.

I have two good nurses dear. My day nurse is the very best in the whole world I am sure. I just love her dearly and she is good to me and the baby. I hope you approve of baby's name—Madeleine L'Engle. I think it just suits her and thought you would want her named Madeleine. I must stop now dear. I'm pretty wobbly but. very very happy

Will try to write again in a day or so.

Your loving wife, Madeleine

When Charles wrote to Mado, on December 5, he explains:

I am more and more uneasy about you. I kept them informed of my address while I was away, so that any cables might be sent on to me. None came, and I did hope to find one here last night, but there's no word yet, and I'll probably be out of here in two or three days, and heaven knows when I'll hear then. I do so long for some word, dear. I can think of little else. Even now you may be in the hospital. I wonder if you can imagine something of the state of my mind.

In his *A History of the 305th Field Artillery* he does write about another young father in his battalion:

The Supply Company had its first casualties at Fismes. While at Chéry, two men were killed and three were gassed. Wagoner Jackob Jackson was one of the men killed. About five minutes before he was hit, he was cleaning his harness in front of his wagon. The company clerk who was passing was stopped and shown a letter that Jackson had just received from his wife. The glad tidings that he was the father of a little boy was conveyed in the letter. Jackson asked the clerk to get him the additional allowance from the government because of the birth of the child, adding "my wife says that things are very high in the States and she needs the money." The clerk promised to attend to it immediately and then jumped into Jackson's wagon to look for some candy that the K of C [Knights of Columbus] had distributed the day before. It wasn't a minute after that that the Hun commenced shelling again, the second one hitting and breaking within five feet of the wagon. On investigation, it was found that Jackson was hit in the back of the head. He died on his way to the hospital.

I have to admire his editorial eye in this passage. Wagoner Jakob Jackson's story has many layers, and the narrator's perspective remains consistent.

In that same letter of December 5, Charles writes of some professional disappointment: a promised promotion has never materialized, likely the same one Mado alludes to in her letter to him.

I found a letter from Homer. He said he expects to go home soon. He said also that Jack Hogan was now a major, and that Barry had been made a first Lieutenant. I heard this morning, too, indirectly that Homer had been made a major. I found last night that my own recommendation for promotion had come back with the endorsement that on orders from the war department no more promotions would be made at present. It was not a disappointment, because I had heard everywhere that there would be no more promotions. Homer must have got his at the last minute. I'm doing my best not to be envious. I think the thing that hurts most is the fear that people will think it queer. I can't go about showing my record book to everybody. And it's hard to have had all your friends so much more materially successful than yourself. Those are two unworthy thoughts that I must put out of my mind. Because I know I've done my job, and I know I've done more valuable service than lots of men who have higher rank. And it's been so much a matter of chance. I wish, for your sake, that I had something more to show for it. And if you are not unhappy about it, I will seriously try not to be.

I quote this at length because I am moved by Charles's voice and by the intimacy of his confession to his wife. I don't think Madeleine (my grandmother) ever read this letter, or the other ones that Mado kept in various bundles that ended up in a suitcase, stashed at Crosswicks until recently. And yet his belief that his complaint about being less successful than his friends was "unworthy" is similar to a journal entry of Madeleine's about her and her husband's professional difficulties as they watch certain of their friends gain renown and prosper. I wonder if Charles was successful in putting those "unworthy thoughts" out of his mind, or if they reared their head again when he stumbled professionally after he came home.

As I read his books, I also looked for a record of Charles's exposure to chemical weapons during his service. He recounts numerous incidents involving different kinds of gasses used in the war, but because he employs third-person narration, it's unclear when and how he was exposed. For a long while, I doubted he had been, and I thought it was just an alibi for his long downward spiral. I understand now that the use of gas was so ubiquitous that there is no doubt he suffered from its effects.

He may also have been exposed during his reporting assignments. For instance, in *War's Dark Frame*, he recounts being given a gas mask as a precaution.

I examined the workings of my gas mask. It was designed to cover the head completely and to be buttoned into one's coat collar. Goggles were fastened to the brown cloth and beneath them was a wooden tube (with an elastic band) that was to be placed between the lips for breathing out. Through the chemical-soaked meshes of the cloth itself, sufficient air filtered for breathing in. It was an unlovely, uncomfortable, and odorous contrivance. We were careful

to keep ours slung over our shoulders—as Williams carried his—
and every officer and man near the fighting line carried one. The
necessity for such a precaution revolted your sense of decency,
aroused a sort of anger.

Later on in the book, he describes seeing a platoon of soldiers
with their masks on: "The brown faces were featureless save for
round, staring goggles. They retained no individuality, no human
semblance. These hideous figures might have been visitors from a far
planet, or monstrosities escaped from this earth, too violently
disturbed." Although extraterrestrial life has been speculated about
since antiquity, his imagination pricks my own about conversations
he and Madeleine might have had that could have primed her for,
and planted early seeds of, the story that became A Wrinkle in Time.

This joint publication of Charles's two books is a project of the
pandemic. Much like my grandparents, who had moved from New
York City to Connecticut in the 1950s, in part to avoid a feared nuclear
blast, I, too, moved to the same house with my family in March 2020,
to have space and fresh air. (That first pandemic spring and summer,
Crosswicks sheltered as many as twelve of us at any given time.)
Family history was more present than ever in the old house, and I
took that opportunity to return to Charles's books. My compassion
for him had increased over the years, as had my interest in my grandmother's relationship with her parents, both resulting from my own
shifting relationships with my own parents and children. I wanted to
do something for Charles, and bringing his writings back into print
felt appropriate. Additionally, these two books about war still have
something to teach us today about conflict, our shared history, and
the violence that people are capable of.

Together, these two books show the United States at a decision
point: just when it seemed unlikely that the country could create a
national army from its patchwork citizens and send this army overseas to enter a conflict that was at a brutal stalemate, the United
States committed to its allies and did exactly that. I love War's Dark
Frame for its definite point of view on the war and what a moral

response to a terrible conflict looks like. It's not an enthusiastic call to action; rather, it's a sober accounting of atrocities and an agonizing assessment that America cannot just standby. On the other hand, *A History of the 305th Field Artillery* is written after victory is won, and is a firsthand account of the creation of the US Army and its deployment, and it honors the soldiers Charles fought with. That he cared to document not only the loss of life but also the intramural sporting events makes me smile and reminds me that yes, those things matter, too.

And what of his influence on his daughter Madeleine, who grew up to write one of the classics of American literature? She absorbed a lot from him, and family photos show that there was a great deal of affection between them. She admired him and from a very young age wanted to be a writer like him; but then, as she began to become aware of his own struggles and demons, she feared that she was too much like him. When reading the book that made her a "famous writer"—*A Wrinkle in Time*—with an understanding of her father's experiences and influence, new layers of meaning are uncovered in the story: it is about families and who saves who; the importance of taking chances for the people you love; and parents and children letting each other go.

Charles's involvement in World War I was formative for him, and it was formative for his daughter Madeleine, as well. It appears that she was aware of her father's war experience from a very young age. In an unpublished fragment about her memory of Christmas in New York, when she heard Christmas carols played on the street outside her apartment on East 81st Street (on the edge of Yorkville, a predominantly German neighborhood at that time), she writes:

[T]here were brass bands playing loudly, slightly out of tune, all the old favorites and I saw nothing ironic in Silent Night played fortissimo. The musicians were usually German, victims one way or other of the first World War, something I simply accepted as a child. Often I would be given pennies wrapped in tissue paper; the front room windows would be opened, and I would be allowed to toss the pennies down, and blessings, in German, would be wafted up.

"Poor things," the Irish cook would say of the German players.

I didn't understand the ways of nations then any better than I do now, but I took the ambiguities far more for granted, and enjoyed tossing down the pennies. My father was ultimately to die of injuries sustained in that war, but he, too, seemed to think the musicians were "poor things," and gave me extra pennies to throw down.

In *The Summer of the Great-Grandmother*, Madeleine tells another story of how her parents were dining at a restaurant, years after the war, when suddenly:

Father got up from the table in great excitement and rushed across the dining room to a man who, in his turn, was hurrying to greet Father. The two men embraced warmly, and Father brought his friend over to the table to meet Mother: the man was a German; he had been an officer in the Kaiser's army; he and Father had fought against each other at the front.

The Camp family moved to France in 1931, impelled by both the stock market crash and Charles's need for cleaner air for his weakened lungs. Madeleine wrote extensively, in both her fiction and nonfiction, about her childhood and the formative time in France, when she was deposited, without any warning, at a boarding school a month into term. She came home from boarding school for the Christmas holidays and saw her parents in a whole new way. She began to understand that her parents were their own persons, with troubles and joys that had nothing to do with her. Having been unhappy at school, she had looked forward to being home again with her parents. As she explained in *The Summer of the Great-Grandmother* "I wanted to balance the pain of school with comfort, safety, changelessness, but I found pain, discovery, change." She discovered that her mother was unhappy and her father was drinking a good deal. Madeleine recalls that upon finding her mother sprawled on her bed, her face contorted in grief, she slowly tiptoed out of the room. A short story she wrote in college, "Gilberte Must Play Bach," is, I believe, an autobiographical story from this visit. In it she recounts, from the perspective of a thirteen-year-old girl, a similar scenario: an unhappy mother who plays Bach on the piano when she is sad; a child who escorts her father to the local bar, knowing that his drinking will make matters worse; and the realization of the child that she is somehow responsible for both of her parents and is inevitably going to fail them by being unable to make them happy.

That winter in France was the first time she understood her parents as separate people, who had their own worries and pains and concerns. That's a real turning point for any child, and she went back to that feeling many times in her writing—sometimes with pathos, sometimes with defensiveness, and sometimes with compassion—always with and for her parents, and with little, it seemed to me, for herself.

She did reflect on her own childhood in many of her nonfiction books, and even as she acknowledged her terror of war stemming from childhood, she doesn't connect it solidly with her father's experience. Again, in *The Summer of the Great-Grandmother*, she writes:

> The lions I feared during my childhood were the lions of war. I was born after the Armistice, and yet the specter of another war after the War to End Wars was always with me, not only because of Father's coughing, but because of my own terror of war; I am not sure where this terror came from, but it was always with me. Quite often I would anxiously ask one parent or the other, "Is there going to be another war?"
>
> They paid me the honor of not trying to comfort me with false promises—though I doubt if they foresaw the enormity of change to come in the ways of waging war.

The family moved back to the United States in 1933, this time to Florida, where her mother's family lived. They did so in part to take care of Mado's mother, but also because Charles and Mado were fearful about rising fascist sentiments that Hitler's appointment to the chancellorship of Germany forewarned. I don't think the family trusted the peace and relative order that the Armistice had brought. Madeleine was again at boarding school, this time in South Carolina, when she heard that Mussolini had taken troops into Ethiopia. "I will never forget the leap of terror in my stomach," she writes in *The Summer of the Great-Grandmother*, "followed by a dull ache of acceptance: this was the beginning of the war about which I had been having nightmares since I was a small child."

If Madeleine inherited her father's war trauma, as I think she clearly did even if she wouldn't have put it that way, she also inherited his love of writing. She wrote in multiple formats and genres, and so had Charles. (Her first professional ambition as an adult was to be a playwright, not a novelist or children's book writer). As a critic, novelist, and journalist, Charles had no trouble finding work before the war, though his income was erratic, as freelance writing so typically is. After the war, he struggled as a writer, and it's unclear whether his drinking and illness were "cause or effect" from his service. But he still wrote, and Madeleine benefited from both his example of a writer's discipline as well as her mother's daily piano practice. I see her use of humor and her belief in the importance of laughter in grim times as coming from her father, too. She writes about what she inherited from him in a slim journal, from her years at Smith College, including in this entry from January 24, 1938:

> He had a quick temper that flared up and burnt out like a match. He never bore grudges. I've got that from him. He lacked the strength to push himself along and make himself work. He had famous friends. He could have been famous too if he had stuck to things harder. I'm afraid I've inherited that too, but for his sake I've got to stick—only it's hard because to do it it takes that very thing that you're trying to get. Which is very vague. Which means I've got to work to find words to clothe my ideas. . . . I wish I had inherited his intellect—but I must use what I have and succeed with my writing for his sake. It meant so much to him.

This is a rare instance of Madeleine being critical of her father, as she suggests that his difficulties with writing were a result of not working hard enough. She worried that she, too, was prone to depression. When she wrote or spoke about him, she usually did mention his drinking and smoking, but she did not connect that to his overall health or cite it as a contributing factor to his recurring illnesses and death. Her son, Bion, would die of end-stage alcoholism when he was just forty-seven, but she blamed his liver failure on allergy medication. Her denial of the causes of my uncle Bion's death has reinforced my suspicions that she wasn't honest about her own father's illness and death.

Her father's death changed her, and changed her writing. The consequences of his death might have been quite different if her mother had not insisted, against the mores of her social class, that she attend college. Her relationship with her mother is a subject for another time, but she always seemed to be yearning for child-like intimacy and naivete with her, and also resistance to parental interference. Charles's death left them both lonely and despairing and unsure of how to support each other. Her writing, however, began to deepen and mature. In Summer of the Great grandmother (again) she says: "Until then my writing was poetry, which wrote me, rather than vice versa, and stories of wish-fulfillment and wild and improbable fantasy. But the year that Father died, writing began to push me."

But her father's memory continued to be a double-edged sword. The college-journal entry above shows Madeleine's fear that she, like her father, might not have the strength to do the necessary hard work to become a writer. (Or become "famous," "important," "successful," adjectives she would continue to struggle to define.) Her early promise, again so much like her father's, was seen by her parents and her professors, and while she was young, it was easy to not be scared of what might be on the other side of youthful promise. When she and her husband, the actor Hugh Franklin, and growing family— including my mother Josephine, born in 1947; Bion, born in 1952; and seven-year-old Maria, adopted in 1956, when her parents, longtime family friends, died within six months of each other—moved from

New York to rural Connecticut, in the early 1950s, she began to doubt her continued success. The move was motivated, in part, by fear of a nuclear attack on New York City, and also by Hugh's disillusionment of his own acting career that never grew as hoped. In *The Summer of the Great-Grandmother*, she relates:

My first novel was a success, and it was not until we were living in Crosswicks and I began to receive what seemed an interminable stream of rejection slips (for nearly a decade I could sell nothing I wrote) that I began to understand what the failure of his last years must have been like for Father.

Let's be clear: she published one novel—*A Winter's Love*—and a couple of short stories in that time, so while it "seemed an interminable stream of rejection slips" to her, in fact she didn't sell "nothing" at all. But there was one novel that she worked and worked on, creating at least six full drafts with significant rewrites, which she sent to publishers in various iterations. The fact that she wasn't able to find an editor willing to work with her on it began to crush her. It is hard to describe the novel—never published, it was called alternately *Rachel Benson* and *Bedroom with a Skylight*—because it went through so many radical revisions, but the consistent thread is the story of a woman who in her mid-thirties experiences a personal crisis and the dissolution of her marriage, which causes her to reflect on her life and past relationships. The replay of her life reveals to her that she has been searching for a father figure who would take care of her, do all her thinking for her, and relieve her from the responsibility of living her own life. That realization leads her to reconcile with her husband. Oh, the irony present here (at least to this twenty-first-century reader)! But, it was rejected multiple times, so perhaps this premise was confusing even to the editors who read it, in the1950's.

When it was rejected again in September 1958, she wrote in her journal:

But I am afraid of the effects of too much failure. I saw what it did to my charming, brilliant father. Oh, father, darling, how well I understand you now. And no one, absolutely no one understood you when you needed it. Mother didn't. She tried, but she couldn't. And you were away from all other artists, from anyone who might have given you a moment of help, of hope.

Madeleine, too, felt isolated from other artists, for she was then living in Goshen, Connecticut, a small, rural, dairy town. The next month, however, she and Hugh decided to sell the general store they had been running since 1953, and move back to New York City, where Hugh would try to get back into acting. Her fear of war was still present, but her and Hugh's dissatisfaction with their life in a small town took precedence. In her journal, she laments that it will be another year before they can afford to move the whole family to the city, and she wonders if they will, in fact, have another year—so acute was her fear of atomic annihilation.

In the spring of 1959, Madeleine and her family went on a camping trip across the United States and back. It was during that trip that she began to work on the story that would become *A Wrinkle in Time*—the story of a young girl who fights to save her family from the forces of evil, and who discovers that while she thought her father could and would fix everything, it is in fact she is the one who must do the hard thing, and that she, in fact is able to.

One of the ways I interpret *A Wrinkle in Time* now is as a story about Madeleine freeing herself from her father. I think she carried a lot of guilt because of the relief she felt at his death. His death meant that she didn't have to do the "debutante coming-out" event that her father was insisting upon, the thought of which made her physically ill. (She spent a lot of time in the bathroom, throwing up, at those parties. Later, she learned that she was allergic to the bivalves that were in heavy rotation at the events she was forced to attend.) His death might have meant that she would have to stay in Florida and take care of her widowed mother, but instead her mother insisted that Madeleine go to college and she supported (though anxiously and ambivalently) her daughter's early career in the theater in New York City. As she expressed in *The Summer of the Great-Grandmother*,

Madeleine also felt some guilt for using L'Engle and not Camp as her professional name.

> When I wrote my first published stories and had to decide on a writing name, it was with a wrench that I decided to use my baptismal name, Madeleine L'Engle.
>
> Mother said, "It's as though you're rejecting your father."
>
> "I'm not!" And I reminded her that it had taken her a year to decide whether or not she could go through life as Mrs. Camp. Add to this the fact that many publishers at that time were friends and contemporaries of my father, and I wanted to be a writer on my own. And I'm sure that Father would have been the first to agree that Madeleine L'Engle is a more felicitous name for a writer than Madeleine Camp.

There was a "Madeleine L'Engle" in another branch of the family who was miffed that her cousin had usurped her name, but they never spoke.

Charles is present in *A Wrinkle in Time* in other ways, as well. Madeleine dedicated the book to him, and to her husband's father, Wallace Franklin, and named a main character after both of them: "Charles Wallace." Even as *A Wrinkle in Time* can be read as a daughter's release from the expectations and fears of a powerful father, the influence of Charles on Madeleine never disappears completely. It changes and shifts, but continues to glow and shimmer.

I've gone back to some of the passages in Madeleine's nonfiction written about her parents and have found my memory of them has been wrong, or at least incomplete. She was usually balanced and clear-eyed about her father's shortcomings. What I had interpreted earlier as an increasing unwillingness to tell the truth about her father, displaying unprocessed trauma, now reads more like a writer's choice, motivated by different psychological imperatives. The stories she wanted to tell changed. I realized that just as I had read Charles's work looking for the things that interested me, I read her work, too, for things that confirmed the story I was already telling myself. She left a very long record, and it's not surprising that the stories that mattered to her changed over the decades she was writing. The stories I look for, the things that I find interesting, the stories I want to tell, have changed, too.

Yet, it is still through the short story "Gilberte Must Play Bach" that a fuller picture of what little Madeleine's life was like emerges. I return to it in order to understand her and see how she absorbed her parents' pain. Her nonfiction, on the other hand, has a strict sense of distance from the painful or formative moments of her childhood— and so these writings are a bit cooler, more detached and analytical, reasonable. But something raw and emotional, broken and sharp, shines through the short story, and it breaks my heart every time I think about it.

I have lost track of that red leather box, lined in velvet and holding the pin and the glass shards from Reims Cathedral. There are still places to look, so I'm not worried that it is gone forever. That's another thing that has changed for me: I am no longer burdened so completely by a feeling of responsibility for both the things and the lives. I believe both the daughter and father, Madeleine and Charles, will be fine. I don't have to search franticly for that box. I know that it will appear in good time, and when I do find it again, I imagine that when I slip off the lid to reveal the contents, I will still feel awe and curiosity but will no longer be overwhelmed. My stories have changed.

Acknowledgments

This was a labor of love, a gesture of thanks to both my grandmother, Madeleine L'Engle, and to my great-grandfather, her father, Charles Wadsworth Camp. It is also an experiment in self-publishing, and I have learned a lot and will be learning more as people find and buy this book outside of traditional book publishing.

Many thanks to my family, who encouraged me, and to Jonathan Bratten for agreeing to write the Introduction (thank you Molly Cantrell-Kraig for introducing us on social media). The manuscript was meticulously copy edited by Margaret Moore Booker. Jaime Green edited my Afterword and was a joy to work with and learn from. I benefited from words of wisdom and encouragement from Barbara Braver, Lisa Ann Cockrel, Jessica Kantrowitz, Jana Reiss, Abigail Santamaria, and Sara Zarr. Renata DiBiase designed the cover.

— Charlotte Jones Voiklis

About Charles Wadsworth Camp

Charles Wadsworth Camp (1879-1936) was a journalist, critic, playwright, novelist, and soldier. He was married to Madeleine Barnett Camp and they were the parents of a daughter Madeleine, who would grow up to become an author of more than 60 books, including the classic *A Wrinkle in Time*.

About the Contributors

CHARLOTTE JONES VOIKLIS is the great-granddaughter of Charles Wadsworth Camp and manages the literary business of her grandmother, the author Madeleine L'Engle (*A Wrinkle in Time*). She is the author of *Becoming Madeleine: A Biography of the Author of* A Wrinkle in Time *by Her Granddaughters* (with Léna Roy) and *A Book, Too, Can Be a Star: The Story of Madeleine L'Engle and the Making of* A Wrinkle in Time (with Jennifer Adams, illustrations by Adelina Lirius). She has a Ph.D. in Comparative Literature and lives in Connecticut and New York City.

JONATHAN D. BRATTEN has been the Maine National Guard Command Historian since 2014 and was the Army Center of Military History's first Scholar in Residence at the U.S. Military Academy at West Point. In 2021 his book *To the Last Man: A National Guard Regiment in the Great War, 1917-1919* received the Army Historical Foundation's award for best unit history. He is a National Guard officer and a veteran of Afghanistan.

Further Reading

Novels by Charles Wadsworth Camp

Sinister Island
The Abandoned Room
The Guarded Heights
The Gray Mask

Selected Work by Madeleine L'Engle

A Wrinkle in Time
A Winter's Love
A Circle of Quiet
Summer of the Great Grandmother
Two Part Invention

Books about Madeleine L'Engle

A Light So Lovely: The Spiritual Legacy of Madeleine L'Engle (by Sarah Arthur)

Becoming Madeleine: A Biography of the Author of *A Wrinkle in Time* by Her Granddaughters (by Charlotte Jones Voiklis and Léna Roy)

A Book, Too, Can Be a Star: The Story of Madeleine L'Engle and the Making of *A Wrinkle in Time* (by Charlotte Jones Voiklis and Jennifer Adams, illustrations by Adelina Lirius)

Listening for Madeleine: A Portrait of Madeleine L'Engle in Many Voices (by Leonard Marcus)